David IZABEL

What is Space-Time Made of?

A Nice Addition to Looking at the Dynamics of General Relativity through the Window of Elasticity

Science Press | edp sciences

EDP Sciences – ISBN(print): 978-2-7598-2573-8 – ISBN(ebook): 978-2-7598-2574-5
DOI: 10.1051/978-2-7598-2573-8

Preamble

"It is absolutely possible that beyond what our senses perceive, unsuspected worlds are hidden" – Albert Einstein. And if on this point, Einstein was still right. Indeed, when we look at photos of galaxies (see figures 1a and 1b), a vortex formed on the sea (see figures 2a and 2b) or a photo taken of a vortex at sea during the tsunami that occurred after the earthquake of March 11, 2011 in Japan or a hurricane on earth (see figures 3a and 3b), how not to be struck by the similarity of spiral shapes and textures, despite the difference in environment (respectively, space vacuum, water and air)! In all cases, these spiral structures seem to bathe in a more or less visible fluid, a sort of elastic continuous medium {171, 172}.

Just as the vortex in sea water is carried by a fluid filling all the space, or the clouds of the hurricane carried by the Earth's atmosphere, is not the cosmos also made up of a substance invisible to our eyes and to our technology today in which galaxies are immersed (see figures 1a and 1b) Indeed, it is well known that astronauts train in swimming pools to simulate the behavior in weightlessness, a proof that in space we feel like floating in a fluid. The analogy of the equations between the gravitomagnetism (Lense-Thirring effect), the electromagnetism (dipole moment effect), the fluid mechanics (rotational fluid dragging a solid sphere immersed in fluid) is also interesting (see Wikipedia GravitoElectroMagnetism). In addition, we only know 5% of the universe that constitutes visible matter. What is the remaining 95% made of? Dark matter or dark energy?

In addition, the eye of the hurricane and the vortex in the spiraling water are undoubtedly reminiscent of the supermassive black hole located at the center of each galaxy.

In view of all this, is there not therefore either an elastic material filling all the space with a sort of elastic jelly as proposed by T. Damour in his lectures on gravitational waves [4] or in his book "If Einstein was told me" {136}? Or, a material that would be like a fluid at low speed allowing massive objects to move in it and becoming an elastic solid ultra-resistant to the speed of light like the skier floating on the surface of the water during a game of sky nautical at high speed and sinking when stationary?

DOI: 10.1051/978-2-7598-2573-8.c901

FIG. 1a – Photo of a galaxy.
https://pixnio.com/fr/espace-fr/galaxie-profond-espace.

FIG. 1b – Photo of a galaxy by Guillermo Ferla on Unsplash.
https://pixnio.com/fr/espace-fr/galaxie-profond-espace.

FIG. 2a – Photo of a whirlpool in water – Photo by Jeremy Bishop on Unsplash.

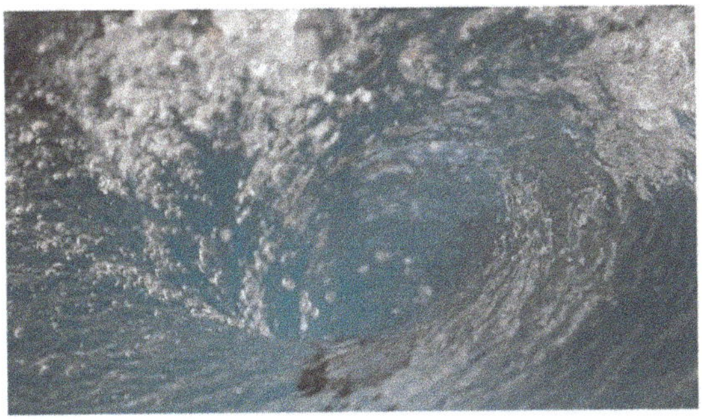

FIG. 2b – Photo of a whirlpool in water – Photo by Enrique Ortega Miranda on Unsplash.

FIG. 3a – Photo of a hurricane.
https://pixnio.com/fr/espace-fr/ouragan-espace-satellite.

FIG. 3b – Photo of hurricane by Nasa on Unsplash.
https://pixnio.com/fr/espace-fr/ouragan-espace-satellite.

This is the initial idea and the common thread that guided me (see appendix A) in an analysis of general relativity from the angle of the continuum mechanics {1–17} and {57} at {59}. This led me to an extraordinary discovery concerning the gravitational constant G and finally to the publication of a scientific article in the peer-reviewed physics journal Pramana[1,2] published on August 13, 2020 by the Indian Academy of Sciences.

Albert Einstein dared to question the Great Isaac Newton based on proven physical and logical facts. Gravity is not instantaneous and bends light in a deformable space-time contrary to what Newton thought in his time. I did the same by pushing Einstein's idea of the deformability of space-time to the end, that is to say of associating the geometric deformation of space-time with a real deformable elastic medium and study its properties according to the principles of elasticity theory by comparing them to what is observed (curvature of star light beam near the sun, torsional deformation of space around the earth in rotation, generalized swelling of space since the Big Bang, and deformation of space during the passage of a gravitational wave). To say that the geometry of space is deformed under the effect of the masses-energies present within it is good, to discover the mechanical properties of the associated elastic physical object which is deformed is better! Especially since the space is supposed to be empty, of classical matter, but could well be filled with an unknown matter which escapes us until now. The quantum vacuum is there to remind us. Emptiness is not nothingness. We can quote Casimir's force joining two plates in a vacuum resulting from quantum field theory which has been measured and verified. The Higgs scalar field and its boson {124–129} which reappear from a vacuum when sufficient energy is supplied to it, as in the CERN accelerator so that it reappears in 2012 because the energies present are approaching from those of the big bang… There is the key to the unification of general relativity with quantum field theory. We will, therefore, conclude this introduction by quoting one of the first great scientists Alhazen Ibn al-Haytham (965–1040) who said:

> The search for truth is arduous, the road to it is fraught with pitfalls, to find the truth, it is necessary to leave aside one's opinions and not to trust the writings of the elders. You must question them and submit each of their statements to your critical mind. Rely only on logic and experimentation, never on the affirmation of each other, for every human being is subject to all kinds of imperfections; in our quest for truth, we must also question our own theories, each of our researches to avoid succumbing to prejudice and intellectual laziness. Do this and the truth will be revealed to you.

Practical information: References in {x} are bibliographic corresponding mainly to the first chapter summarizing physics.

References in [x] are the relevant ones that were used to develop/justify the theories of this book.

---

[1]PRAMANA Journal of physics springer Published by the Indian Academy of Sciences (impact factor 1688 2019) editor in Chief Umesh V Waghmare.

[2]David I. (2020) Mechanical conversion of the gravitational Einstein's constant, *Pramana – J. Phys.* **94**, 119.Online August 13, 2020. https://doi.org/10.1007/s12043-020-01954-5.

Pramāṇa (devanāgarī: प्रमाण) (valid means of knowledge) is a term of Indian philosophy) source Wikipedia.

Contents

CHAPTER 7

CHAPTER 8

CHAPTER 9

CHAPTER 10

Symbology

L	Span of a beam
b	Width of a beam section
S	Section of a beam
F	Force
δ	Displacement
H or h or $2h$	Height of a beam
R	Radius of curvature
R_x, R_y, R_{xy}	Bending curvatures of a plate
S	Section
$M_{(x)}$	Bending moment
M_e	Elastic bending moment
M_p	Plastic bending moment
χ	Curvature $(1/R)$
χ_e	Elastic curvature $(1/R_e)$
$y_{(x)}$	Deflection of a beam
$w_{(x,y)}$	Deflection of a plate
N	Normal effort (tension or compression)
T or M_T	Torsion/twist moment
I	Bending inertia
I_t	Torsion inertia
E or Y	Young's modulus
ν	Poisson's ratio
ρ	Density or radius
τ	Shear stress
σ	Normal stress
$T_{\mu\nu}$	Stress energy tensor
$G_{\mu\nu} = R_{\mu\nu} - \frac{1}{2} g_{\mu\nu} R$	Einstein's tensor
$R^{\alpha}_{\mu\nu\lambda}$	Riemann curvature tensor

$R^{\lambda}_{\mu\nu\lambda}$	Contraction on λ of Riemann's tensor
$\boldsymbol{R}_{\mu\nu}$	Ricci's tensor
$\boldsymbol{g}_{\mu\nu}$	Metric tensor (unknown to Einstein's gravitational field equation)
$\boldsymbol{\eta}_{\mu\nu}$	Minkowski's flat metric tensor
$\boldsymbol{h}_{\mu\nu}$	Perturbation of the metric tensor
$\bar{\boldsymbol{h}}_{\mu\nu}$	Perturbation of the metric tensor corrected
$\boldsymbol{\varepsilon}_{\mu\nu}$	Strain tensor
$\boldsymbol{\sigma}_{\mu\nu}$	Stress tensor
$\boldsymbol{\Gamma}^{\theta}_{\mu\nu}$	Christoffel's symbol
$\boldsymbol{\Lambda}$	Cosmological constant
$\boldsymbol{R}$	Scalar curvature
$\boldsymbol{r}$	Radius of a cylinder, radius of the fibers of the space fabrics
$\boldsymbol{c}$	Speed of light
$\boldsymbol{G}$	Gravitational constant or shear modulus
$\boldsymbol{u}$	Displacement
$\boldsymbol{h}$	Planck's constant or beam height, or trace of $h_{\mu\nu}$
$\bar{\boldsymbol{\varepsilon}}$	Trace of the strain tensor
$\boldsymbol{m}$	Masse
$\boldsymbol{\omega}$	Circular frequency
$\boldsymbol{f}$	Frequency
$\boldsymbol{V}$	Volume
$\boldsymbol{\hbar} = \frac{h}{2\pi}$	Reduced Planck's constant
$\boldsymbol{\kappa} = \frac{8\pi G}{c^4}$	Gravitational Einstein's constant (topic of this book)
$\boldsymbol{k}$	Spring rigidity
$\boldsymbol{t}$	Time
$\boldsymbol{K}$	Rigidity ($F = K\delta$)
$\boldsymbol{g}$	Acceleration of Earth's gravity 9.81 m/s^2
$\boldsymbol{\theta}$	Rotation
$\boldsymbol{\phi}$	Gravitational potential
$\boldsymbol{W}_{\text{ext}}$	External work
$\boldsymbol{W}_{\text{int}}$	Internal work
$\boldsymbol{U}$	Strain energy
$\boldsymbol{U}_c$	Kinetic energy
$\boldsymbol{E}_v$	Vacuum energy
$\boldsymbol{\rho}_v$	Density of emptiness
$\boldsymbol{\delta L}$ or $\boldsymbol{\Delta L}$	Length variation
$\boldsymbol{u}$	Four speed
$\boldsymbol{u}_i$	Displacement
$\boldsymbol{v}$	Speed
$\boldsymbol{J}$	Moment of inertia in rotation of a cylinder
$\boldsymbol{\mu}, \boldsymbol{\lambda}$	Lamé's coefficients
$\boldsymbol{\delta}_{ij}$	Kronecker's symbole

$\boldsymbol{\Delta}$	Laplacian operator
$\Box$	Dalembertian operator
$\overrightarrow{\nabla}$	Gradient operator
div	Divergence operator
$\dot{\boldsymbol{x}}$	Velocity as the first derivative with respect to the time of a position
$\ddot{\boldsymbol{x}}$	Acceleration as a second derivative with respect to the time of a speed
γ	Shear strain – angular distortion
ε_{00}	Elastic time strain
ε_{xx}	Longitudinal elastic strain according to x
ε_{yy}	Longitudinal elastic strain according to y
$\boldsymbol{\Delta t'}$	Time variation
t_q	Quantum time
β	Rotation, angle
θ	Rotation
S_r	Reduced shear section
H_0	Hubble's constant
α	Thermal expansion coefficient, angle
μ	Shear modulus $Y/(2(1 + \nu))$
$\mu\nu$	Index 0, 1, 2, 3 (space-time)
ij	Index 1, 2, 3 (space)

Introduction

What is the Problem with 21st Century Physics?

After completing a publication on Hazard/random draw [1] in mathematics in 2015, I was hit, amazed by current physics that fails to reconcile general relativity and quantum mechanics (quantum field theory), the two pillars of our current knowledge of the world of the infinitely large (the universe) and the infinitely small (atoms and their constituents, electrons, quarks,...). Indeed, their reconciliation is essential to understand the end of the big bang or black holes that necessarily mix these two theories (infinitely small and infinitely hot and dense in energy and so in mass with $E = mc^2$).

Indeed, how to do when one (general relativity) is deterministic and the other (quantum mechanics) is probabilistic? General relativity makes it possible to know the speeds and trajectories of objects moving along geodesics (shortest path on a curved surface) in a malleable space-time elastic by these said objects and continuous (except for singularities; black holes), characterized by a limit speed of cause and effect stuck on the speedometer at 299 792 458 km/s (c, the speed of light), where gravity reigns (illusion of force but in fact a deformation of space-time) and the work of a single man A. Einstein in 1915/1916. While the other, the work of a generation of physicists in the 1920s/1930s (Planck, Einstein, Schrödinger, De Broglie, Heisenberg, Bohr, Dirac, etc.) is discontinuous and quantified, proceeds by jumps, bundles of energy E at frequencies v given *via* a constant of proportionality h: ($E = hv$), or by exchanges of virtual particles between quantum fields (bosons, for example virtual photon vector of transmission of the electromagnetic force), it does not allow to know with precision speeds and positions of its quantum objects (Heisenberg principle). Moreover, according to this theory, the same quantum objects sometimes seem to be particles, sometimes waves (*e.g.* light) without being in fact neither (experience of Young's slits), being entangled giving the impression of communicating faster than the speed of light (quantum entanglement)! What to

DOI: 10.1051/978-2-7598-2573-8.c902

think of this theory or physics and associated mathematics can only predict the probabilities of the presence of these quantum objects in certain places and time (wave function and probability density, virtual particle fields) and all this in a space-time rigid non-deformable whereas that of general relativity is precisely deformed by its content.

When we think that each theory is verified in its field *via* ultra-precise physics measurements (for example for gravitational waves we measure space deformations of 10^{-21}) at orders of magnitude of 13 digits after the comma and have each leading to our current technology which works well (GPS for general relativity, in today's digital world for quantum physics), there is enough to lose its Latin and Its Greek or break your teeth. No doubt there is something at the fundamental level that we need to revisit somewhere that we need to change to reconcile these two worlds because physics is schizophrenic.

Moreover, why is the speed of light limited to what it is? is it in connection with the nature, the density of an elastic medium filling the empty space?

Einstein gave us, when we analyze with hindsight his work, a way to do it: first try to understand how the world works and then adapt the associated mathematical tool, not the other way around. He placed himself by thought in situations at the extremes of our daily life: He dared to question what seemed indestructible to our reason, to our senses: time is not absolute as Newton thought, it is not the same for everyone, it actually manifests itself when we move at speeds close to the speed of light and we are hidden at our daily pace, the space is malleable and charged to the extreme it tears (black holes). Time and space are intertwined to form space-time. Mass and energy are connected by the speed of light $E = mc^2$ (mass is energy in stock as frozen in some way), in the particle accelerators of the moving energy turns into mass; gravity is not a force (illusion) objects move in the universe following the curvatures of space distorted by the objects found there, the luminiferous ether does not exist, light propagates in a vacuum without vibrating a mysterious medium. However, a new kind of ether seems to reappear when we consider space as a deformable physical object. We will see this throughout this book.

Einstein thus transformed a simplified formula (Newton's gravitational formula) into a larger tensor formulation, more general, independent of the reference frame of calculation, remaining valid for very massive objects where gravitational forces are enormous as for massless objects (light). And when we think that all this started from an inaccuracy of 43 s of arc in the delay of mercury perihelion, that Newton's formulation was unable to explain while Einstein's formula found it exactly. The devil hides in detail says the popular adage. Hence the ultra-precise measurements that today's science is required to perform because in physics reality hides at scales that are billions and billions of times what happens on our scale whether in the infinitely large as in the infinitely small for that matter!

Another way, developed by Kaluza–Klein, is to consider an additional dimension to the 4 currently used to study space-time to unify general relativity and electromagnetism. A 5th dimension could be a way to effectively unify quantum mechanics and general relativity and maybe to explain why gravity is so low in our 4 dimensions (because acting in a 5th dimension). This 5th dimension would be invisible to us because rolled up on itself like a thread appearing to us in one dimension instead of three. An other image help to understand this. In a two-dimensional world (ant world), a flat surface covered with snow for example, our steps in the snow are quantified regularly by the width of our steps (h = about 90 cm). But in a three-dimensional world we understand that the quantification of these steps is an illusion since it is a man (or a woman) in 3D that creates these steps at first discontinuous in 2D.

Theories still in the works are trying to solve this dilemma. String theory, but there are 10^{500} possibilities, it takes a space of 10 or even 11 dimensions, and it involves measurements at orders of magnitude that we cannot reach (Planck time, Planck measurement); Loop quantum gravity or time no longer exists at the quantum level! and where space is quantified (ultimate grains of space of Planck length). However these new worlds are *antipodes* of our daily lives and are still inaccessible to our means of measurement current.

Trying to see clearly and provide a clue to cure this schizophrenia in physics is one of the objectives of this book. This implies first of all to understand what has been done to date and then to start again from 0 to see more clearly, to look for the clue that escaped us and to try to disentangle or rather to intertwine these two beautifully demonstrated theories to reconcile them – the process leading to this 9-year book from 2011 to 2020. The main stages are given in appendix A. The path to truth is often winding!

Rather than attacking these two theories head-on at the same time, to the extent that one of the big differences is that space-time is elastic and deformable in Einstein while it is rigid and indeformable in the context of quantum field theory, the author preferred to go through an intermediate theory whose role is precisely to describe and study the behavior of deformations, the stresses of the medium: the elasticity theory and its simplifications in one dimension that is the beam theory and two dimensions that is plate theory {1–14} and {57–59}. In this book, we will see the different advantages of using its various theories to understand, or even reinterpret, general relativity under this window of elasticity theory and how one naturally falls on knowledge relating to graviton on the one hand and the gravitational constant G on the other.

Chapter 1

Where is Physics Today? – Synthetic Overview of the State of the Art of Physics Today

1.1 Introduction

In order to try to go beyond current physics, it is therefore necessary to begin by fully understanding where current physics has arrived and therefore to study it. This is the aim of this chapter. Not having been trained in the mold of theoretical physics of the XXth and XXIst centuries but in mechanics (continuum mechanics, elasticity theory, strength of materials, Eurocodes), but having learned to learn, I then studied for 8 years from 2011 to 2019 as an autodidact by all available means and in particular internet, these two pillars of physics from their foundations until today. So I studied the online lectures in physics – electromagnetism, special relativity, general relativity, cosmology – of R. Taillet teacher-researcher at the University of Savoie Mont Blanc 2014, the e-lecture of L. Susskind of the university of Stanford on gravitation, the conferences of T. Damour and K. Thorne on gravitational waves, black holes and geometrodynamic, the quantum mechanics lectures of Polytechnique X by P. Grangier and J. Dalibard in 2013, on physics, gravitation, time, the history and philosophy conferences of physics with E. Klein, quantum field theory with the lectures of P. Kumar Tripathy from the Indian Institute of Technology in Madras and the YouTube science channel Viascience, conferences on the quantum entanglement of A. Aspect, the quantum gravity lectures of C. Rovelli, the cosmology conferences on dark matter and dark energy of F. Combes etc.). So here is a summary of the ideas that seem to me to be key in physics with the main key formulations and associated dimensional equations. The reader will find in appendix C a summary of the historical dates and principal formulas of physics. The appendicies D and E will give for the reader some main results of general relativity; the scalar curvature R of a sphrere and the Friedmann–Lermaître equations at the base of the cosmology.

DOI: 10.1051/978-2-7598-2573-8.c001

1.2 Newton's Gravitation

Regarding gravity, this is the first of the 4 forces of nature. First there is Newton's approach, which in his book Philosophy naturalis principa Mathematica published in 1687 considers that 3-dimensional space is absolute, rigidly indeformable and where time is universal and immutable. According to him, two bodies of mass M and m distant from a distance r attract each other by an instantaneous attractive force dependent in particular on a universal gravitational constant G that will be the source of many of the discussions in this book. How it acts remotely instantly remains a mystery. But his approach is correct in weak gravity field (solar system excepted Mercury). In particular, it allowed U. le Verrier to discover by calculation the planet Neptune in 1846.

Newton's gravitational force:

$F = G\frac{M.m}{r^2}$ with F a force in kg.m/s^2.

G gravitational constant 6.674 30(15) × 10^{-11} (m^3/kg.s^2).

r the distance between the two massive objects (m).

M and m two masses (kg) that attract to each other.

In the case of the earth we have thus for each mass m_i a force F(N) which is applied to the center of gravity these masses attracting it towards the center of the earth of radius R_T (m) and mass M_T (kg) with an intensity F in (N) of:

$$F = G\frac{M_T}{{R_T}^2}m_i = g \times m_i = \frac{6.6743015 \times 10^{-11} \times 5.97210^{24}}{6\,371\,000^2}m_i = 9.81m_i$$

The acceleration of gravity in m/s^2 is generally noted

$$g = G\frac{M_T}{{R_T}^2}$$

Source: Wikipedia.

1.3 Electromagnetism/GravitoElectroMagnetism

J.C. Maxwell in 1864/1865 unified the equations of electricity and magnetism into a series of four differential equations that allowed us to discover electromagnetic waves. Our eyes see only a part of this spectrum (this is the visible spectrum), but there is a multitude of electromagnetic waves that only human-built instruments can measure (radio wave, X-ray, gamma ray etc.). Mathematics and physics allowed us to discover what our senses did not see.

These equations demonstrating the propagation of electromagnetic waves at a speed c = 299 792 458 m/s have enabled us to understand in particular that light is an electromagnetic wave moving without support in empty space unlike acoustic waves which need a support, the air, to transmit their vibrations or the swell which needs sea water. We will see much later that the vacuum is not as empty as that... It is Maxwell's equations which by uniting the electric field and the magnetic field, like two sides of the same coin, give rise to the equations of electromagnetic waves. Later, Einstein showed that sometimes light also behaves as if it were made of particles! This is the particle-wave duality (sometimes behaves like a wave sometimes like a particle depending on the other). It is interesting to note the similarities of these Maxwell's equations with a rotating gravity field (GravitoElectroMagnetism-GEM). It was done by O. Heaviside in 1893.

J.C. Maxwell's equations/O. Heaviside Gravitoelectromagnetism (formula noted with index *g*)

The divergence of the electric field:

$$\operatorname{div}\vec{E} = \frac{\partial E_x}{\partial x} + \frac{\partial E_y}{\partial y} + \frac{\partial E_z}{\partial z} = \frac{\rho}{\varepsilon_0} \rightarrow \operatorname{div}\vec{E}_g = \frac{\partial E_{gx}}{\partial x} + \frac{\partial E_{gy}}{\partial y} + \frac{\partial E_{gz}}{\partial z} = -4\pi G \rho_g$$

$$\frac{1}{\text{m}}\left(\frac{\text{kgm}}{\text{s}^2\text{A.s}}\right) = \frac{\text{A.s}}{\text{m}^3 \times \frac{\text{A}^2.\text{s}^4}{\text{kg.m}^3}} = \frac{\text{kg}}{\text{A.s}^3} \rightarrow \frac{\text{m}}{\text{s}^2} \times \frac{1}{\text{m}} = \frac{\text{m}^3}{\text{kgs}^2} \times \frac{\text{kg}}{\text{m}^3}$$

The rotational of the electric field:

$$\overrightarrow{\text{rot}}\,\vec{E} = \begin{pmatrix} \frac{\partial E_z}{\partial y} - \frac{\partial E_y}{\partial z} \\ \frac{\partial E_x}{\partial z} - \frac{\partial E_z}{\partial x} \\ \frac{\partial E_y}{\partial x} - \frac{\partial E_x}{\partial y} \end{pmatrix} = -\frac{\partial \vec{B}}{\partial t} \rightarrow \overrightarrow{\text{rot}}\,\vec{E}_g = \begin{pmatrix} \frac{\partial E_{gz}}{\partial y} - \frac{\partial E_{gy}}{\partial z} \\ \frac{\partial E_{gx}}{\partial z} - \frac{\partial E_{gz}}{\partial x} \\ \frac{\partial E_{gy}}{\partial x} - \frac{\partial E_{gx}}{\partial y} \end{pmatrix} = -\frac{\partial \vec{B}_g}{\partial t}$$

$$\frac{1}{\text{m}}\left(\frac{\text{kgm}}{\text{s}^2\text{A.s}}\right) = \frac{\text{kg}}{\text{s}^2\text{A}} \times \frac{1}{\text{s}} \rightarrow \frac{\text{m}}{\text{s}^2} \times \frac{1}{\text{m}} = \frac{1}{\text{s}} \times \frac{1}{\text{s}}$$

The divergence of the magnetic field:

$$\operatorname{div}\vec{B} = \frac{\partial B_x}{\partial x} + \frac{\partial B_y}{\partial y} + \frac{\partial B_z}{\partial z} = 0 \rightarrow \operatorname{div}\vec{B}_g = \frac{\partial B_{gx}}{\partial x} + \frac{\partial B_{gy}}{\partial y} + \frac{\partial B_{gz}}{\partial z} = 0$$

$$\frac{1}{\text{m}} \times \frac{\text{kg}}{\text{s}^2\text{A}} = 0 \rightarrow \frac{1}{\text{s}} \times \frac{1}{\text{m}} = \frac{1}{\text{s}} \times \frac{1}{\text{m}} = 0$$

The rotational magnetic field:

$$\overrightarrow{\text{rot}}\,\vec{B} = \begin{pmatrix} \frac{\partial B_z}{\partial y} - \frac{\partial B_y}{\partial z} \\ \frac{\partial B_x}{\partial z} - \frac{\partial B_z}{\partial x} \\ \frac{\partial B_y}{\partial x} - \frac{\partial B_x}{\partial y} \end{pmatrix} = \mu_0 \vec{j} + \varepsilon_0 \mu_0 \frac{\partial \vec{E}}{\partial t} = \mu_0 \begin{pmatrix} j_x \\ j_y \\ j_z \end{pmatrix} + \varepsilon_0 \mu_0 \begin{pmatrix} \frac{\partial E_x}{\partial t} \\ \frac{\partial E_y}{\partial t} \\ \frac{\partial E_z}{\partial t} \end{pmatrix} \rightarrow \overrightarrow{\text{rot}}\,\vec{B}_g$$

$$= \begin{pmatrix} \frac{\partial B_{gz}}{\partial y} - \frac{\partial B_{gy}}{\partial z} \\ \frac{\partial B_{gx}}{\partial z} - \frac{\partial B_{gz}}{\partial x} \\ \frac{\partial B_{gy}}{\partial x} - \frac{\partial B_{gx}}{\partial y} \end{pmatrix} = \frac{-4\pi G}{c^2}\vec{j}_g + \frac{1}{c^2}\frac{\partial \vec{E}_g}{\partial t}$$

$$\frac{1}{\mathrm{m}} \times \frac{\mathrm{kg}}{\mathrm{s^2A}} = \frac{\mathrm{kg.m}}{\mathrm{A^2s^2}} \times \frac{\mathrm{A}}{\mathrm{m^2}} + \frac{\mathrm{A^2.s^4}}{\mathrm{kg.m^3}} \times \frac{\mathrm{kg.m}}{\mathrm{A^2s^2}} \times \left(\frac{\mathrm{kgm}}{\mathrm{s^2As}}\right) \times \frac{1}{\mathrm{s}} \rightarrow \frac{1}{\mathrm{m.s}}$$
$$= \frac{\mathrm{m^3}}{\mathrm{kg.s^2}} \times \frac{\mathrm{s^2}}{\mathrm{m^2}} \times \frac{\mathrm{kg}}{\mathrm{s.m^2}} + \frac{\mathrm{s^2}}{\mathrm{m^2}} \times \frac{\mathrm{m}}{\mathrm{s^2}} \times \frac{1}{\mathrm{s}}$$

The electromagnetic wave equation from the electric field that is obtained by mixing the equations above together (unification of the electric field with the magnetic field = electromagnetic wave) or with the GravitoElectroMagnetism, the gravitational wave:

$$\Delta\vec{E} = \varepsilon_0\mu_0 \frac{\partial^2 \vec{E}}{\partial t^2} \rightarrow \Box\bar{h}_{\mu\nu} = 0$$

Source: Online course of Richard Taillet teacher-researcher at Savoie Mont blanc university "introduction to electromagnetism".

This speed depends on the permittivity ε_0 and the magnetic constant μ_0 of the vacuum. The vacuum, therefore, has electrical and magnetic properties. It is a physical object anyway. The vacuum (void) is no longer empty.

The vacuum is therefore characterized like any material by these two parameters ε_0 and μ_0

$\varepsilon_0 = 8.85418782 \times 10^{-12}\ \frac{\mathrm{A^2.s^4}}{\mathrm{kg.m^3}}$

$\mu_0 = 4\pi \times 10^{-7}\ \frac{\mathrm{kg.m}}{\mathrm{A^2s^2}}$

$$c = \frac{1}{\sqrt[2]{\varepsilon_0.\mu_0}} \rightarrow \frac{\mathrm{m}}{\mathrm{s}} = \frac{1}{\left(\frac{\mathrm{A^2.s^4}}{\mathrm{kg.m^3}} \times \frac{\mathrm{kg.m}}{\mathrm{A^2s^2}}\right)^{1/2}}$$

$$c = \frac{1}{\sqrt{8.85418782 \times 10^{-12} \times 4\pi \times 10^{-7}}} = 299\,792\,458\,\mathrm{m/s}$$

The electromagnetic wave equation is written:

$$\Delta\vec{E} = \frac{1}{c^2} \times \frac{\partial^2 \vec{E}}{\partial t^2}$$

Source: Online e-lecture of Richard Taillet teacher-researcher at Savoie Mont blanc university "introduction to electromagnetism".

Note: At that time it was believed that light like all waves needed a medium to propagate. He had been called (the supposedly immobile luminiferous Ether). In 1887, Michelson and Morley built an interferometer that bears their names today. Their idea was to measure whether the speed of light varies according to whether the earth moves in the direction of the ether or in the opposite direction. Their result is that whatever is the situation, the speed of light remains constant. The speed of light vector does not add (or subtract) from the speed of the moving body from which it is measured. The speed of light is an absolute constant, which at the time was a serious contradiction with Newton's mechanics and the law of composition of velocities. Indeed, if you walk at 10 km/h in a train moving at 100 km per hour, you are traveling at a total speed of 110 km/h. Well, this principle is no longer true if you move at the speed of light in the train! Your speed is no longer equal to the speed of the train + the speed of light, but remains frozen at the speed of light... How is that possible? Einstein then found the solution. A speed being a distance divided by a time, this implies in order to keep the speed of light constant that time slows down or distances contract. They are no longer absolute as Newton considered them to be, they depend on the observer's point of view. Special relativity was born.

1.4 Special Relativity

Thus A. Einstein in 1905 taught us that space and time are linked to create a space-time and that in this space-time nothing can go faster than light c (see Michelson and Morley's experiment at search for luminiferous ether {131–140}). Consequently when observers placed in a train which moves at speeds close to or not negligible compared to this speed c or on the platform compare their points of view on apparent simultaneities of events (sum of a position in the space and time), depending on the point of view in which they place themselves (on the platform or in the train), they do not reach the same conclusions on the notion of simultaneity of events. Their conclusions relate to their movement. Since we cannot exceed this speed c, since the speed of the moving observer cannot add up to the speed of light, then distances must contract and time slow down. At the same time, Einstein considers at this time that he no longer needs ether. Light is an electromagnetic wave which moves itself without support. He will come back to this idea by considering in 1920, during a conference in Leiden, and following discussions with Lorentz (1916), that there must still exist a kind of ether whose general relativity, in a way, calculates the geometric deformations *via* the metric $g_{\mu\nu}$ (see [27, 28]) but which, however, has nothing to do with the original luminiferous ether. This distortion of space-time then becomes gravity because it pushes the objects that follow this distorted metric to approach each other giving the illusion of an attractive force. But it is no longer the luminiferous ether that carries the light (it moves on its own, see J. Maxwell's equations) it is something else. Professor and physicist Thibault Damour gives this definition; this excellent image I quote: "space-time is an elastic structure which is deformed by the presence within it of mass energy" see [5]. He also explains very well in this same talk about the speed of light and gravitational waves I quote "the speed of sound for the elastic deformation of space in Einstein's theory is the speed of light; gravitational waves are waves of elastic deformations of space" see [5] and appendix G. We will see later why these two sentences are fundamental for what will follow in the unification of the two pillars of physics.

So, each has its own time relative to the other. Newton's Universal time is an illusion. He does not exist. When taking very precise measurements (*e.g.* atomic clock), the position of the clock in relation to the earth's gravity field (on earth or in space) is enough to desynchronize two identical and synchronized clocks (the one on earth slowed down compared to the one in space). It is the same if one clock is placed on an airplane and the other remains on earth. On return, they no longer indicate the same time. This is special relativity. So when we move at very high speeds (20% and more of the speed of light for example to feel something), the natural time t' of the one who moves slows down (expands) relatively to the natural time t of the one who does not move. The distances for the one who is moving appear to contract compared to the one who is at rest. So, time seems to have an elastic behavior! It expands.

$$\Delta t'(\text{in movement}) = \frac{1}{\sqrt{1-\left(\frac{v}{c}\right)^2}}\Delta t(\text{at rest}) => \text{Time dilation}$$

$$\Delta x'(\text{in movement}) = \sqrt{1-\left(\frac{v}{c}\right)^2}\,\Delta x(\text{at rest}) => \text{contraction of distances}$$

Source: references from this book {70, 71}.

Two events (characterizing what happens at a given time and at a given place of space, therefore at a given point in space-time) can begin together and end together, but not have the same duration. This is contrary to our intuition and our daily life: when we go with our colleague to the cafeteria, leaving and returning together, we are convinced that we have spent the same amount of time together (1 h for example). But this is only an illusion, if we measure time precisely with atomic clocks on the one hand, and if we move fast enough for a given period of time (in an airplane for example) on the other hand, ultra-precise clocks get out of sync. And this is a fact measured experimentally by physicists today. This is also the case with the paradox of Langevin's twins presented in 1911 where two twin brothers leave each other on a given date, one remaining on earth and the other going on a space to do a trip of 1 year for him in his rocket at a non-negligible speed compared to the speed of light then turns around and returns to earth where he finds his brother. As he moved at the speed of light, his own time slowed down, so he delayed his aging compared to that of his brother to stay on earth and so it was for example 1 year for him in space while 2 years have passed for his brother on Earth or time flies. The astronaut is therefore younger than the earthling. A resounding consequence of this is that if we superimpose the durations of the brother who remained on earth (2 years) and the brother who went to space (1 years) on the departure date, the time did not elapse in the same way, there are several times somehow. The durations for each of the astronauts are no longer equal. So, absolute time becomes an illusion. Special relativity nevertheless respects causality, the arrow of time. The cause has an effect which always occurs after the cause and not before. Thus, nothing propagates faster than the speed limit of light. If the sun goes out, we will not see and feel the effects until 8 min 20 after it is gone. The speed of light is not just about the speed of light. It turns out that this speed limit was first discovered in the case of light, but in fact any phenomenon whatsoever cannot propagate faster than the speed of light. The only apparent paradox in this is quantum entanglement where entangled quantum particles interact at great distance without exchanging information instantly, so *a priori* faster than the speed of light. But in fact, there again we cannot know the information of what happens on the second particle instantly if a time corresponding to that of a transfer of this information to our knowledge from the particle to the second does not exceed the speed of light. From special relativity arises the four-vector positions, the four-velocity vectors, the four-momentum vectors and finally Einstein's famous equation relating the energy to the mass of an object at rest $E = \gamma mc^2$. This equation which can be read from left to right translates that energy can be concentrated into a small amount of mass. Read from right to left that mass can be transformed into a tremendous amount of energy. This equation is verified in the large particle accelerators at CERN or by projecting protons at speeds close to the speed of light, from the shock born of new particles from the energy of their shocks that have produced mass. We have more mass after the shock than the initial mass that crashed into it! Energy of movement is transformed into mass! So, we do live in space-time and special relativity is just as strange as it is; the shifting of atomic clocks, the calibration of GPS in satellites revolving around the earth and the creation of particles in CERN shocks are there to prove it to us.

Special relativity

$$v(\mathrm{m/s}) = \frac{d(\mathrm{m})}{t(\mathrm{s})}$$

The speed of light is constant whatever the frame of reference considered.

You cannot move faster than this speed c.

No event can be transmitted faster than the speed of light c.

A simultaneous event in one frame of reference is not necessarily in another in movement with respect to the first.

Time expands at the speed of light so have aged less quickly $t' = \gamma t$.

Distances contract at the speed of light $L' = L/\gamma$.

This results in the Lorentz transformation to go from a stationary frame R' to a moving frame R in translation at the speed v.

Lorentz transformations for a following motion: x:

$$\begin{cases} ct' = \gamma\left(ct - \frac{vx}{c}\right) \\ x' = \gamma(x - vt) \\ y' = y \\ z' = z \end{cases} => \begin{pmatrix} ct' \\ x' \\ y' \\ z' \end{pmatrix} = \begin{pmatrix} \gamma & -\beta\gamma & 0 & 0 \\ -\beta\gamma & \gamma & 0 & 0 \\ 0 & 0 & 1 & 0 \\ 0 & 0 & 0 & 1 \end{pmatrix} \begin{pmatrix} ct \\ x \\ y \\ z \end{pmatrix}$$

$$\gamma = \frac{1}{\sqrt{1 - \left(\frac{v}{c}\right)^2}}$$

$$\beta = \frac{v}{c}$$

You have to make space-time diagrams with universe lines and cones of light to solve the problems.

Causality: $t_B - t_A > \frac{x_B - x_A}{c}$ will know what happened to A when the cone of light hits it; that is, when the light has traveled the distance between A and B. $t_B = (x_B - x_A)/c + t_A$.

We must consider the space-time => four vector position, speed, acceleration, momentum.

It follows from the relativistic invariants $ds^2 = c^2dt^2 - dx^2 - dy^2 - dz^2$.

$E = \gamma mc^2$ appears at the level of the time line of the four vectors.

Source: On line e-lecture of Richard Taillet teacher-researcher at Savoie Mont blanc university "special relativity".

An illustration of space-time is given in figure 1.1. The proof that gravitational waves move in a 4-dimensional space has been shown by {181}.

1.5 General Relativity

Newton's equation of gravitation not being satisfactory in Einstein's eyes since it assumes an instantaneous effect of the force of gravity as we said in the introduction, which is not possible according to special relativity where nothing can propagate

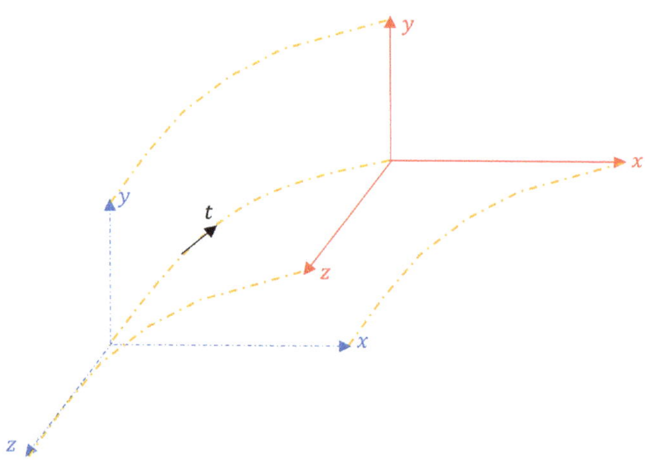

FIG. 1.1 – Space-time.

faster than the speed of light c, he proposes a new approach unifying the effects of gravity with special relativity: general relativity {18–31} and {52–56}. In 1915/1916, [8, 9] he postulated a tensor equation describing gravity which is no longer a force of gravity as seen by Newton, but the manifestation of the deformation of space-time (see appendix B for the measurements of deformations of this space which have been carried out for a century). This equation is valid whatever the frame of reference (use of tensors in the mathematical sense of the term) and comes in a multitude of differential equations making it hyper complex to understand and solve when starting from 0 (see an example of resolution in cosmology in appendix E). However, it can be summarized in a simple principle. Space-time is a physical object that is curved, distorted by the matter/energy in it. According to Wheeler J.A. "mass and energy tell space-time how to bend, and the curvature of space-time tells matter how to behave." Suddenly the objects move straight in space, but follow the curvatures or what amounts to the same deformations of this one. If the curve, called geodesic, forms a kind of cone (2-dimensional image), for example, when a very heavy object is placed in the center of a trampoline, a "pétanque" ball for example, and small balls are thrown on the trampoline deformed, we then have the impression that they revolve around the central object which created this curvature. If the speed of the objects slows down, they dive inward in a cone giving the illusion of a force which attracts them towards the massive object placed in the center of the trampoline. So in Einstein's approach, there is no more force (we can therefore wonder why scientists today from a fundamental point of view continue to consider that there are 4 forces in nature since gravitation is not one!!!) but a distorted space-time!!

General relativity or the incorporation of special relativity to gravitation

Curvature – angle/surface of space = $\kappa \times$ mass density/energy

$$G_{\mu\nu} = -\frac{8\pi G}{c^4} T_{\mu\nu}$$

$$\frac{1}{\mathrm{m}^2} = \frac{\mathrm{s}^2}{\mathrm{kg.m}} \times \left(\frac{\frac{\mathrm{kg.m}^2}{\mathrm{s}^2}}{\mathrm{m}^3} \right)$$

$$\frac{8\pi G}{c^4} = \frac{8 \times \pi \times 6.6742 \times 10^{-11}}{299\,792\,458^4} = 2.0766 \times 10^{-43} \frac{s^2}{\mathrm{kg.m}} = 2.0766 \times 10^{-43}\,\mathrm{N}^{-1}$$

$G_{\mu\nu} = R_{\mu\nu} - \frac{1}{2} g_{\mu\nu} R$ Einstein's tensor that reflects the curvature of space-time ($1/\mathrm{m}^2$).

$R_{\mu\nu}$ Ricci's tensor which is a tensorial contraction of Riemann's curvature tensor. Riemann's curvature tensor consists of the off-diagonal Weyl tensor and Ricci's tensor.

$g_{\mu\nu}$ metric (unknown to the equation and also gravitation).

R scalar curvature, a tensorial contraction of $R_{\mu\nu}$.

$\kappa = \frac{8\pi G}{c^4}$ Einstein's constant in s^2/Kgm or N^{-1} (subject of this book and of the paper by D. Izabel published in Pramana).

$T_{\mu\nu}$ the strain energy tensor (energy density) which translates the energy mass that curves space-time into $(kgm^2/s^2)/m^3$.

In multilinear algebra, contraction tensor is a computation process on tensors involving duality. It is possible to contract a unique tensor of rank p into a tensor of rank p-2, for example, by calculating the trace of a matrix. "Wikipedia".

Source: online e-lecture of Richard Taillet teacher-researcher at Savoie Mont blanc university "introduction to general relativity".

1.6 Black Holes

When the mass present in one place in space-time becomes extremely large and is concentrated in an extremely small volume, the curvature of space-time becomes extremely large, then we get a black hole. K. Schwarzschild was the first in 1916 {182} to find an exact solution to Einstein's equation. A visualization of what a black hole is can be done *via* beam theory (see appendix F).

Schwarzschild's solution gives the metric around a spherical star of mass M and radius r without rotation from the metric below. The metric is the set of terms that are in front of dt^2, dr^2, $d\theta^2$ and $d\varphi^2$ in the expression for the squared space-time interval ds^2. It is these terms that Einstein's equation can calculate the famous $g_{\mu\nu}$ or its inverse $g^{\mu\nu}$. The first image of the massive central black hole M87*of the galaxy M87 located at 53 millions light years from Earth, was established by the Event Horizon Telescope and presented on april, 10 2019. The magntic field of this same balck hole was presented on March, 24 2021 with this same telescope. The general relativity is again confirmed.

Schwarzschild's metric and the first non-rotating black hole:

$$ds^2 = B_{(r)}dt^2 - A_{(r)}dr^2 - r^2\left(d\theta^2 + \sin^2\theta d\varphi^2\right)$$

$$ds^2 = c^2\left(1 - \frac{2GM}{rc^2}\right)dt^2 - \frac{1}{\left(1 - \frac{2GM}{rc^2}\right)}dr^2 - r^2\left(d\theta^2 + \sin^2\theta d\varphi^2\right)$$

Or by posing as Schwarzschild's radius: $a = \frac{2GM}{c^2}$

$$ds^2 = c^2\left(1 - \frac{a}{r}\right)dt^2 - \frac{1}{\left(1 - \frac{a}{r}\right)}dr^2 - r^2\left(d\theta^2 + \sin^2\theta d\varphi^2\right)$$

Or we see that if $a = r$ you get a singularity or a black hole. Time stops and disappears, and space is torn apart. The diameter of the black hole becomes larger than its circumference (see [5, 7]).

Source: online e-lecture of Richard Taillet teacher-researcher at Savoie Mont blanc university "introduction to general relativity".

R. Kerr in 1963 subsequently found the formula of a black hole rotating moment kinetic J.

Boyer–Lindquist's coordinates

$$ds^2 = -\left(1-\frac{r_s r}{\Sigma}\right)c^2dt^2 + \frac{\Sigma}{\Delta}dr^2 + \Sigma d\theta^2 + \left(r^2+a^2+\frac{r_s r a^2}{\Sigma}\sin^2\theta\right)\sin^2\theta d\phi^2 - \frac{2r_s r a \sin^2\theta}{\Sigma}cdtd\phi$$

where

$$r_s = \frac{2GM}{c^2}$$

$$\Sigma = r^2 + a^2\cos^2\theta$$

$$\Delta = r^2 - \frac{2GM}{c^2}r + a^2$$

$$ds^2 = -\left(1-\frac{\frac{2GM}{c^2}r}{r^2+a^2\cos^2\theta}\right)c^2dt^2 + \frac{r^2+a^2\cos^2\theta}{r^2-\frac{2GM}{c^2}r+a^2}dr^2 + \left(r^2+a^2\cos^2\theta\right)d\theta^2 + \left(r^2+a^2+\frac{\frac{2GM}{c^2}ra^2}{r^2+a^2\cos^2\theta}\sin^2\theta\right)\sin^2\theta d\phi^2 - \frac{2\frac{2GM}{c^2}ra\sin^2\theta}{r^2+a^2\cos^2\theta}cdtd\phi$$

$$ds^2 = -\left(1-\frac{2GMr}{c^2(r^2+a^2\cos^2\theta)}\right)c^2dt^2 + \frac{(r^2+a^2\cos^2\theta)}{r^2-\frac{2GM}{c^2}r+a^2}dr^2 + (r^2+a^2\cos^2\theta)d\theta^2 + \left(r^2+a^2+\frac{2GMra^2\sin^2\theta}{c^2(r^2+a^2\cos^2\theta)}\right)\sin^2\theta d\phi^2 - \frac{4GMra\sin^2\theta}{c^2(r^2+a^2\cos^2\theta)}cdtd\phi$$

$$\mathrm{m}^2 = \left(\frac{\frac{\mathrm{m}^3}{\mathrm{kgs}^2}\mathrm{kgm}}{\frac{\mathrm{m}^2}{\mathrm{s}^2}(\mathrm{m}^2+\mathrm{m}^2)}\right)\frac{\mathrm{m}^2}{\mathrm{s}^2}\mathrm{s}^2 + \left(\frac{\mathrm{m}^2}{\mathrm{m}^2+\frac{\frac{\mathrm{m}^3}{\mathrm{kgs}^2}\mathrm{kgm}}{\frac{\mathrm{m}^2}{\mathrm{s}^2}}+\mathrm{m}^2}\right)\mathrm{m}^2 + (\mathrm{m}^2+\mathrm{m}^2) + \left(\mathrm{m}^2+\mathrm{m}^2+\frac{\frac{\mathrm{m}^3}{\mathrm{kgs}^2}\mathrm{kgmm}^2}{\frac{m^2}{s^2}(\mathrm{m}^2+\mathrm{m}^2)}\right) + \left(\frac{\frac{\mathrm{m}^3}{\mathrm{kgs}^2}\mathrm{kgmm}}{\frac{\mathrm{m}^2}{\mathrm{s}^2}(\mathrm{m}^2+\mathrm{m}^2)}\right)\frac{\mathrm{m}}{\mathrm{s}}\mathrm{s}$$

$$a = \frac{\frac{\mathrm{kgm}^2}{\mathrm{s}}}{\mathrm{kg}\frac{\mathrm{m}}{\mathrm{s}}} = \mathrm{m}$$

The kinetic moment per unit of mass:

$$j = \frac{J}{M} = \frac{\frac{\text{kgm}^2}{\text{s}}}{\text{kg}} = \frac{\text{m}^2}{\text{s}}$$

$$\bar{a} = \frac{c^2}{G} \times \frac{a}{M} = \frac{\frac{\text{m}^2}{\text{s}^2}}{\frac{\text{m}^3}{\text{kgs}^2}} \times \frac{\text{m}}{\text{kg}} = 1$$

$$\frac{\bar{a}GM}{c^2} = a = \frac{j}{c} = \frac{J}{Mc}$$

a is the kinetic moment of the black hole per unit of mass and per unit of speed:

$$\boldsymbol{a} = \frac{\boldsymbol{J}}{\boldsymbol{Mc}} = \frac{\frac{\textbf{kgm}^2}{\textbf{s}}}{\textbf{kg}\frac{\textbf{m}}{\textbf{s}}} = \textbf{m}$$

$$\frac{\boldsymbol{\bar{a}GM}}{\boldsymbol{c^2}} = \boldsymbol{a} = \frac{\boldsymbol{j}}{\boldsymbol{c}} = \frac{\boldsymbol{J}}{\boldsymbol{Mc}}$$

Source: Moore general relativity translated into French by Richard Taillet teacher-researcher at Savoie Mont blanc university edition deboeck, Wikipedia, conference T. Damour enigmatic black hole IHES November 7, 2019.

1.7 Gravitational Waves

Finally, in 1918, [25] Einstein linearized his equation by writing that the metric of space-time $g_{\mu\nu}$ is equal to Minkowski's flat metric $\boldsymbol{\eta_{\mu\nu}}$ corrected by a small perturbation $\boldsymbol{h_{\mu\nu}}$: $\boldsymbol{g_{\mu\nu} = \eta_{\mu\nu} + h_{\mu\nu}}$. He then showed that space-time is subject to gravitational waves such as the swell on the surface of the sea when extremely massive objects move there (coalescence of two black holes or two neutron stars for example).

Gravitational waves or elastic vibrations of space-time:

$$\boldsymbol{\bar{h}_{\mu\nu} = h_{\mu\nu} + \frac{1}{2}\eta_{\mu\nu}\bar{h}}$$

The linearized expression of Einstein's gravitational field equation is:

$$\partial^\lambda \partial_\lambda \bar{h}_{\mu\nu} = \Box \bar{h}_{\mu\nu} = -\frac{16\pi G}{c^4} T_{\mu\nu}$$

$\square$: The d'Alembertian

In the vacuum we have the equation of the gravitational wave:

$$\partial^{\lambda}\partial_{\lambda}\overline{h}_{\mu\nu} = \square\overline{h}_{\mu\nu} = 0$$

The solution is:

$$\overline{h}_{\mu\nu} = A_{\mu\nu}\cos(k_{\sigma}x^{\sigma})$$

$$\boldsymbol{\overline{h}_{\mu\nu} = A_{\mu\nu}e^{(ik_{\sigma}k^{\sigma})}}$$

k^{σ} The four-wave vector $\left(\frac{\omega}{c};\overrightarrow{k}\right)$

$$\|\overrightarrow{k}\|^{2} = \frac{\omega^{2}}{c^{2}};\, x^{\sigma} = (ct,\, x,\, y,\, z)$$

$\omega = 2\pi f$ the circular frequency of the wave

$$A_{\mu\nu} = A_{+}\begin{bmatrix} 0 & 0 & 0 & 0 \\ 0 & +1 & 0 & 0 \\ 0 & 0 & -1 & 0 \\ 0 & 0 & 0 & 0 \end{bmatrix} + A_{\times}\begin{bmatrix} 0 & 0 & 0 & 0 \\ 0 & 0 & +1 & 0 \\ 0 & +1 & 0 & 0 \\ 0 & 0 & 0 & 0 \end{bmatrix}$$

$$\eta_{\mu\nu} = \begin{bmatrix} 1 & 0 & 0 & 0 \\ 0 & -1 & 0 & 0 \\ 0 & 0 & -1 & 0 \\ 0 & 0 & 0 & -1 \end{bmatrix}$$

h: the trace of $h_{\mu\nu}$ with $\overline{h} = -h$

Source: e-lecture on line of Richard Taillet teacher-researcher at Savoie Mont blanc university "introduction to general relativity".

Gravitational waves were first measured on Ligo interferometers in the US on September 14, 2015 and the discovery was announced on February 11, 2016. The strains were in the range of $\pm 10^{-21}$ and corresponded to the coalescence of two black holes [23, 24].

1.8 Quantum Mechanics

The second, pillar, quantum mechanics {32–42} concerning the infinitely small, aims for its part to give differential equations making it possible to understand the quantified energy jumps of particles and the probabilities of presence of these wave/particle objects in a given volume. Indeed, it was first discovered that quantum particles, infinitely small, proceed by energy leaps, quantas, hence the name quantum mechanics.

In 1900, the study of the radiation of black bodies and the behavior of the ultraviolet allowed M. Planck to discover Planck's constant h (sometimes noted) $\hbar = \boldsymbol{h}/2\pi$).

U_ν is the energy of an oscillator (J).
u_ν is the energy density of the electromagnetic field (J.s/m^3).
k is Boltzmann's constant: $1.38064852 \times 10^{-23}$ J/K.
c the speed of light (m/s).
h the Planck constant $6.62607004 \times 10^{-34}$ kg m^2/s = (J.s).
ν frequency of the electromagnetic wave in (1/s = Hz).

$$u_\nu = \frac{8\pi\nu^2}{c^3} U_\nu$$

$$u_\nu = \frac{8\pi h}{c^3} \times \frac{\nu^3}{e^{h\nu/kT} - 1}$$

Source: The Great Ideas of Science – Max Planck and Quantum Physics Collection presented by E. Klein 2014.

In 1905, Einstein showed that the light which we took for a wave, only a wave, could behave in the photoelectric effect like particles, quanta of energy which we will call later in 1926 of photons. For this he received the Nobel Prize in 1921. The behavior of quantum particles is therefore not continuous but proceeds in small jumps according to their frequency.

The energy of quantum particles is quantified:

$$E = hf = h\nu = \hbar\omega$$

$$\mathrm{J} = \mathrm{J.s} \times \frac{1}{\mathrm{s}}$$

where:

Planck's constant $h = 6.62607004 \times 10^{-34}$ (kgm^2/s = J.s).
The frequency $f = \nu$ in s^{-1}.
The circular frequency of the wave in $\omega = 2\pi f(\frac{\mathrm{rad}}{s})$.
The energy $E = \mathrm{kgm^2/s^2}$ = (J).
Quantum particles also have a momentum:

$$p = \frac{hv}{c}$$

$$\frac{\frac{\text{kgm}^2}{\text{s}^2}.\text{s}\frac{1}{\text{s}}}{\frac{\text{m}}{\text{s}}} = \text{kg}\frac{\text{m}}{\text{s}}$$

p momentum in J.s/m.
c the speed of light in m/s.
f frequency ($f = v$) en s^{-1} (Hz).
The relativistic energy squared of a particle is therefore:

$$E^2 = m^2c^4 + p^2c^2$$

The first term is mass energy at rest, the second term energy related to the movement of the particle.

Source: see reference of this book {37–51} and J. Dalibard's lectures in quantum mechanics given in the 1st year of polytechnic in 2013.

In 1909 Einstein highlights this wave-particle duality {47–51} Light is both a wave and a particle. See A. Aspect's excellent lecture: "the photon wave or particle quantum strangeness brought to light" [6].

The thermal fluctuation radiation (Einstein on the present status of the radiation problem 15 March 1909 Physikalische Zeitschrift 185–193).

Planck's law is:
E energy (J = kgm^2/s^2).
h Planck's constant (J.s).
k Boltzman's constant (J/K).
V volume (m^3).
T température (Kelvin).
v frequency (1/s).

$$E = \frac{8\pi hv^3}{c^3} \times \frac{1}{e^{hv/kT}} Vdv$$

$$\text{J} = \frac{\text{J.s}}{\frac{\text{m}^3}{\text{s}^3}} \times \frac{1}{\text{s}^3} \times \frac{1}{e^{\frac{\text{J.s}\frac{1}{\text{s}}}{\frac{\text{J}}{\text{K}}\text{K}}}} \times \text{m}^3 \times \frac{1}{\text{s}}$$

The fluctuation of the thermal radiation $\overline{\varepsilon^2}$ in J^2 is:

$$\overline{\varepsilon^2} = hvE + \frac{c^3E^2}{8\pi v^2 Vdv} = kT^2\frac{\partial E}{\partial T}$$

$$\frac{\overline{\varepsilon^2}}{Vd\nu} = \frac{h\nu E}{Vd\nu} + \frac{c^3 E^2}{8\pi\nu^2 V^2 (d\nu)^2}$$

We define the spectral radiation density (J.s/m^3):

$$\rho = \frac{E}{Vd\nu}$$

$$\overline{\varepsilon^2} = \left(h\nu\rho + \frac{c^3\rho^2}{8\pi\nu^2}\right)Vd\nu$$

$$\mathrm{J}^2 = \left(\frac{\mathrm{J.s}}{\mathrm{s}} \times \frac{\mathrm{J.s}}{\mathrm{m}^3} + \frac{\frac{\mathrm{m}^3}{\mathrm{s}^3}}{\frac{1}{\mathrm{s}^2}} \times \frac{\mathrm{J}^2\mathrm{s}^2}{\mathrm{m}^6}\right)\mathrm{m}^3 \times \frac{1}{\mathrm{s}}$$

The first term is related to the particle aspect the second is related to the wave aspect.

Source: conference A. Aspect the photon waves or particles quantum strangeness highlighted (2019) friends of the IHES and {183–186}.

In 1913 N. Bohr proposed a new model of the atom. There is of course a nucleus, but the electrons orbiting around have privileged orbits with precise energies, they jump towards an outer orbit with quantities of quantified energy when there is an increase in energy and descend to lower levels when there is de-excitation of the atom. Each jump is made with an emission of photon and therefore light. Each jump is an integer multiple of Planck's constant h. We thus find the atomic spectra.

Bohr's Postulate

The orbits of the electrons are quantified

The circumference of the circle R should contain an entire number of times n the wavelength λ:

$$2\pi R = n\lambda$$

But the wavelength is written in function of the momentum p:

$$\lambda = \frac{h}{p}$$

So:

$$2\pi R = n\frac{h}{p}$$

But the momentum p is written:

$$p = mv$$

Hence the formula:

$$mvR = pR = n\frac{h}{2\pi} = n\hbar$$

where

R the orbit radius of the electron.
m electron mass in kg.
h Planck's constant.
n integers numbers.
v a speed.

The radius of the Bohr atom:

ε_0 The permittivity of the vacuum: $8.85418782 \times 10^{-12}$ m^{-3} kg^{-1} s^4 A^2.
m the electron mass (kg).
e the electrical charge of the electron in Coulomb: -1.6×10^{-19} C (A s).

$$R = n^2\frac{h^2\varepsilon_0}{\pi m e^2} = a_0 n^2$$

a_0 is called the Bohr radius for the hydrogen atom.

Numerical application:

$h = 6.62 \times 10^{-34}$ J.s (=kgm^2/s)
$\varepsilon_0 = 8.854 \times 10^{-12}$ m^{-3} kg^{-1} s^4 A^2
$e = q = 1.610 \times 10^{-10}$ A s
$m = 9.109 \times 10^{-31}$ kg
$a_0 = 5.29 \times 10^{-11}$ m or 0.529 Å Angstrom.

a_0 is called the first Bohr radius for the hydrogen atom since $R = a_0$ for $n = 1$ and $Z = 1$. A practical unit: The Angstrom 1 Å $= 10^{-10}$ m.

Bohr's energy

The energy variation between each orbital jump is written:

$$E_T = -\frac{E_0}{n^2} = -\left(\frac{e^4 m}{8\varepsilon_0^2 h^2}\right)\frac{1}{n^2} \quad E_0 = \left(\frac{e^4 m}{8\varepsilon_0^2 h^2}\right) = 13.6\,\text{eV} = 2.1710^{-18}\,\text{J}$$

Why do we see rays in the electromagnetic spectrum of hydrogen?

Between two orbits, two energy jumps there is a quantity, a quantum of energy $h\nu$:

$$|\Delta E_{n,p}| = |E_n - E_p| = h\nu$$

$$|\Delta E_{n,p}| = \left|-\frac{E_0}{n^2} + \frac{E_0}{p^2}\right| = E_0\left|\frac{1}{p^2} - \frac{1}{n^2}\right| = h\nu = h\frac{c}{\lambda}$$

So, the different wavelengths λ are with n and p integer numbers:

$$\frac{1}{\lambda} = \frac{E_0}{hc}\left|\frac{1}{p^2} - \frac{1}{n^2}\right| = R_H\left|\frac{1}{p^2} - \frac{1}{n^2}\right| \text{ with } n > p$$

Source: Lecture of T. Brierre L1 CHIM 110 atoms and molecules USTHB info.

In 1922 Stern and Gerlach have shown that quantum particles had a kind of internal kinetic moment a small internal magnetic field, a spin that caused quantum particles immersed in a magnetic field to sometimes go up and down depending on the sign of their spin.

It has been found that the bosons (force vector) have in whole spins:

- The potential graviton would be vector of gravitational force (spin 2, speed c) [17] which will later be found in mechanics *via* Mohr's circle.
- The Higgs boson is spin 0.
- Photons are vectors of electromagnetic interaction (spin 1).
- The eight gluons are vectors of strong interaction (spin 1).
- The Z0, W^{-0} and W^{-0} bosons are vectors of weak interaction (spin 1).

The fermions (electrons, protons, neutrons), on the other hand, are whole half spin (the electron has a spin of 1/2).

Then Young's slit experiment was performed with electrons pulled one by one through two slits. We noticed that after a large number of shots, the electrons were deposited on the screen placed behind the 2 slits according to interference fringes, like light or water does through two slits. Particles behaved like waves. As if the electrons interacted with them even through the two slits, such is the interpretation of Copenhagen in 1927. Another interpretation exists but is subject to discussion, the electrons would follow a pilot wave (approach of L. DeBroglie, D. Bohm) and would strike the screen at will or the pilot wave would have taken them. It looks a bit like a balloon (the electron) is floating on water passing through one of the slits and going to hit the screen as the waves move. We will come back to this approach much later in our book. This approach may not be all that wrong. Perhaps, a space-time material, made up of or interacting with gravitons, could influence the shots of particles on this scale. This wave particle duality [6] has been characterized by mathematical expressions of particles as wave packets. Both waves and particles are actually neither one nor the other.

De Broglie in 1923/1924 showed that particles such as electrons can actually behave like waves. The birth of matter waves. This was demonstrated in 1927 by Davisson and Germer on the one hand, then by G.P. Thomson and Ponte.

De Broglie matter wave approach

To any particle of m mass and velocity v, has a momentum $p = mv$, one can associate a wave of vector wave k:

$$\vec{k} = \frac{\vec{p}}{\hbar}$$

$$\frac{\text{kgm/s}}{\frac{\text{kgm}^2}{\text{s}^2\text{rad}}s} = \frac{\text{rad}}{\text{m}}$$

A wavelength λ in meters is also written:

$$\lambda = \frac{2\pi}{k} = \frac{2\pi\hbar}{p} = \frac{h}{p} = \frac{h}{mv}$$

Sources: see references of this book {37–51} and J. Dalibard's lectures in quantum mechanics given in the 1st year of polytechnic in 2013.

In 1925 W. Pauli demonstrated the principle of exclusion: two fermions (spin half integers like leptons (electrons, muons, tauons; neutrinos) and quarks) cannot occupy the same quantum state. On the contrary, the force vector bosons (whole spin, photons etc.) can (example of the laser).

In 1925 W. Heisenberg, M. Born, P. Jordan gave a matrix description of quantum mechanics.

If we have waves, or more exactly waves of probability of the presence of particles, then we need a differential equation reflecting the behavior of this wave. It was E. Schrödinger who discovered it in 1925/1926. His goal was to find functions describing the probabilities of finding quantum objects in rigid space that sometimes manifest as waves, sometimes as particles, but which are in fact neither. This equation predicted the different energy levels of the basic atom of the universe, the initial building block that is the hydrogen atom. It allows to find the wave function $\psi_{(\vec{r},t)}$ for a particle of mass m, of momentum p located at a radius $\vec{r} = (x_{(t)}, y_{(t)}, z_{(t)})$ at time t immersed in a potential (an energy) U: This differential equation is complex in the mathematical sense of the term because it involves $i^2 = -1$ (see X polytechnic lectures by J. Dalibard). It was therefore necessary to create wave functions whose mathematical formalism (Fourier series in exponential form) simultaneously translates this wave-particle duality.

The foundations of quantum mechanics

The wave function must also satisfy the Schrodinger equation:

$$i\hbar\frac{\partial}{\partial t}\psi_{(r,t)} = \hat{H}\psi_{(r,t)} = E_m\psi_{(r,t)} = \left[\frac{p^2}{2m} + U_{(r,t)}\right]\psi_{(r,t)}$$

$$= \left[-\left(\frac{\hbar^2}{2m}\right)\frac{\partial^2}{\partial r^2} + U_{(r,t)}\right]\psi_{(r,t)} = -\frac{\hbar^2}{2m}\frac{\partial^2\psi_{(r,t)}}{\partial r^2} + U_{(r,t)}\psi_{(r,t)}$$

The dimensional equation gives:

$$\frac{\mathrm{Js}}{\mathrm{s}} = \mathrm{J} = \mathrm{J} = \frac{\left(\mathrm{kg}\,\frac{\mathrm{m}}{\mathrm{s}}\right)^2}{\mathrm{kg}} + \mathrm{J} = \frac{\left(\frac{\mathrm{kgm}^2}{\mathrm{s}^2}\,\mathrm{s}\right)^2}{\mathrm{kgm}^2} + \mathrm{J}$$

The form of the function is for example in a problem has one dimension:

$$\psi_{(x,t)} = Ae^{i(kx+t)} = Ae^{i\left(\frac{p}{\hbar}x+\frac{E}{\hbar}t\right)} = A(\cos(kx+t) + i\sin(kx+\omega t))$$

In this expression of the wave function (whose square gives the density of probability of presence at a point of space and time), we do have a wave with the cos and sin structure and a particle having a certain energy in the term $\left(\frac{p}{\hbar}x + \frac{E}{\hbar}t\right)$.

Note: For a particle in a potential well in 1, 2 dimensions, the mathematical expression of the solution of the Schrödinger equation is identical to the strength of material formulas of a vibrating beam or plate (Eigen frequencies un eigen modes) (see {57}).

Sources: see references of this book {37–51} and J. Dalibard's lectures in quantum mechanics given in the 1st year of polytechnic in 2013.

In 1926, O. Klein and W. Gordon tried to introduce special relativity into quantum mechanics (what happens when quantum particles go at the speed of light?). He finds the Klein–Gordon equation. We will see later in quantum field theory that this equation governs particles of spin 0 of the scalar fields of the vacuum, the famous Higgs field Ψ which is the unknown of the problem becomes in quantum field theory a state vector belonging to at a Fock's space.

Klein–Gordon's equation is written:

$$\hbar^2 c^2\left(\Delta\psi_{(r,t)} - \frac{1}{c^2}\frac{\partial^2\psi_{(r,t)}}{\partial t^2}\right) - m^2c^4\psi_{(r,t)} = 0$$

$$\left(\Delta\psi_{(r,t)} - \frac{1}{c^2}\frac{\partial^2\psi_{(r,t)}}{\partial t^2}\right) = m \times \frac{mc^2}{\hbar^2}\psi_{(r,t)}$$

Dimensional equation:

$$\frac{1}{\mathrm{m}^2} = \frac{\mathrm{kg}^2 \times \frac{\mathrm{m}^2}{\mathrm{s}^2}}{\left(\frac{\mathrm{kgm}^2}{\mathrm{s}^2} \times \mathrm{s}\right)^2} = \frac{\mathrm{s}^2 \times \mathrm{kg}\,\frac{\mathrm{m}^2}{\mathrm{s}^2}}{\mathrm{kgm}^4} = \frac{\mathrm{s}^2}{\mathrm{kgm}^4} \times \mathrm{kg}\frac{\mathrm{m}^2}{\mathrm{s}^2} = \frac{\mathrm{s}^2}{\mathrm{kgm}} \times \frac{1}{\mathrm{m}^3} \times \mathrm{kg}\frac{\mathrm{m}^2}{\mathrm{s}^2}$$

Note: It is interesting to see that this equation has the dimension of a curvature and somehow translates that the curvature of the field (of the probability wave) depends on the masse/energy density present.

By posing the d'Alembertian $\square$:

$$\frac{1}{c^2} \times \frac{\partial^2}{\partial t^2} - \left(\frac{\partial^2}{\partial x^2} + \frac{\partial^2}{\partial y^2} + \frac{\partial^2}{\partial z^2}\right) = \frac{1}{c^2} \times \frac{\partial^2}{\partial t^2} - \Delta$$

Klein–Gordon's relativistic equation is written:

$$\hbar^2 c^2 \left(\Delta \psi_{(r,t)} - \frac{1}{c^2} \frac{\partial^2 \psi_{(r,t)}}{\partial t^2}\right) - m^2 c^4 \psi_{(r,t)} = 0$$

$$\left(-\hbar^2 c^2 \square - m^2 c^4\right) \psi_{(r,t)} = 0$$

Source: Wikipedia Klein–Gordon equation and reference of this book {43–51}.

In 1927 Mr. Born gave an interpretation of what the wave function was:

The probability P to find the quantum object at the moment t in a volume $d^3 r$ surrounding the point $\vec{r}$ *is* :

$$d^3 P = \left|\psi_{(\vec{r},t)}\right|^2 d^3 r$$

We are talking about density of probability (probality P divided by a volume V):

$$\frac{d^3 P}{d^3 r} = \left|\psi_{(\vec{r},t)}\right|^2$$

$$\iiint_{-a;-b;-c}^{a;b;c} \left|\psi_{(\vec{r},t)}\right|^2 d^3 r = 1$$

The integral on the volume of the wave function standardized square $\left|\psi_{(\vec{r},t)}\right|^2$ is worth 1: the particle is necessarily somewhere by in the radius sphere r.

So the goal of quantum mechanics is to determine the probabilities of the presence of a quantum object in Newtonian space and time and the associated energy levels.

Sources: see references of this book {37–51} and J. Dalibard's lectures in quantum mechanics given in the 1st year of polytechnic in 2013.

Heisenberg in 1927 demonstrated that it is not possible to know precisely the speed and position of a quantum object. This is Heisenberg's famous principle:

Heisenberg's principle:
It is impossible to know precisely the position $\boldsymbol{x}$ and the momentum p:

$$\Delta x \Delta p \geq \frac{\hbar}{2}$$

It is impossible to know precisely the position x and speed v:

$$\Delta x \Delta v \geq \frac{\hbar}{2m}$$

It is impossible to know precisely the energy E and the time t or we have this energy:

$$\Delta E \Delta t \geq \frac{\hbar}{2}$$

Sources: see references of this book {37–51} and J. Dalibard's lectures in quantum mechanics given in the 1st year of polytechnic in 2013.

Then in 1927, we wanted to know how quantum particles behaved at speeds close to the speed of light, relativistic quantum mechanics was born with P.A. Dirac. By doing this, its equation highlights the spins of particles on the one hand, and positive and negative energies on the other hand. He discovers antimatter equal to matter, but of opposite electrical sign. The positon (positive electron) was measured in cosmic rays by K.D. Anderson in 1932/1933.

$$i\hbar \frac{\partial \psi_{(x,t)}}{\partial t} = \left(\alpha_0 mc^2 - i\hbar c \sum_{j=1}^{3} \alpha_j \frac{\partial}{\partial x_j} \right) \psi_{(x,t)}$$

$$i\hbar \frac{\partial \psi_{(\vec{x},t)}}{\partial t} = \left[\alpha_0 mc^2 - i\hbar c \left(\alpha_1 \frac{\partial}{\partial x^1} + \alpha_2 \frac{\partial}{\partial x^2} + \alpha_3 \frac{\partial}{\partial x^3} \right) \right] \psi_{(\vec{x},t)}$$

where

$$\alpha_0 = \begin{pmatrix} 1 & 0 \\ 0 & -1 \end{pmatrix}; \vec{\alpha} = \begin{pmatrix} 0 & \vec{\sigma} \\ \vec{\sigma} & 0 \end{pmatrix}$$

where $\vec{\sigma}$ the Pauli's Matrix.

Source: "E. Wikipedia" Dirac equation.

In 1927, N. Borh, W. Heisenberg and M. Born defined the interpretation of Copenhagen.

In 1934, E. Fermi discovered the low source force of radioactivity (short distance action).

In 1935, Yukawa discovered strong force (short distance action) (linkage of nucleus constituents).

Then comes the quantum field theory developed between 1920 and 1950 where quantum particles are replaced by quantum fields which interact with each other *via* Boson vectors a bit like an exchange of balls (bosons) between two boats on a pond (the field), ends up pushing them back. This theory is based on the Lagrangian (difference between kinetic and potential energy) which is minimized, principle of least action, the shortest path that the particle will travel to spend as little energy as possible. The Hamiltonian is the associated energy.

We thus write the Lagrangian of the standard model which is the sum of the Lagrangians of each particle of force: Higgs + electromagnetic fields (Klein Gordon, scalar fields) (photon fields)+... Below, you will find:

– An Example with the Klein–Gordon equation found from a quantum field Lagrangian.

– An Example with the equation of Dirac that can be found from an other quantum field Lagrangian.

Either a ψ scalar field whose Lagrangian is written

$$\mathcal{L} = \frac{1}{2}(\partial_\mu \psi)(\partial^\mu \psi) - \frac{1}{2}\frac{m^2c^2}{\hbar^2}\psi^2$$

Dimensional equation:

$$\frac{1}{\mathrm{m}^2} - \frac{\mathrm{kg}^2\left(\frac{\mathrm{m}^2}{\mathrm{s}^2}\right)}{\left(\frac{\mathrm{kgm}^2}{\mathrm{s}^2}\mathrm{s}\right)\times\left(\frac{\mathrm{kgm}^2}{\mathrm{s}^2}\mathrm{s}\right)} = \frac{1}{\mathrm{m}^2}$$

$$\{\partial^\mu\} = \frac{\partial}{\partial x_\mu} = \begin{pmatrix} \frac{1}{c}\frac{\partial}{\partial t} \\ -\frac{\partial}{\partial x} \\ -\frac{\partial}{\partial y} \\ -\frac{\partial}{\partial z} \end{pmatrix} = \begin{pmatrix} \frac{1}{c}\frac{\partial}{\partial t} \\ -\vec{\nabla} \end{pmatrix}$$

$$\{\partial_\mu\} = \frac{\partial}{\partial x^\mu} = \begin{pmatrix} \frac{1}{c}\frac{\partial}{\partial t} \\ \frac{\partial}{\partial x} \\ \frac{\partial}{\partial y} \\ \frac{\partial}{\partial z} \end{pmatrix} = \begin{pmatrix} \frac{1}{c}\frac{\partial}{\partial t} \\ \vec{\nabla} \end{pmatrix}$$

$$\{\partial_\mu\} = \left(\frac{1}{c}\frac{\partial}{\partial t}, \frac{\partial}{\partial x}, \frac{\partial}{\partial y}, \frac{\partial}{\partial z}\right) = \left(\frac{1}{c}\frac{\partial}{\partial t}; \vec{\nabla}\right)$$

$$\{\partial_\mu\}\{\partial^\mu\} = \left(\frac{1}{c^2}\frac{\partial^2}{\partial t^2} - \frac{\partial^2}{\partial x^2} - \frac{\partial^2}{\partial y^2} - \frac{\partial^2}{\partial z^2}\right) = \left(\frac{1}{c^2}\frac{\partial^2}{\partial t^2} - \Delta\right) = \square$$

$$\mathcal{L} = \frac{1}{2}\left[\left(\frac{\partial\psi}{\partial x_0}\frac{\partial\psi}{\partial x_0}\right) - \sum_{\mu=1}^{3}\frac{\partial\psi}{\partial x_\mu}\frac{\partial\psi}{\partial x_\mu}\right] - \frac{1}{2}\frac{m^2c^2}{\hbar^2}\psi^2$$

To find the equation of movement (the shortest path between two known situations), we use the Euler–Lagrange equation (principle of least action):

$$\frac{\partial}{\partial x_\mu}\left(\frac{\partial\mathcal{L}}{\partial\left(\frac{\partial\psi}{\partial x_\mu}\right)}\right) - \frac{\partial\mathcal{L}}{\partial\psi} = 0$$

Or noted:

$$\frac{\partial}{\partial_\mu}\left(\frac{\partial\mathcal{L}}{\partial(\partial_\mu\psi)}\right) - \frac{\partial\mathcal{L}}{\partial\psi} = 0$$

We calculate the derivatives:
We define: $X_i = \frac{\partial\psi}{\partial x_\mu}$
We define: $(X_i^2)' = \frac{\partial\psi}{\partial x_\mu}\frac{\partial\psi}{\partial x_\mu} = 2X_i \times 1$

$$\left(\frac{\partial\mathcal{L}}{\partial(\partial_\mu\psi)}\right) = \left(\frac{\partial\mathcal{L}}{\partial\left(\frac{\partial\psi}{\partial x_\mu}\right)}\right) = \frac{1}{2} \times 2(\partial^\mu\psi) = (\partial^\mu\psi)$$

We calculate the derivative:

$$\frac{\partial}{\partial_\mu}\left(\frac{\partial\mathcal{L}}{\partial(\partial_\mu\psi)}\right) = \partial_\mu(\partial^\mu\psi)$$

We calculate the derivative:

$$-\frac{\partial\mathcal{L}}{\partial\psi} = \frac{m^2c^2}{\hbar^2}\psi$$

In the end:

$$\frac{\partial}{\partial_\mu}\left(\frac{\partial\mathcal{L}}{\partial(\partial_\mu\psi)}\right) - \frac{\partial\mathcal{L}}{\partial\psi} = 0$$

$$\partial_\mu(\partial^\mu\psi) + \frac{m^2c^2}{\hbar^2}\psi = 0$$

$$\frac{1}{c^2}\frac{\partial^2\psi_{(t,x,y,z)}}{\partial t^2} - \Delta\psi_{(t,x,y,z)} + \frac{m^2c^2}{\hbar^2}\psi_{(t,x,y,z)} = 0$$

$$\Box\psi_{(t,x,y,z)} + \frac{m^2c^2}{\hbar^2}\psi_{(t,x,y,z)} = 0$$

Which is the Klein–Gordon equation.
The solution is with

$$k_\mu = \pm k = \begin{pmatrix} \pm\omega/c \\ \overrightarrow{k} \end{pmatrix} = \begin{pmatrix} k_0 = \pm\omega_k/c \\ k_1 \\ k_2 \\ k_3 \end{pmatrix}$$

$$\psi_{(\vec{x},t)} = e^{-ik_\mu x^\mu} = e^{-i(k_0x^0+k_1x^1+k_2x^2+k_3x^3)} = e^{-i(k_0ct+k_1x+k_2y+k_3z)} = e^{-i\left(\omega t+k_xx+k_yy+k_zz\right)}$$
$$= e^{-i\left(\omega t+k_jx^j\right)} = e^{-i\left(\omega t+\overrightarrow{k}\,.\,\overrightarrow{x}\right)}$$

Note: The mathematic structure of this solution is identical at the structure of the gravitational wave equation (see $\bar{h}_{\mu\nu}$). The difference is that the Klein–Gordon equation admits a scalar solution (wave function or quantum field) for a massive particle of spin 0 (ex Higgs Boson) while the linearized solution of Einstein's field equation, which leads to gravitational waves with two polarizations, is tensorial type for a hypothetical gravitational boson of spin 2 (graviton) without mass (theoretically) since moving at the speed of light.

Reference sources see references of this book {72–88}.

Either the Lagrangian associated with Dirac's equation:

$$\mathcal{L} = i\hbar c\left[\overline{\psi}\gamma^\mu(\partial_\mu\psi)\right] - mc^2\overline{\psi}\psi$$

To find the equation of movement (the shortest path between two known situations), we use the equation of Euler–Lagrange:

$$\frac{\partial}{\partial x_\mu}\left(\frac{\partial\mathcal{L}}{\partial\left(\frac{\partial\overline{\psi}}{\partial x_\mu}\right)}\right) - \frac{\partial\mathcal{L}}{\partial\overline{\psi}} = 0$$

We calculate the derivative:

$$\left(\frac{\partial\mathcal{L}}{\partial(\partial_\mu\overline{\psi})}\right) = \left(\frac{\partial\mathcal{L}}{\partial\left(\frac{\partial\overline{\psi}}{\partial x_\mu}\right)}\right) = 0$$

We calculate the derivative:

$$\frac{\partial}{\partial_\mu}\left(\frac{\partial \mathcal{L}}{\partial(\partial_\mu \overline{\psi})}\right) = 0$$

We calculate the derivative:

$$-\frac{\partial \mathcal{L}}{\partial \overline{\psi}} = -i\hbar c\left[\gamma^\mu(\partial_\mu \psi)\right] + mc^2\psi$$

In the end we obtain:

$$\frac{\partial}{\partial_\mu}\left(\frac{\partial \mathcal{L}}{\partial(\partial_\mu \psi)}\right) - \frac{\partial \mathcal{L}}{\partial \psi} = 0$$

We can find Dirac's equation:

$$-i\hbar c\left[\gamma^\mu(\partial_\mu \psi)\right] + mc^2\psi = 0$$

$$\boldsymbol{ic\gamma^\mu \partial_\mu \psi - \frac{mc^2}{\hbar}\psi = 0}$$

$$\boldsymbol{i\gamma^\mu \partial_\mu \psi - \frac{mc}{\hbar}\psi = 0}$$

Which is the Dirac equation.
The dimensional equation is:

$$\boldsymbol{\frac{1}{\mathrm{m}} = \frac{\mathrm{kg} \times \frac{\mathrm{m}}{\mathrm{s}}}{\frac{\mathrm{kgm}^2}{\mathrm{s}}}}$$

See source see references of this book {72–92}.

The 3 forces approach in quantum field theory is very well explained in the conferences of E. Klein on the Higgs boson. A summary is given below:

Each quantum object (neither particle nor wave but manifesting itself according to one or the other according to physics experiments) is therefore represented by a quantum field whose mathematical expression is its own Lagrangian. Once we know the Lagrangian associated with this quantum field everywhere present in space (scalar (Higgs, Klein Gordon) or vectorial or tensor (graviton)), then it becomes possible to know all the effects, energies and forces that is to say all the interactions that this field can exert *via* force vector particles which are specific to it.

Example when we have the Lagrangian of the electromagnetic field, it is associated with the photon which is the vector of the repulsion interaction force of the field consisting of electrons for example.

The scalar Lagrangian of the Higgs field, at the time of the big bang, interacted with the particles initially massless to give them more or less mass *via* the Higgs boson.

We know that we have the right Lagrangian representative of the field and its associated force vector particles when, when a local gauge transformation is applied to it, it remains invariant; that is, unchanged. To remain invariant, the field must have interactions, exchanges of information with its environment *via* particles mediating its information or forces, the famous force vector bosons.

To build this kind of invariant Lagrangian by local gauge transformation, it is necessary to take a non-invariant starting Lagrangian, apply a local gauge transformation to it, see the residual term in addition which appears during this gauge transformation, and compensate exactly for this term by adding to it a new mathematical term an interaction which one calls a gauge field (and the associated particles) which will compensate, cancel this complementary term, thus making the Lagrangian invariant. It is this complementary term which is the field of the mediating particle vector of the force of interaction.

This complementary term is associated with a gauge group or symmetry group. For the electromagnetic force, it is the SU (1) group. For the weak nuclear interaction, it is the SU (2) group a 2*2 matrix and for the strong nuclear interaction, it is the SU (3) group a matrix 3*3. Interactions between particles therefore take place through exchanges of particles called bosons, much like exchanges of balls between two boats on a pond allowing the direction of the boats to be changed by energy transfers at each reception and throwing of balls.

The same approach, however, fails with the field of the graviton vector of the force of gravity.

In 1948 R. Feynman solves Quantum Electro Dynamics (QED) developed in 1920 with P. Dirac. R. Feynman in 1950/1965 studies the interaction between electric and magnetic fields within the framework of quantum mechanics associated with special relativity where the photon the grain of light becomes a boson vector of the electromagnetic force. He develops the integrals of the path built on the principle of least action. When two particles (the electrons here) bump into each other, they produce several other types or possibilities of Hazard particles. They transmit their energy by photons (boson vector of the electromagnetic force). We then establish a Feynman diagram by the type of result that can give this collision of two electrons. We try to find all these diagrams to characterize the collision. Each diagram is associated with a mathematical expression characteristic of what the collision could be. After adding the contributions of all the diagrams, it is possible to calculate the probability of each outcome that may result from the collision and compare it to the experimental data. In fact, in colliders there are thousands of collisions and we know that such a phenomenon occurs x times out of a total of y collisions obtained, which is the probability of obtaining x/y of having this collision mode.

In the quantum vacuum, the Heisenberg principle allows that there are creations and annihilations of a very short duration of particles that are called virtual (they have an energy), but it lasts only a fraction of a second to respect the principle. All his creations and annihilations of particles lead to phenomena that can be measured as Casimir's force in 1948 {60–68}, which brings two metal plates closer to one of the other in a vacuum (more particle and equivalent wave form – particle wave duality – outside the plates than between the plates, suddenly the shocks of the more numerous external particles make them approach) it is the same thing if we put two glass plates in water and cause waves. As the plates in a vacuum get closer, we measured it, to be sure the vacuum is not empty. It is indeed filled with all these virtual particles.

The unification of quantum electrodynamics with weak force leads to the electroweak force explaining beta radioactivity in action within the sun.

Then comes (1965–1975) the strong force studied *via* the Quantum Chromo Dynamics (QCD) in 1973 which is the force which ensures the binding of the quarks constituting the ultimate (to date!) of the protons and neutrons. The standard particle model was born. In this world of quantum field, we discover that the interstellar vacuum is not empty, that there is a non-zero ground state of energy made up of virtual particles that create and cancel each other out.

The vacuum can be seen as a harmonic oscillator:

$$E_n = \hbar\omega\left(n + \frac{1}{2}\right)$$

Not-zero fundamental state, when $n = 0$

$$E_0 = \frac{1}{2}\hbar\omega$$

Sources: see references from this book {37–39}.

In 1964 there is also a Higgs scalar field {43–56} which at the very beginning of the universe gave mass to particles, it is the famous Higgs Boson discovered in 2012. Thus, it is the interaction of particles with this Higgs field that gives them mass.

Either a Lagrangian density of the form:

$L = T - U_{(\phi)} = \frac{1}{2}(\partial_\mu \phi^*)(\partial^\mu \phi) - \mu^2 \phi^* \phi - \lambda(\phi^* \phi)^2$

In this expression, μ takes the same definition as in the Lagrangian equation that allows to obtain Klein–Gordon's equation:

Lagrangian:

$$\boldsymbol{\mathcal{L} = \frac{1}{2}(\partial_\mu \psi)(\partial^\mu \psi) - \frac{1}{2}\frac{m^2c^2}{\hbar^2}\psi^2}$$

Klein Gordon's Equation:

$$-\Delta\psi + \frac{1}{c^2}\frac{\partial^2\psi}{\partial t^2} + \mu^2\psi = 0$$

where

For the light $\mu = 0$.
For particles of matter $\mu = \frac{2\pi mc}{h} = \frac{mc}{\hbar}$.
μ^2 is therefore a mass term related to the mass of the field.
λ represents the (positive) coupling constant in relation to its self-interaction.
In this expression the field ϕ is the Higgs field and is worth:

$$\phi = \begin{pmatrix} \phi^+ \\ \phi^0 \end{pmatrix} = \frac{1}{\sqrt{2}} \begin{cases} \phi_a + i\phi_b \\ \phi_c + i\phi_d \end{cases} = \begin{cases} R\phi^+ + iI\phi^+ \\ R\phi^0 + iI\phi^0 \end{cases}$$

R is the real part and i the imaginary part of the field.
ϕ^* is the conjugate complex field of ϕ:

$$\phi^* = \begin{cases} R\phi^+ - iI\phi^+ \\ R\phi^0 - iI\phi^0 \end{cases}$$

As well:

$$\phi^*\phi = (R\phi^+ + iI\phi^+) \times (R\phi^+ - iI\phi^+) = (R\phi^+)^2 + (I\phi^+)^2 = |\phi|^2$$

So, we can rewrite the potential energy in the form:

$$U_{(\phi)} = \mu^2|\phi|^2 + \lambda|\phi|^4$$

And so, the Lagrangian in the form:
$L = T - U_{(\phi)} = \frac{1}{2}(\partial_\mu\phi^*)(\partial^\mu\phi) - \mu^2|\phi|^2 - \lambda|\phi|^4$
So, there is invariance of the Lagrangian:
You get the same result if you replace ϕ with $-\phi$.
The potential is therefore invariant by symmetry.
It is noted that this form is similar to that established for the buckling of the post, a column in compression in strength of material (see J. Iliopoulos book the origin of mass):

$$E = C_0 + C_1\|\vec{\delta} \cdot \vec{\delta}\| + C_2\|\vec{\delta} \cdot \vec{\delta}\|^2 + \cdots$$

$$U_{(\phi)} = \phi^*\phi - \lambda(\phi^*\phi)^2$$

$$U_{(\phi)} = \mu^2|\phi|^2 + \lambda|\phi|^4$$

The minimum of potential is achieved by looking for when the derivative of the potential cancels out:

$$\frac{dU_{(\phi)}}{d|\phi|} = 2\mu^2|\phi| + 4\lambda|\phi|^3 = 0$$

$$\frac{dU_{(\phi)}}{d|\phi|} = |\phi|\left(\mu^2 + 2\lambda|\phi|^2\right) = 0$$

where $|\phi| = 0$

$$|\phi|^2 = -\frac{\mu^2}{2\lambda}$$

We ask:

$$|\phi|^2 = -\frac{\mu^2}{2\lambda} = \frac{v^2}{2}$$

It is necessary that:

$$v^2 = -\frac{\mu^2}{\lambda} > 0$$

The minimum of the potential ϕ noted is therefore obtained for: ϕ^0

$$\phi_{\text{minimum}} \rightarrow v = \pm\sqrt{-\frac{\mu^2}{\lambda}}$$

$$|\phi|^0 = \frac{v}{\sqrt{2}}$$

v being the minimum potential is associated with the vacuum:
We draw this potential: $U_{(\phi)}$

$$U_{(\phi)} = U_{(\phi)} = \mu^2|\phi|^2 + \lambda|\phi|^4$$

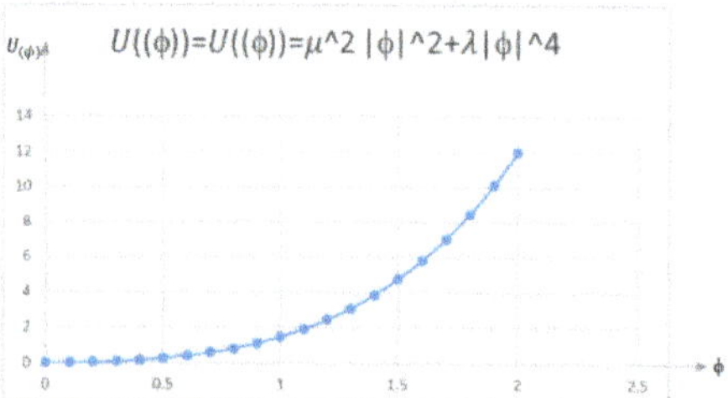

Potential curve $U_{(\phi)}$ for $\mu^2>0$ - stable post condition -

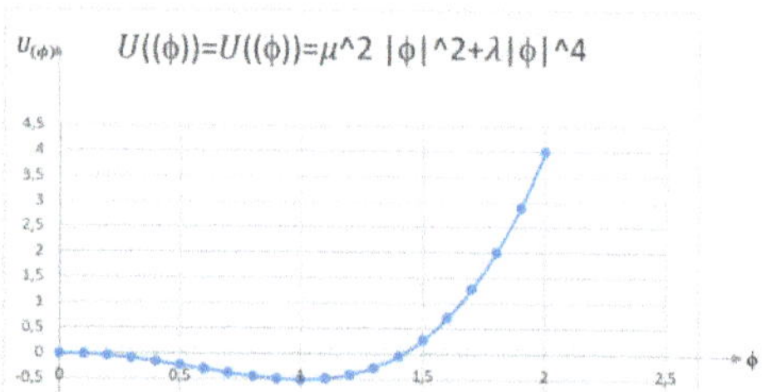

Potential curve $U_{(\phi)}$ for $\mu^2<0$ - unstable post condition -

When μ^2 is >0 then it is shaped like a glass of champagne. In this case the vacuum takes on a stable value in $v = 0$.

When $\mu^2 < 0$ then it has the shape of a Mexican hat.

The minimum of potential is achieved with:

$$\phi_{\text{minimum}} \rightarrow v = \pm\sqrt{-\frac{\mu^2}{\lambda}}$$

So, terms $v^2\lambda = \mu^2$ are mass terms.

$$|\phi|^0 = \frac{v}{\sqrt{2}}$$

v being the minimum potential is associated with the vacuum (average value between creations and particle annihilation). So, we note:

$$\phi^0 = \frac{1}{\sqrt{2}}\begin{pmatrix} 0 \\ v \end{pmatrix} \rightarrow \frac{v}{\sqrt{2}}$$

v is the minimum of potential.

So, the Higgs field is the sum of the fundamental field of vacuum and a complementary field:

$$\phi_{(x)} = \phi^0 + \beta_{(x)}$$

That we can write again:

$$\phi = \begin{pmatrix} \phi^+ \\ \phi^0 \end{pmatrix} = \frac{1}{\sqrt{2}}\begin{cases} \phi_a + i\phi_b \\ \phi_c + i\phi_d \end{cases} = \frac{1}{\sqrt{2}}\begin{cases} 0 \\ v + \beta_{(x)} + ih_{(X)} \end{cases}$$

So ultimately is a complex field of form: $\phi_{(X)}$

$$\phi_{(X)} = \frac{1}{\sqrt{2}}[\text{Real} + i\text{Imag}] = \frac{1}{\sqrt{2}}\left[\left(v + \beta_{(X)}\right) + i\left(h_{(X)}\right)\right]$$

$$\phi^*{}_{(X)} = \frac{1}{\sqrt{2}}[\text{Real} - i\text{Imag}] = \frac{1}{\sqrt{2}}\left[\left(v + \beta_{(X)}\right) - i\left(h_{(X)}\right)\right]$$

$$X \rightarrow (x, y, z, t)$$

$\beta_{(X)}$ and $h_{(X)}$ are two infinitesimal perturbations of the field to the vicinity of v. v is the minimum of potential and is an independent constant X.

As the field is symmetrical, we could also have written:

$$\phi_{(X)} = \frac{1}{\sqrt{2}}[\text{Real} + i\text{Imag}] = \frac{1}{\sqrt{2}}\left[-\left(v + \beta_{(X)}\right) + i\left(h_{(X)}\right)\right]$$

To estimate the effect of Spontaneous symmetry breaking (equivalence with a buckling of a column in strength of material), one must study what happens in the vicinity of v. The Lagrangian is written:

$$L = T - U_{(\phi)} = \frac{1}{2}(\partial_\mu\phi^*)(\partial^\mu\phi) - \mu^2\phi^*\phi - \lambda(\phi^*\phi)^2$$

$$L = T - U_{(\phi)}$$
$$= \frac{1}{2}(\partial_\mu \beta)(\partial^\mu \beta) + \frac{1}{2}(\partial_\mu h)(\partial^\mu h) - \mu^2\left(\frac{1}{\sqrt{2}}[(v+\beta)+i(h)]\frac{1}{\sqrt{2}}[(v+\beta)-i(h)]\right)$$
$$- \lambda\left(\frac{1}{\sqrt{2}}[(v+\beta)+i(h)]\frac{1}{\sqrt{2}}[(v+\beta)-i(h)]\right)^2$$

$$L = T - U_{(\phi)}$$
$$= \frac{1}{2}(\partial_\mu \beta)(\partial^\mu \beta) + \frac{1}{2}(\partial_\mu h)(\partial^\mu h) - \frac{\mu^2}{2}[(v+\beta)+i(h)][(v+\beta)-i(h)]$$
$$- \frac{\lambda}{4}([(v+\beta)+i(h)][(v+\beta)-i(h)])^2$$

$$L = T - U_{(\phi)}$$
$$= \frac{1}{2}(\partial_\mu \beta)(\partial^\mu \beta) + \frac{1}{2}(\partial_\mu h)(\partial^\mu h) - \frac{\mu^2}{2}\left[v^2 + 2v\beta + \beta^2 + h^2\right]$$
$$- \frac{\lambda}{4}\left(\left[(v^2 + 2v\beta + \beta^2) + h^2\right]\right)^2$$

If we are interested in the h-free part: We get:

$$-\frac{\mu^2}{2}\left[v^2 + 2v\beta + \beta^2\right] - \frac{\lambda}{4}(v^4 + 4v^3\beta + 6v^2\beta^2 + 4v\beta^3 + 2v\beta + \beta^4)$$

$$-\left[\frac{v^2}{2}\left(\mu^2 + \frac{1}{2}v^2\lambda\right) + \beta v(\mu^2 + \lambda v^2) + \frac{\beta^2}{2}(\mu^2 + 3\lambda v^2) + \lambda v\beta^3 + \frac{1}{2}\lambda\beta v + \frac{1}{4}\lambda\beta^4\right]$$

The following terms are interactions:

$$\lambda v\beta^3 + \frac{1}{2}\lambda\beta v + \frac{1}{4}\lambda\beta^4$$

And there is a mass term:

$$\frac{\beta^2}{2}(\mu^2 + 3\lambda v^2)$$

In addition, there are the terms $v^2\lambda = \mu^2$ in which are terms of mass. The interaction of the vacuum with the Higgs boson creates mass.

Sources: see references from this book {124–130}.

In 1974 S. Hawking discovered the Hawking radiation leading to the evaporation of black holes when two virtual particles of matter and antimatter appear in a vacuum (the vacuum is zero on average but is not empty as such), if one of them is too close to the event horizon of a black hole (the sphere of radius r = Schwarzschild radius) and penetrates into the other remains and seems to emerge to radiate from the black hole.

Hawking Radiation

$$T_H = \frac{\hbar c^3}{8\pi k_B G M} = \frac{\frac{\mathrm{kgm}^2}{\mathrm{s}^2} \times \mathrm{s} \times \frac{\mathrm{m}^3}{\mathrm{s}^3}}{\frac{\mathrm{kgm}^2}{\mathrm{Ks}^2} \times \frac{\mathrm{m}^3}{\mathrm{kgs}^2} \times \mathrm{kg}} = K$$

T_H Hawking temperature (K).
M the mass of the black hole (kg).
k_B Boltzmann's constant $\mathrm{m^2kg/(K.s^2)}$.
G constant gravitation $\mathrm{m^3/(kg.s^2)}$.
c speed of light (m/s).

Source: Wikipedia Hawking radiation.

1.9 Highlighting the Differences between the Two Pillars of Physics

Having made this summary, we better understand the differences between general relativity on the one hand, the quantum world on the other. General relativity assumes a deformable space-time, while it is rigid in the quantum world (see appendix B to see the space deformations that have been measured). The universe seems to vibrate like a crystal according to {100–106} and is traversed by gravitational waves {93, 94} as any elastic solid [5].

Gravity is not a force, it is a geometric and therefore physical deformation of space whereas in quantum field theory there are 3 forces, magnetic force, electroweak force and strong force that results in virtual particle exchanges Bosons. An alternative is graviton that would convey the force of gravity. But this Boson of spin 2 is hypothetical now because not measured.

The world is continuous in general relativity and proceeds by quanta, by discontinuous leap in quantum field theory.

The position and velocity of mass or energy in general relativity is precisely predicted, whereas in quantum mechanics, the result is a field equation giving the probability of the presence of particles in a particular place sometimes even in several places at the same time. Particles in quantum mechanics can interact remotely when they have been entangled together without any exchange of information, which is impossible in general relativity!

To this is added the knowledge brought by general relativity when applied to the whole universe by considering it homogeneous and isotropic. It is Friedmann–Lemaitre's equations (see appendix E) that lead to the big bang and radiation of the cosmological diffuse background. The space is dynamic.

Then Hubble's law that shows us that space is expanding {141}. Then to the acceleration of the expansion of the universe requiring a dark energy and a possible cosmological constant Λ to add to Einstein's equation of the general relativity. This constant can be considered to the left of Einstein's equation, in this case it should be an additional constant curvature. The other assumption is that this constant must be considered on the right side, in this case it will be an additional repulsive energy (opposite to gravitation).

The speed of rotation of stars on the periphery of galaxies should decrease so as not to expel said stars by rotation. We measure constant (and therefore high) velocities along the diameter of the galaxy disk. The stars therefore seem to be held in place by a gravity stronger than predicted by Newton's law unless one is mistaken about the present mass and that Newton's gravity is exercised not only by the mass of visible stars, gas, and dusts, but also by the mass of an unknown dark matter (see figures 1.2a and 1.2b).

The other hypothesis is that Newton's law changes on a very large scale (see MoND theory Modified Newtonian dynamics).

The last assumption that it remains is that G is not a constant (variation in the time? Other ($G_{(\rho)}$ with $\rho_{(R)}$ [2]?) see figure 1.2b). But it is not demonstrated and proven.

The observation of the universe on a large scale (figure 1.3) resembles a kind of plastic foam made up of closed cells with filaments and organizations of billions of galaxies in structure (see photos of the Hubble deep field and see Max Planck Institute for Astrophysics, Millennium Simulation Project) which is possible again only if this unknown dark matter condenses galaxies according to these filaments and structure.

Interestingly, recent research has found a lot of similarities between the human brain frame and the large scale universe frame (see {193}).

Then we end up with moments (big bang) or objects (Black holes) in the universe where its two pillars must be united:

The big bang or the masses, temperatures and energies are gigantic but concentrated in an infinitesimal point. The Cosmic microwave background {96–99} and {115–117} thus gives us this image of the universe 280 000 years after the big bang or quantum fluctuations of this field in line with dark matter seem to have given us

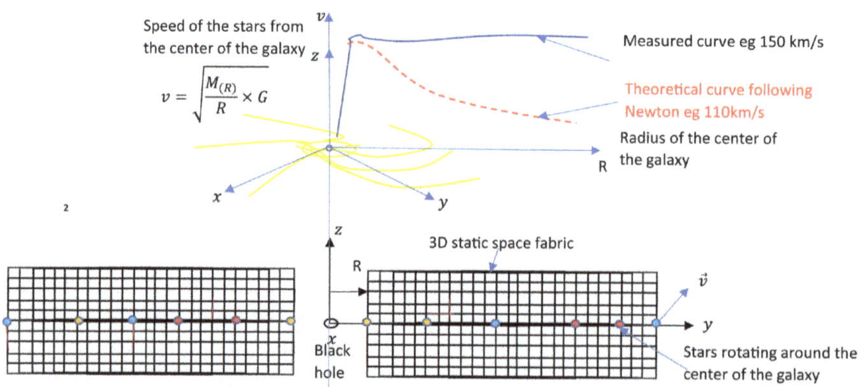

FIG. 1.2a – Typical theoretical and measured curves of the stars velocities of a galaxy. Source (https://www.futura-sciences.com/sciences/definitions/astronomie-theorie-mond-12229/).

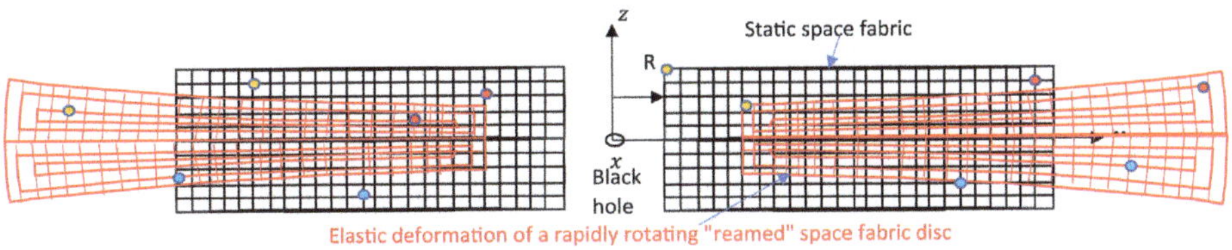

FIG. 1.2b – Deformation of an elastic galactic disk rotating around the axe z [2, 64].

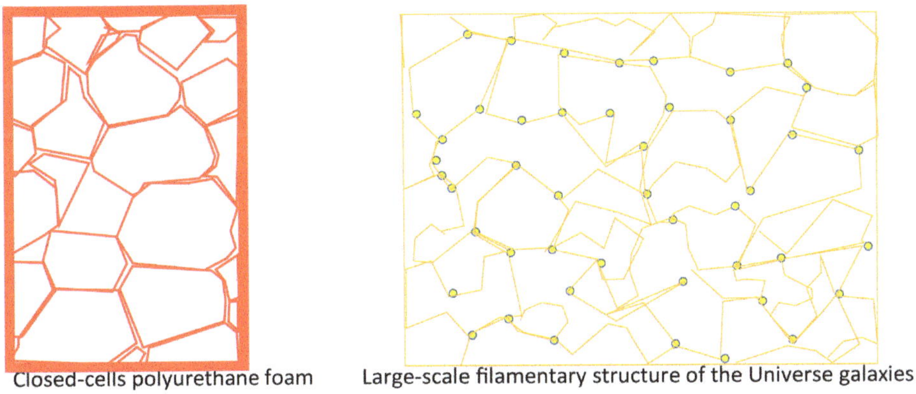

FIG. 1.3 – Comparison of the texture of a polyurethane foam and the texture of the universe on a large scale.

the galaxies of today. Friedmann–Lemaitre's equations (see appendix E) allow us to estimate the age of the universe at 13.7 billion years {67, 68}.

In summary, how do we reconcile these two pillars of physics (see appendix C for the synthesis of key formulations)? This is the quest of physicists today [55].

String theory {107–114}, or quantum gravity, is an avenue of solutions. In this book, we will propose another, simpler, more pragmatic approach guided by a single principle: Understanding how it works, how all the pieces of this puzzle can intertwine, and whether a new ether different from that of the luminiferous ether can be born {131–133}.

Others try to introduce torsion into the equations of general relativity it is the theory of Einstein–Cartan or to take into account the torsion of space-time by the rotation of the earth (satellite Probe B in 2011) {142–170}.

Nature plunges us into multiple illusions constantly, and physics is there to allow us to see clearly.

1.10 Nature Plays with Our Senses

One point is fundamental. Nature does not stop disturbing our senses and delving into illusions that must be constantly foiled.

- It was believed that the bodies fell the faster they were dense. A kg of feather falls to earth slower than a kg of lead. But this is only an illusion (well shown on the moon during an Apollo 15 mission in 1971 where the hammer and the feather hit the ground at the same time). All bodies fall at the same speed in a field of pure gravity in a vacuum to an accuracy of 15 decimal places as shown by the microscope mission. But when they fall in the air, the friction of the air slows down the feather, giving us the impression that the more massive an object is, the faster it falls. Galileo understood this.
- When we suck water from a glass through a straw, we are convinced that we make the water rise by vacuum suction, but in fact it is the pressure of the atmosphere on the surface of the water in the glass which pushes the water upwards in our mouth, balances the pressures.
- Electricity and magnetism were thought to be different phenomena, but in fact they only form two visions, two different points of view of the same phenomenon: electromagnetism.
- We thought there were only light waves, but in fact the field of electromagnetic waves is much larger, it was Maxwell who taught us.
- Newton believed that there was a force of gravity that made objects attract each other. But in fact, Einstein showed that objects follow the curvature of the field of gravity giving us the illusion of being attracted by a force. There is actually no force.
- We were convinced that the particles had masses since that is what we measure. But in fact, they have no mass but interact with a Higgs field whose intensity of this interaction more or less slows them down, giving them the impression of having mass.
- We see a blue object, but in fact it looks blue to us because it absorbs all wavelengths except the blue it reflects. Our brain reconstructs an image of reality.
- We see a Star in the distance in the sky and are convinced that it exists but if it is very far away, its light may have taken billions of years to reach us and therefore perhaps it no longer exists today. Looking far into the universe, we look at the past and not the present.
- We were convinced that there is an absolute elapsed time, but special relativity has shown us that this is not the case, the measurement of atomic clocks has shown us.
- We would never have imagined that gravity would slow down time; two atomic clocks initially synchronized on a table become out of sync if one is brought to the ground and the other remains on the table. The one on the ground slowed down its atomic ticking compared to the one on the table. The greater the gravity, the longer the time slows down.
- We thought of mass and energy as two different physical objects, they are in fact connected and one and the other can transform into one or the other with $E = mc^2$.
- We thought the universe was static on our scale but we know it is expanding rapidly.

- We thought that black holes did not exist, we measured the gravitational waves resulting from their coalescences. They were thought to be emitting nothing and S. Hawking has shown this not to be the case, by his Hawking radiation.
- It was thought that light needed an ether to propagate but it is not, the electric field and the electromagnetic field intermingle and self-stimulate.
- We thought the space vacuum was empty and in fact it is full of quantum fluctuations (Casimir force) of a dark energy, of a Higgs field. It is in a way empty on average, but the constituents making it possible to calculate this average are not zero. Emptiness is not nothingness.
- The moon always presents its same face to us so it does not seem to turn on itself, and indeed not it turns but has a synchronous rotation with the rotation of the earth giving us this illusion of absence of rotation. We thought that everything is at rest around us in our flat, but in fact all the objects move at the speed of rotation of the earth, which revolve around the sun, which revolve around the center of the Milky way our galaxy, which moves also in an galaxy clusters, etc.
- Earth was thought to be the center of everything, but it is just a small planet, revolving around a middle star, in a classic spiral galaxy lost in the midst of billions of others. The only marvelous originality of size, it is home to the life and author of this book, but then again, there is no originality and life merges in the universe. Moreover, if statistically there is at least one existence of life per galaxy (this is the case for the Milky Way: us), it seems logical that there is at least one per galaxy, and as there is billions of galaxies in the universe...
- Etc. etc. etc.

1.11 How to Reconcile the Two Physics

In order to advance physics and our knowledge, we must therefore compare for a given phenomenon what we measure as accurately as possible with what the candidate theories predict. The good theory will be the one whose predictions are in agreement, without exceptions, with what is measured for the phenomenon studied. This is very well explained by E. Klein in his various conferences and in particular that of 2018 "the fundamental structure of matter" under thinker view, of which we take up the main lines below:

When there is disagreement between theory and measurement:

- Either the measurements are false due to an instrumental and experimental problem or an external disturbing effect (there is no problem in fact of law Just a problem of measurement):

  This was the case with the original gravitational wave measurements of the big bang which did not take any in 2014 bicep experiment 2.

- Either the mathematical and physical theory is correct but the interpretation of the measurements is incorrect or an object has been forgotten in the analysis: (ontological modification, we keep the law but we must add something in the study system to solve the problem):

This case occurred while studying the motion of Uranus. The trajectory obtained with Newton's equations did not correspond to reality. It was therefore necessary to mathematically add a planet that we did not know, Neptune (by a calculation by Le Verrier in 1846) so that we find the measured disturbance of Uranus. J.G. Galle then observed this planet using indications obtained from Le Verrier exactly where he had predicted it would be by applying Newton.

- Either the theory is false or is just a simplification of a larger theory and it is the measures that are correct (legislative change, we change the law to solve the problem):

  This was the case with gravitation: Newton's equations work well in the solar system, they found the planet Neptune, but could not predict the delay of the perihelion of Mercure (infinitesimal angle of 43 arc second that could have been considered a measurement problem rather than a problem in the law), *i.e.* in gravity field very close to the sun. Thus, Einstein will change the law by producing the general relativity that will predict this delay exactly and describe much more precisely in accordance with the special relativity what gravity is.

- Either we have to make a mixture of the two (ontological and legislative modification):

  This was the case with quantum field theory which works extremely well to explain the Standard Model, but which implies that the masses of the particles are all zero, which is contrary to what we measure since hundreds of particles have masses! It was therefore believed that the model based on quantum field theory was wrong. Three physicists in 1964 (R. Brout, F. Englert, and P. Higgs) assumed that it was good, but that it is this principle that particles are necessarily born with a mass (mass = matter) that had to be questioned (legislative amendment). They then hypothesized a new scalar field in the Standard Model (ontological modification), the Higgs field that existed at the time of the big bang. The particles appeared massless and by interacting more or less with this field *via* the Higgs Boson acquired more or less mass. The Higgs Boson was discovered in 2012 at the LHC thus confirming their approach and questioning what was thought to be acquired, mass is a secondary property of particles. The particles initially had no mass and time of their own. By interacting with this field, they have acquired more or less mass and we begin to have our own time. So, particles move at the speed of light (when they have no mass) or slower than the speed of light (when they have mass) by interacting with this Higgs field *via* this boson of Higgs.

This gravitation is a distortion, a warped of space-time by matter/energy found there. Newton's universal law was merely a simplistic approach to reality that was much more mathematically complex (tensing); simpler at a time: a single principle curvature of space = $K\times$ the density of present energy and without any mysterious action (remote action without connection between objects and not instantaneous because the setting of its action is limited to the speed of light).

The scientific method is based on the following principles (see E. Klein conferences):

- Reducing the complex to the elemental.
- Understanding how the elemental works, finding operating laws.
- Deduce the laws of the complex from those of the elemental (so we go back from the elementary to the complex).
- Do not hesitate to demonstrate the real by the impossible.
- Do not forget that it is in the details that the devil hides.

With all these reminders of our knowledge, having learned from the elders how to work (understand how things work before putting mathematics), trusting mathematics when we push beyond what we know the laws we have found, wary of the appearances we perceive with our limited senses, having understood that it is in the details or the 15th decimals that reality is found, we can now attach ourselves to our objective. Trying to reconcile these two physics: finding the flaw that has eluded us so far to reconcile them. The author's actual journey is given in appendix A. It was made didactic by the order of the chapters of this book to facilitate its understanding. As E. Klein indicates, in his lecture "le gout du réel/passion of reality" at the Sorbonne on October 2, 2020, science (which is a body of knowledge taken for granted) should not be confused with research (theory that remains to confirm). This book synthesizes the science in chapters 1 and 2 and is research in the rest of the other chapters. As C. Rovelli points out at the start of his quantum gravity lectures posted online in 2018, science has always progressed in a safe way based on a solid foundation never starting from 0. So that is what did:

- Newton by putting together the physics of Galileo (all bodies fall together at the same speed in a vacuum), and Kepler (motion of the planets) (this gave the law of universal gravitation and made it possible to discover Neptune).
- Maxwell by combining electricity and magnetism (this made it possible to discover electromagnetic waves).
- Einstein by putting together Newton's physics and special relativity (General Relativity gives Newton's theory in a weak field, space has become a deformable physical object, we have discovered black holes and gravitational waves).
- Dirac by putting together quantum mechanics and special relativity (this gave birth to antiparticles).

This is what we are going to do in this book as well. Indeed, we will put together general relativity and the theory of elasticity. We would rely in a solid way on the elastic strain measurements of space established with certainty *via* gravitational waves on the one hand, and taking into account the vacuum data on the other hand, to try to express Gravitational Einstein's constant κ and Newton gravitational constant's G, as a function of mechanical parameters like Young's modulus of space-time Y. We will calibrate all our results in order to find all the values of the known universal constants. Authors in 2018 such as T.G. Tenev, M.F. Horstemeyer, M.R. Beau, R. Weiss, A.C. Melissinos, K.T. Mcdonald were also interested in this question, so we will compare our approach with their results in the conclusion in chapter 22.

Chapter 2

First Ask the Right Question

2.1 What is the State of the Art and the Issues that Arise from It?

Let us go back to the starting point, that is to say, to what we have learned since Einstein:

(a) Since gravity is not a force, but the materialization of the geometric deformation of space-time by the massive objects (or energy) that follow its curvature according to Einstein [8–10] and his general relativity.
(b) Since space-time is not rigid but deformable and elastic (see appendix B):

- The earth goes straight and follows the curvature of space created by the sun.
- The light of the stars around the sun is deflected (seen during an eclipse by Eddington) [11] when the sun positions between them and us in our target field, and then resumes theirs initials position when the sun leaves the target area.
- Space vibrates and transmits gravitational waves (coalescence of black holes or neutron stars) [4, 5] and [15, 16].
- Galaxies are "fixed" but move away from us all the faster as they are far away in a space that expands (Hubble Law) [21].
- The Probe B satellite mission, with gyroscopes, measured the twisting deformation of space around the earth [22].
- The Ligo and Virgo interferometers measure space deformations (10^{-21}) for example from the coalescence of 2 black holes [23, 24].
- Space seems to sound like a crystal (Ringermacher) [26].

The fundamental questions that then lead to which Einstein and his successors did not answer, are as follows:

What is space-time made of? What concretely deforms, twists, vibrates, propagates elastic waves of gravitation? Is there a space-time material? Is it a quantum fluid? Is it a new sort of deformable dark ether?

DOI: 10.1051/978-2-7598-2573-8.c002

2.2 What is the Nature of Space-Time?

Indeed, is space not made up of a super fluid without viscosity (see approach by Stefano Liberati and Luca Maccione) at low speed allowing the stars to move so much and does not it become like a hyper resistant solid at the speed of light? much like water skiing or skis sink at low speed and lean on water at high speed)? T. Damour showed that the Navier Stokes equations (fluid mechanics) are hidden in the equations of general relativity when studying black holes [5] and that general relativity can be interpreted and understood as a Hooke's law in elasticity [69–71]. Newton showed by a simple formulation that gravity could be seen as an instantaneous force pulling objects together, but was unable to say what it was and how this force materialized between objects attracting each other at an unrelated distance between them. Its whole demonstration depends on a mysterious constant G whose dimension equation is bizarre (not direct) as seeming to result from the assembly of other elements: $m^3/(kg.s^2)$. And why this value? Some authors consider this idea of a constant consisting of a density multiplied by a squared frequency [13].

2.3 Can Einstein's Equation be Reconstructed without Passing through Newton's Weak Field Limits? Without Using G?

Einstein showed in 1915 that gravity was not a force but a deformation of the Space giving us the illusion of a force [12]. But he did not say what this space-time was made of. He found a mathematical model, geometric and physical behaving like the real Space (a model is a mathematical and scientific construction that behaves like reality but that is not reality much like the computer finite element model of a bending plate is not the plate itself see figures 1.2a and 1.2b) but he did not say how his model was made in reality. In the two-dimensional images (so simplified because it assumes that objects rest on space, when in reality objects are drowned in a 3 dimensions gangue of space, volume deformation) where space-time is considered as a fabric plan of a trampoline, it is possible by replacing the sun with a "pétanque ball" place in the center, to visualize the curvature of the fabric. Planets that are replaced by small Balls that one launches with a certain speed then begin to describe orbits around the ball of "pétanque", following the curvature of the fabric, and when they lose speed they end up falling towards the ball of "pétanque" giving the illusion that they are attracted to it *via* a mysterious force (see figure 2.1a). But if mathematics can describe, to find the continuous functions predicting curvature and thus the geometry of this fabric and the movement of the marbles on the fabric, the fabric Elastic exists at the start! It is not "nothing" that bends. It is made of an elastic material.

Note: The image of figure 2.1a presents well the notions of curvature, but failing to suggest that space-time is deformed by something placed on it, the reality is rather that space-time is deformed in its thickness by the objects therein (a bit like in figure F.5).

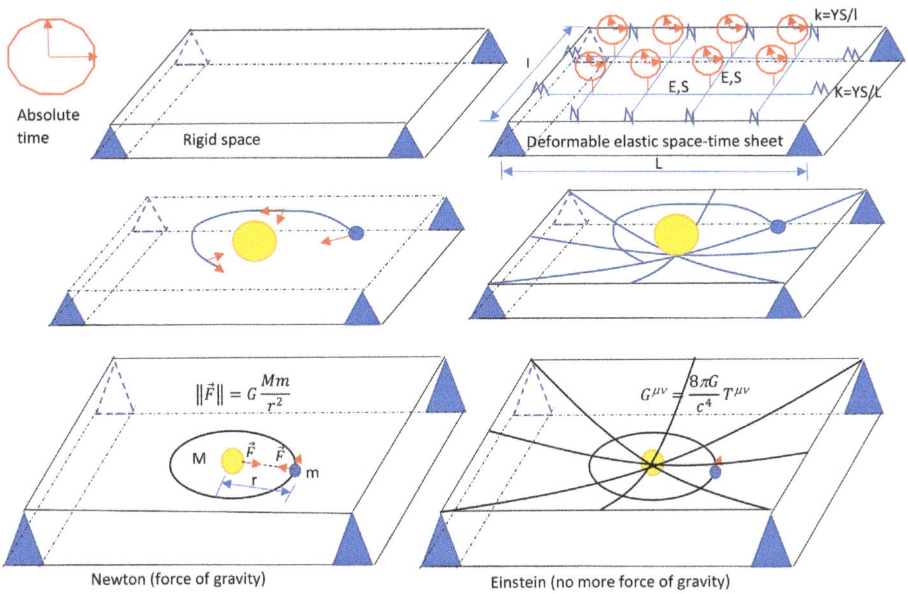

FIG. 2.1a – Different ways of seeing gravitation (left according to Newton *via* the force concept) on the right according to Einstein (*via* warped space-time). Source DrphysicsA of Bob Eagle completed by D. Izabel.

Einstein calibrated his theory by a passage to the limits (weak gravitational fields) where Newton's law works well (use of the 00 component of the metric tensor $g_{00} = 1 + \frac{2\phi}{c^2}$ and of Poisson's law $\Delta\phi = 4\pi G\rho$). It therefore reused G (which appears in κ) using the temporal component of its tensors. It therefore did not use the spatial components of its tensors to calibrate. Its proportionality constant $\kappa = \frac{8\pi G}{c^4}$ between the curvature $G_{\mu\nu}$ and the strain energy spatial tensor $T_{\mu\nu}$. Given the elastic behavior of space, he could have attempted to calibrate his equation by analyzing the spatial behavior of space-time and using its elastic properties according to the literature [29, 30]. But Einstein did not use or even refer to the elasticity theory, which is nevertheless appropriate for this kind of studies of elastic medium and common application for engineers who use it almost systematically to study elastic deformable solids or fluids [69–71]. So, that is what we are going to do in this book. Attempting to recalibrate Einstein's equation not from a boundary switch with Newton using the temporal components of tensors and thus the gravitational constant G, but using the elasticity theory applied to the spatial components of tensors and we will see the interesting consequences that result from it. Because if Newton's equation is only a simplification of the true law of gravitation, why continue to keep G associated with this simplified law in our modern equations of the twenty-first century?

2.4 What Brings Us Contemporary Data of the Vacuum?

The other aspect that has not been used by Einstein, and there he could not do it, all the data accumulated for a century regarding the quantum mechanics, relativistic quantum mechanics and the quantum fields theory. In fact, empty Space is not empty! There is an Energy from emptiness to rest demonstrated through Casimir's force. Virtual particles and antiparticles that appear and disappear instantly or almost thanks to the Heisenberg's principle which when they do near a black hole, for example, generate radiation, Hawking radiation. There are also a scalar field which gives mass to the particles, the famous Higgs's field and reappears in CERN's particle accelerators when collisions bring enough energy to the vacuum to allow it to re-emerge *via* these famous Higgs's bosons discovered in 2012. In summary, the emptiness at rest has a minimum Energy, therefore depending on the special Relativity an equivalent mass ($E = mc^2$) and so dividing this mass by a volume, a density ρ. But according to the Quantum mechanics, at this Emptiness energy corresponds also an Eigen frequency f of vibration ($E = hf$). So, the vacuum is not empty. It is not nothingness. It also has a permittivity. In summary, the space vacuum can be considered as a physical object. It can therefore be considered as a pseudo material that makes up space..., a different material from this which constitutes us, but a material with elastic properties. Finally, according to quantum field theory, there is indeed a force that transmits gravitation through a Boson, the graviton of spin 2 and mass null (speed c following the gravitational waves GW170817 and others) (see figure 2.1b).

We must also consider Michelson and Morley's experiment who tells us that there is no ether at rest, the speed of light is not altered by this ether. What seems to have contradicted with the paragraphs above which implies a certain elastic substance filling all the space. T. Damour speaks of "jelly" the famous English's dessert to illustrate this point of elastic substance filling the space.

How can it be that there is no ether but still some kind of other elastic ether seems to fill the space and warp? We will see later how to solve this paradox. Einstein himself minded on the issue in 1915 (*via* special relativity) where he suppressed the ether and then in 1920 at a conference in Leiden, he rehabilitated a kind of ether [28] or during a letter with Lorentz [27]. Based on its data of general

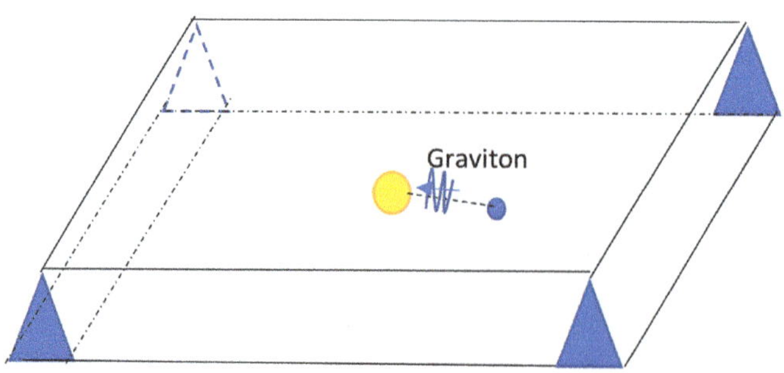

FIG. 2.1b – Spin 2 graviton that transmits gravitational force.

relativity (space is an elastic body), the quantum field theory (the vacuum is not empty but has a minimum energy, a minimal Eigen frequency or there are gravitons), but also the elasticity theory (all elastic body has mechanical characteristics $E = Y$, v), it seems possible to seek by the elasticity theory that it could be the nature and the structure of an elastic material constituting the space itself. This is what we will attempt to do in this book.

2.5 Space-Time as a Physical Object an Elastic Medium

All this seems to indicate that space-time could be seen as an elastic material with mechanical properties (elastic Young's modulus, Eigen vibration frequency) [4, 5, 18, 69–71] {173} at {178} and that the gravitational constant G may not be an ultimate constant, but is in fact calibrated from an assembly of the mechanical and quantum properties of this space! [13]. But as a great physicist of quantum field theory told me when we met in his office where I had presented him with my first ideas "all these words. It has to be proven mathematically and physically, because phenomena can follow identical mathematical laws but have nothing to do with it from a physical point of view". Thus, space-time is warped [7, 19, 20, 23–25], the vacuum strain measurements of Ligo and Virgo show us almost every day now thanks to the coalescences of black holes or other neutron stars that generate gravitational waves. The space deforms very little since synchronized strains are measured $\varepsilon_{xx} = \frac{\Delta L_x}{L_x} = -\varepsilon_{yy} = -\frac{\Delta L_y}{L_y} = \pm 10^{-21}$. It oscillates as it passes through the earth into the xy plane of the perpendicular vacuum arms of the interferometer perpendicular to the propagation direction z of gravitational waves where the advances and retreats of mirrors elongating and shortening the path of the laser beam are measured in the vacuum tubes. But then what is physically warped, deformed in these arms where the near perfect vacuum has been made and in which the lasers come to hit mirrors immersed in an almost perfect void? What constitutes the fabric of this trampoline, its curvatures, what vibrates to transmit at the speed of light gravitational waves? In summary, how is this mysterious space-time made mysterious? What constitutes this elastic jelly as said by T. Damour in his popular lectures on gravitational waves or black holes [2, 3]? How does it work mechanically? Einstein's formulation retains a part since it tells us that something is distorted without telling us what it is made of. This is remid us to a certain extent that of Newton who found a law that allows us to calculate the effects of gravity, but which does not explain how this force is transmitted, mysteriously, without connection between the massive objects on which it acts… This is the question that guided this exciting research of the true. This led me to give a possible way to unify general relativity with the elasticity theory and then with quantum mechanics and quantum field theory *via* graviton and its spin and with the vacuum data. For this I acted where few people had really dug before, that is to say on Einstein's constant of proportionality of the gravitational field equation which relies on an old-fashioned G constant from a simplified formulation of Newton that does not represent the final true theory. This was the starting point of my reflection and this exciting research and here is how…

The following chapters are a development of the author's papers published in Pramana physical peer reviewed journal on August 13, 2020.

Chapter 3

A Strange Analogy between S. Timoshenko's Beam Theory and General Relativity

Chapter Summary

$$\frac{1}{R^2} = \frac{2}{EI}\left(\frac{W_{\text{ext(total)}}}{L}\right) = K\left(\frac{W_{\text{ext(total)}}}{L}\right)$$

$$G^{\mu\nu} = -\frac{8\pi G}{c^4}(T^{\mu\nu}) = -\kappa(T^{\mu\nu})$$

3.1 Introduction

Having not studied general relativity during my engineer lectures, I decided to get to know her self-taught *via* all the means available on internet. Thanks to this fabulous tool that is this search engine on the net, I very quickly fell one evening in January 2011, by typing "general relativity" on my computer keyboard, on an exceptional conference by Thibault Damour No. 183 on Saturday 1st July 2000 "General relativity Thibault Damour". This presentation very well done, was the result of a series of lectures on science and physics: 2000 in France: "university of all knowledges". At that time, the slides and the computer were not currently used and the right old method based on transparencies applied. The conference was dense, it is true that summarized in 1 h all Einstein's work 10 years ago was a real challenge. T. Damour explains very well the special relativity how Einstein unified space and time to create

DOI: 10.1051/978-2-7598-2573-8.c003

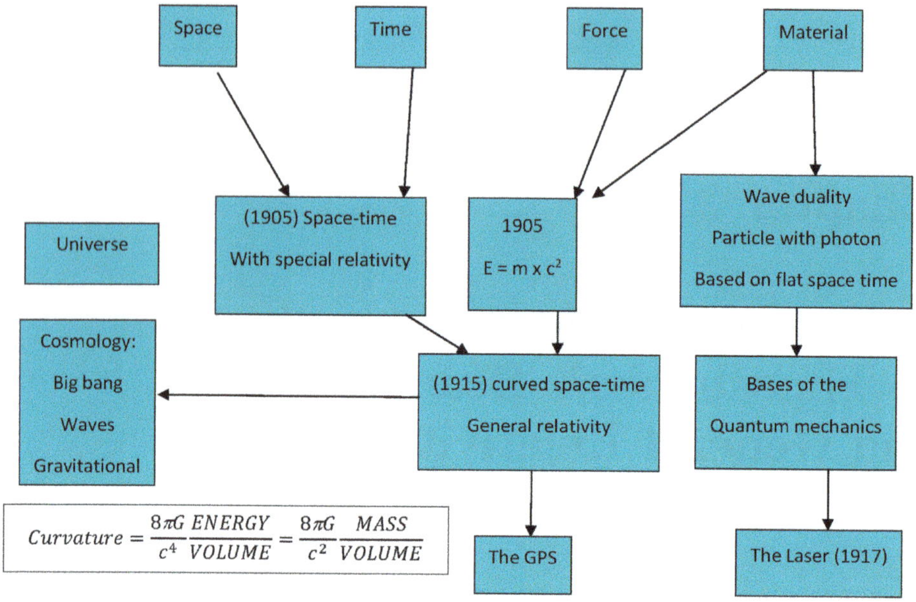

FIG. 3.1a – D. Izabel's presentation of a transparent from the presentation of T. Damour Conference No. 183 on Saturday, July 1, 2000, "General Relativity".

space-time, unifies force and matter with its famous equation $E = mc^2$ and his work on particle wave duality (see figures 3.1a and 3.1b).

3.2 Generalities on General Relativity

T. Damour goes on to show the results of Einstein's work on gravitation that unifies the curvature of space with matter-energy. The great French scientist thus formalizes Einstein's revolutionary work with two simple formulas, which required the German scientist to work hard for more than 10 years from 1905 to 1915: Finally, I begin to understand the great principles of this theory. The presentation of Einstein's formula is presented by T. Damour (see figure 3.1c).

The first equation gives us the definition of curvature and indicates that the presence of mass placed in a triangle distorted this triangle so that the sum of the angles $\alpha + \beta + \gamma$ is no longer 180°.

We note for memory that the unit: degree/m^2.

The second equation creates a bridge between the geometry of the space represented by its curvature as defined above and its energy or mass content, both of which are equivalent to the $E = mc^2$ formula.

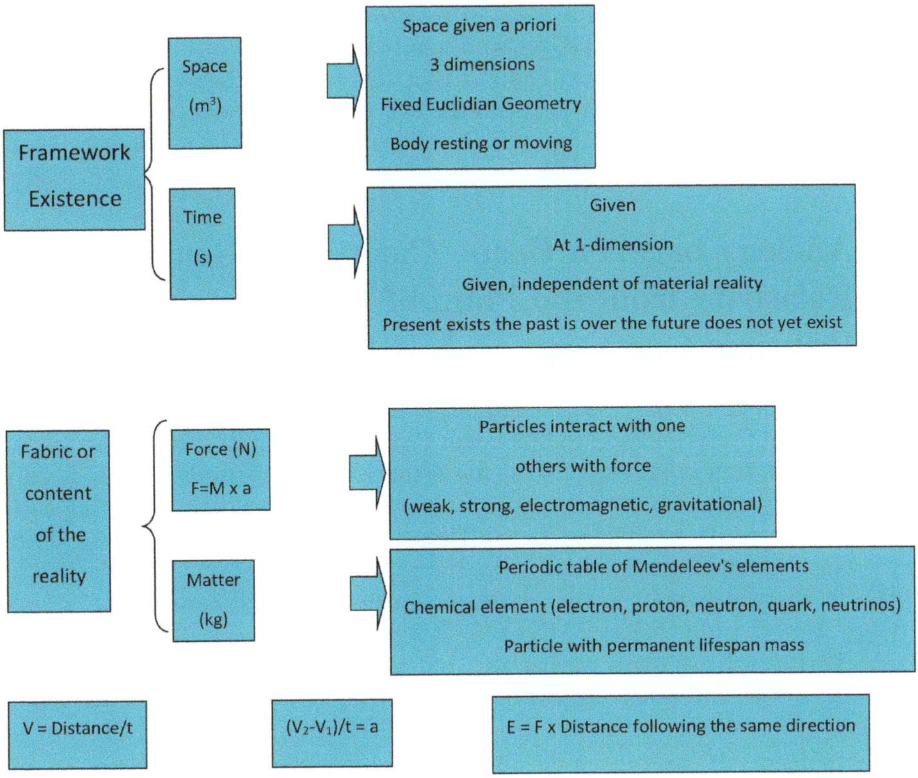

FIG. 3.1b – D. Izabel's presentation of a transparent from the presentation of T. Damour Conference No. 183 on Saturday, July 1, 2000 "General Relativity".

$$\mathit{Curvature} = \frac{\alpha + \beta + \gamma - 180°}{\mathit{Area}}$$

Einstein's Equation:

$$\begin{matrix}\mathit{Mean}\\ \mathit{SPATIAL}\\ \mathit{Curvature}\end{matrix} = \frac{8\pi G}{c^4}\frac{\mathit{ENERGY}}{\mathit{VOLUME}} = \frac{8\pi G}{c^2}\frac{\mathit{MASS}}{\mathit{VOLUME}}$$

FIG. 3.1c – Synthetic expression of general relativity. Source: T. Damour Conference No. 183 of Saturday, July 1, 2000 "General Relativity".

The dimensional equation of this second equation is:

$$\frac{1}{\mathrm{m}^2} = \frac{\frac{\mathrm{m}^3}{\mathrm{kgs}^2}}{\left(\frac{\mathrm{m}}{\mathrm{s}}\right)^4} \times \frac{\frac{\mathrm{kgm}^2}{\mathrm{s}^2}}{\mathrm{m}^3} = \frac{\frac{\mathrm{m}^3}{\mathrm{kgs}^2}}{\left(\frac{\mathrm{m}}{\mathrm{s}}\right)^2} \times \frac{\mathrm{kg}}{\mathrm{m}^3}$$

Reading the above formula, I then felt very excited, it was amazing. Einstein's equation of general relativity used the same notions that I used in strength of material. Indeed, for study beams or plates or shells, it is necessary to use the theory of Stephen Timoshenko (December 11, 1878 – May 29, 1972), considered as one of the founding fathers of the Continuum mechanics from which the strength of materials. We will see in the next chapter that this bending theory is also based on the principle of the relationship between the curvatures of a beam or of a plate and the associated strain energy, related to the external work of the applied forces and moments.

3.3 Analogy between Beam Theory and General Relativity from the Point of View of the General Principle Curvature = K × Energy Density

Considering in a first step the second equation of the T. Damour's presentation (see figure 3.1c). So, let us go back to our beam theory so that you can see the analogy with general relativity that appealed to me so much at the start of my work. So, if we take the analogy of a simple beam on 2 supports in pure bending according to the strength of materials established by S. Timoshenko [29, 30] from a more general theory which is the elasticity theory (see figure 3.2).

This beam has a span L (m), is made up of an elastic material of Young's modulus $Y = E$ (unit MPa = MN/m^2), has a section $S = bh$ (m^2), an inertia $I = bh^3/12$ in (m^4), a mass m (kg/m) and a radius of curvature R (m). The equivalent of the curvature equation of this beam is the equation that defines, according to the elasticity theory, the beam deflection $y_{(x)}$ under the effect of two external moments M (daN.m) applied at each end and which act as masses m_i with ($m_i \gg m$) bending under the effect of the acceleration of earth's gravity $g = G\frac{M_T}{R_T^2}$ applied on the beam (see figure 3.3).

The relationship between the curvature ($1/R$) and the deflections or geodesic of the beam depends on the second derivative of the deflection $y_{(x)}$ (see appendix H) as in general relativity $R_{\mu v}$ that depends on the second derivatives of the metric $g_{\mu v}$:

$$\frac{M}{EI} = \frac{\frac{d^2 y_{(x)}}{dx^2}}{\left(1 + \left(\frac{dy_{(x)}}{dx}\right)^2\right)^{3/2}} = \frac{1}{R}$$

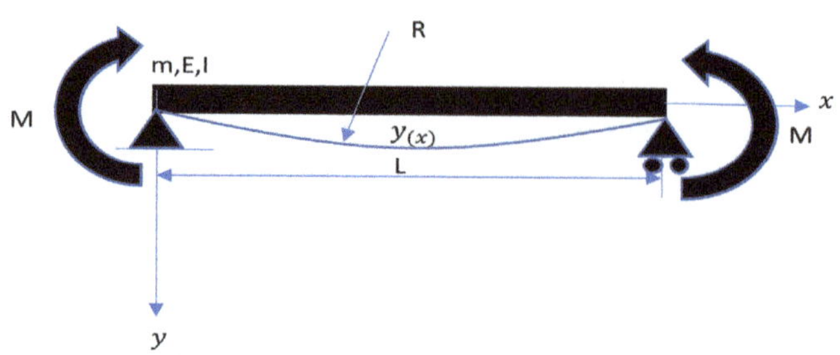

FIG. 3.2 – Radius of curvature R, mechanical parameters EI and mass m of a beam in strength of materials loaded by two bending moments M.

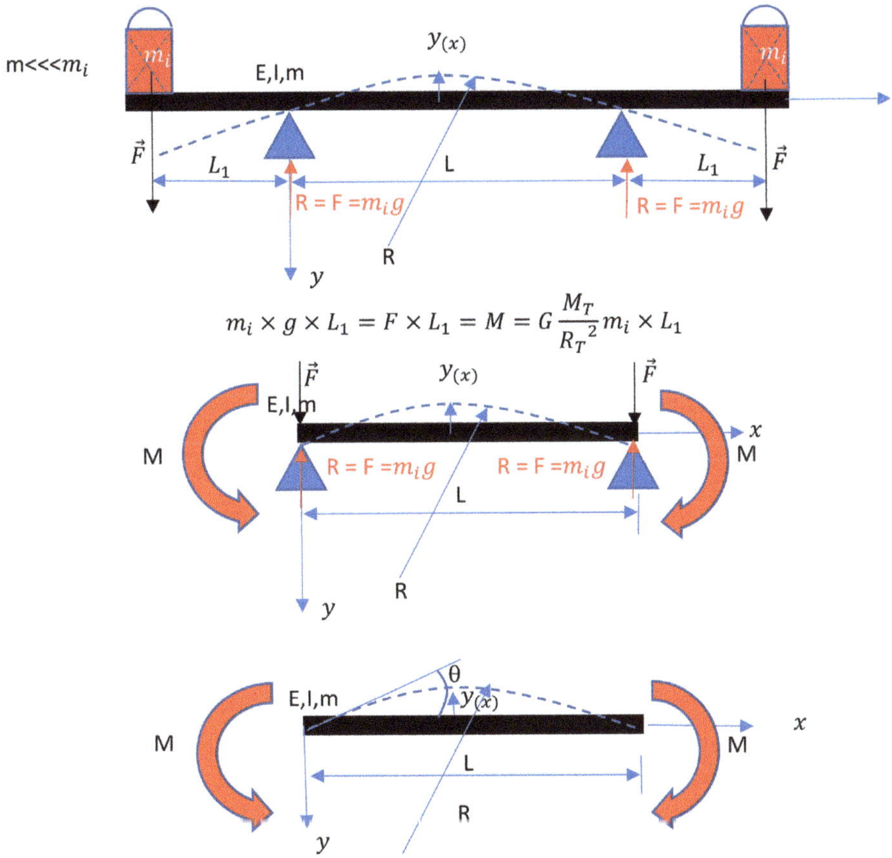

FIG. 3.3 – Equivalent beams in pure bending (zero-support reactions).

For small curvature and for small rotation, we have:

$$\frac{d^2 y_{(x)}}{dx^2} = \frac{d\left(\frac{dy_{(x)}}{dx}\right)}{dx} = \frac{d\theta_{(x)}}{dx} \approx \frac{1}{R}$$

So, in pure bending (case of figures 3.2 and 3.3) we have with a moment $(M = M_{(x)} = cte)$ at each end of the beam:

$$M_{(x)} = M = -\frac{EI}{R}$$

The strain energy U (internal forces work) of the simple bending beam is written since M is constant:

$$U = \frac{1}{2}\int_0^L \frac{\left(M_{(x)}\right)^2}{EI}\,dx = \frac{M^2 L}{2EI}$$

In the above expression, we see the link between the work of the external forces on the right as a function of the moments M applied at each end (themselves attached to the applied masses, see figure 3.3) and the work of the internal forces or strain energy noted U here on the left. This internal work is due to the fibers, to the three-dimensional fabric of the beam which lengthens and shortens in proportion to Young's modulus E of the material constituting it and thus generating the displacement $y_{(x)}$ of the beam and its curvature following a constant radius R here.

By substituting M function of curvature $(1/R)$ in the strain energy U of the beam we obtain:

$$U = \frac{1}{2}\frac{EIL}{R^2}$$

Or:

$$\frac{1}{R^2} = \frac{2}{EI}\left(\frac{U}{L}\right) = \frac{M^2}{(EI)^2}$$

Or to get closer to the general relativity formalism:

$$\frac{1}{R^2} = K\left(\frac{U}{L}\right) = \frac{2}{EI}\left(\frac{U}{L}\right) = \frac{M^2}{(EI)^2}$$

(Strength of material, beam in 1 dimension)

$$\text{Curvature} = K \times \frac{\text{Energy}}{\text{Length}}$$

$$G_{\mu v} = \kappa T_{\mu v} \quad \text{(General Relativity in 4 dimensions)}$$

$$\text{Curvature} = \kappa \times \frac{\text{Energy}}{\text{Volume}}$$

where $K = \frac{2}{EI}$ the coupling constant between the curvature $(1/R^2)$ on the left and the linear energy density (U/L) on the right (logical since we consider here a beam according to a dimension of length).

Note: Constant $1/(EI)$ represents a kind of flexibility of the beam (flexibility $= 1/\text{rigidity}$).

Indeed, in the case of a spring of rigidity k (N/m) we have for a displacement δ (in m) a force F (N) that is written:

$$F = k\delta$$

and in a beam on two supports uniformly loaded with a load p (N/m), L(m) span, $I(\text{m}^4)$ the inertia of the beam section, and E (MPa) the Young's modulus of the material constituting the beam, we have for the deflection f (in m) at mid span:

$$f = \frac{5pL^4}{384EI}$$

So:

$$pL = F = \frac{32}{5}\frac{12EI}{L^3}f$$

The rigidity of the beam is called $\frac{12EI}{L^3}$ in (N/m).

We therefore obtain a formalism in one dimension between curvature of an elastic medium and energy density (or strain energy density) close to that of general relativity in 4 dimensions, hence my great surprise when listening to T. Damour's 2000s lecture.

In addition, we can write the expression of the external work created by the two moments applied to each end of the beam (see figures 3.2 and 3.3).

The expression of the external work due to a moment M is written:

$$W_{\text{ext(1 moment)}} = \frac{1}{2}M\theta$$

The rotation θ at each end of our beam under two constant moments is written:

$$\theta = \frac{ML}{2EI}$$

Using the two expressions below we get a new expression of the work of the external forces of our beam.

$$W_{\text{ext(1 moment)}} = \frac{M^2L}{4EI}$$

We can then re-express M^2 according to the work of the external forces of the beam loaded by two constant moments:

$$W_{\text{ext(total)}} = \frac{M^2L}{2EI}$$

So, we have:

$$M^2 = \frac{2EI}{L}W_{\text{ext(total)}}$$

which we can substitute in the expression giving the work of internal forces U:

$$\frac{1}{R^2} = \frac{2}{EI}\left(\frac{U}{L}\right) = \frac{M^2}{(EI)^2} = \frac{2}{EI}\left(\frac{W_{\text{ext(total)}}}{L}\right)$$

Note: If we now consider a sphere of radius r and calculate its scalar curvature R according to the Riemann curvature tensor in this two-dimensional space (see appendix D) we obtain $2/r^2$ so a function in $1/r^2$ compatible with the curvature $1/R^2$ defined in the formula above of a beam in pure bending. Einstein's gravitational field formulation therefore allows us to find this analogy with the beam theory in one dimension at the level of the curvature. Thus, the curvature of the beam in pure

TAB. 3.1 – Comparison of dimensional equation according to Einstein's general relativity theory and according to S. Timoshenko's beam theory.

Parameter	General relativity (4 dimensions)	Units	Strength of material (1 dimension)	Units
Principle	$G^{\mu\nu} = \kappa T^{\mu\nu}$	$\frac{1}{m^2}$	$\frac{1}{R^2} = \frac{2}{EI}\left(\frac{U}{L}\right) = \frac{2}{EI}\left(\frac{W_{\text{ext(total)}}}{L}\right)$	$\frac{1}{\text{m}^2}$
Curvature	$G^{\mu\nu}$	$\frac{1}{m^2}$	$\frac{1}{R^2}$	$\frac{1}{\text{m}^2}$
Linear strain energy density of the spatial space fabric	Have to be built $M^{\mu\nu}$	$\frac{N.m}{m^3}$	$\left(\frac{U}{L}\right)$	$\frac{\text{N.m}}{\text{m}}$
External work	$T^{\mu\nu}$	$\frac{N.m}{m^3}$	$\frac{W_{\text{ext(total)}}}{L}$	$\frac{\text{N.m}}{\text{m}}$
	$\kappa = \frac{8\pi G}{c^4}$	$\frac{1}{N}$	$K = \frac{2}{EI}$	$\frac{1}{\text{N.m}^2}$

bending in one dimension is indeed an analogy of the equation of the field of gravity in 4 dimensions established by Einstein according to the parallelism given in table 3.1:

$$\frac{1}{R^2} = \frac{2}{EI}\left(\frac{W_{\text{ext(total)}}}{L}\right) = K\left(\frac{W_{\text{ext(total)}}}{L}\right)$$

$$G^{\mu\nu} = R_{\mu\nu} - \frac{1}{2} g_{\mu\nu} R = -\frac{8\pi G}{c^4}(T^{\mu\nu}) = -\kappa(T^{\mu\nu})$$

Indeed, $T_{\mu\nu}$ represents the masses (energy) applied in space-time (the work of external forces in a way) while $G_{\mu\nu}$ represents the curvature of space due to the work of external forces.

The consequences of this analogy is that κ seems related to the flexibility of space structure because $2/EI$ is a flexibility as opposed to a rigidity $(\frac{12EI}{L^3})$.

3.4 Analogy between the Definition of Curvature in Strength of Material and General Relativity

The other expression of T. Damour (the first listed above figure 3.1c in his presentation about the curvature definition) tells us that the curvature in Einstein's sense is an angle divided by an area. We will show this in the case of the same beam loaded this time by a thermal gradient (temperature difference between the lower and upper fibers of the beam $\Delta_T = T_{\text{ext}} - T_{\text{int}}$).

Now consider the case of an identical beam in pure bending undergoing a thermal gradient (see figure 3.4).

The strain energy is always:

$$U = \frac{1}{2}\int_0^L \frac{M^2}{EI}\,dx$$

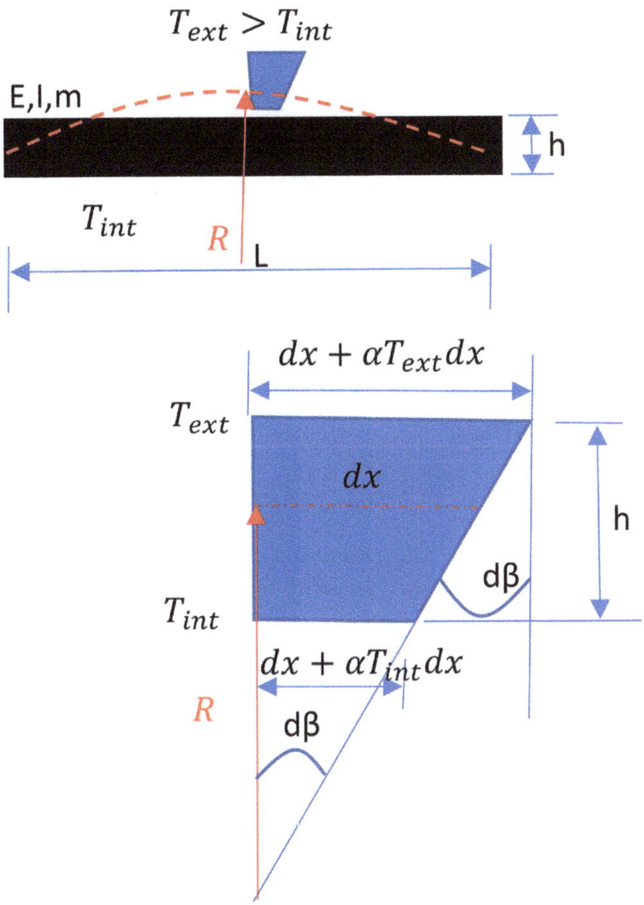

FIG. 3.4 – Deformation of a beam in pure bending under thermal gradient ΔT.

Figure 3.4 allows us to write the following geometric relationships between the angle β (radiant) and the curvature $(1/R)$:

$$\tan(d\beta) = \frac{dx}{R} = \frac{(dx + \alpha T_{\text{ext}}\, dx) - (dx + \alpha T_{\text{int}}\, dx)}{h} = \frac{(\alpha T_{\text{ext}} - \alpha T_{\text{int}})\, dx}{h} = \frac{\alpha \Delta T dx}{h}$$

Given the figure 3.4, the relationship between the curvature of a beam under a thermal gradient and the second derivative of its deflection $y_{(x)}$, we obtain:

$$\frac{d^2 y}{dx^2} = \frac{M}{EI} = \frac{1}{R} = \frac{\tan(d\beta)}{dx} = \frac{d\beta}{dx} = \frac{\alpha \Delta T}{h}$$

We obtain a new expression of the bending moment M:

$$M = EI \frac{d\beta}{dx}$$

By replacing the bending moment M with its above formula in the equation of the strain energy U of the beam, we obtain:

$$U = \frac{1}{2}\int_0^L \frac{\left(EI\frac{d\beta}{dx}\right)^2}{EI}\,dx$$

That is, after simplification if the angle is constant (no variation along the length of the beam, pure bending):

$$U = \frac{EI}{2}\left(\frac{d\beta}{dx}\right)^2 L$$

So, after some mathematics:

$$\left(\frac{\boldsymbol{d\beta}}{\boldsymbol{dx}}\right)^2 = \frac{\mathbf{2}}{\boldsymbol{EI}}\frac{\boldsymbol{U}}{\boldsymbol{L}}$$

which can be compared with the expression already found in the previous paragraph in the case two masses m_i distorted the same beam:

$$\frac{\mathbf{1}}{\boldsymbol{R^2}} = \frac{\mathbf{2}}{\boldsymbol{EI}}\left(\frac{\boldsymbol{U}}{\boldsymbol{L}}\right)$$

So, we have well an angle on a surface as a definition of curvature $(1/R)$ square of the beam.

$$\left(\frac{\boldsymbol{d\beta}}{\boldsymbol{dx}}\right)^2 = \frac{\boldsymbol{d\beta^2}}{\boldsymbol{dx \times dx}} = \frac{\boldsymbol{d\beta^2}}{\mathbf{Area}} = \left(\frac{\mathbf{1}}{\boldsymbol{R}}\right)^2$$

This corroborates the expression of T. Damour:

$$\text{Curvature} = \frac{\alpha + \beta + \gamma - 180^\circ}{\text{Area}}$$

The strain energy becomes for a constant thermal gradient:

$$U_{\Delta T} = \tfrac{EI}{2}\left(\tfrac{d\beta}{dx}\right)^2 L = \tfrac{EI}{2}\left(\tfrac{\alpha\Delta T}{h}\right)^2 L$$
$$\left(\tfrac{\boldsymbol{\alpha\Delta T}}{\boldsymbol{h}}\right)^2 = \tfrac{\mathbf{2}}{\boldsymbol{EI}}\tfrac{\boldsymbol{U_{\Delta T}}}{\boldsymbol{L}}$$

As we are in elasticity, we can superimpose the load cases and thus superimpose the case of the beam in pure bending (due to the moment M at each extremity of the beam) with the case of the beam under thermal gradient (uniform constant ΔT), we then obtain:

$$\frac{\mathbf{1}}{\boldsymbol{R^2}} + \left(\frac{\boldsymbol{\alpha\Delta T}}{\boldsymbol{h}}\right)^2 = \frac{\mathbf{2}}{\boldsymbol{EI}}\left(\frac{\boldsymbol{U_M}}{\boldsymbol{L}}\right) + \frac{\mathbf{2}}{\boldsymbol{EI}}\left(\frac{\boldsymbol{U_{\Delta T}}}{\boldsymbol{L}}\right) = \frac{\mathbf{2}}{\boldsymbol{EI}}\left(\frac{\boldsymbol{W}_{\mathbf{ext(total(}\boldsymbol{M}+\boldsymbol{\Delta T}\mathbf{))}}}{\boldsymbol{L}}\right)$$

For memory, the complete Einstein's formula in general relativity, with the cosmological constant Λ (dark energy effect, space expansion), is:

$$R^{\mu\nu} - \frac{1}{2} g^{\mu\nu} R + \Lambda g^{\mu\nu} = -\frac{8\pi G}{c^4} T^{\mu\nu}$$

Note: According to an article published on October 12, 2020 in the scientific journal Astrophysical journal {179}, contrary to popular belief, the universe is heating up (there are hotspots and colder spots). It would therefore be interesting to study whether there is not a correlation between the expansion of space linked to a possible elastic material with a coefficient of thermal expansion α, to this temperature rise and to a possible associated thermal gradient [2]. If this were the case, the result would be an additional curvature of space-time linked to this rise in temperature, which could itself be linked to the cosmological constant. If it is the case, dark energy could then be correlated with a thermal expansion of space see also {180}. But, for the moment, this is not proven.

The comparison of dimensional equations of Einstein's theory and the beam theory that results from a simplification of the elasticity theory that itself derives from simplification from the Continuum mechanics is also conclusive regarding the parallelism between these two theories (see table 3.2).

3.5 Extension of Curvature to Other Strength of Material Solicitations

We can extend the reasoning done in bending to all types of solicitations that exist in strength of material (see table 3.2).

The various equations and their associated dimensions are given in table 3.3.

Note: If we take our analogy of the one-dimensional beam, we find a link between the curvature and the deformations of the beam.

Indeed, the flexural normal stress is written as a function of the moment M (N.m), of the inertia I (m^4) and of the ordinate v (m) ($v_{\max} = h/2$):

$$\sigma = \frac{M_{(x)} v}{I} \rightarrow \frac{\sigma I}{v} = M_{(x)}$$

By transferring this expression in the energy-curvature formula, we obtain:

$$\left(\frac{d^2 y}{dx^2}\right)^2 = \frac{M^2{}_{(x)}}{E^2 I^2} = \frac{1}{R^2}$$

$$\left(\frac{d^2 y}{dx^2}\right)^2 = \frac{\sigma^2}{E^2 v^2} = \frac{1}{R^2}$$

TAB. 3.2 – Comparison of space curvature equations = K linear energy density for the different stresses present in strength of material.

Solicitations	Relationship solicitation/motion/rotation/deformation	Related displacement/rotation	Curvature/energy formula
Bending moment $M_{(x)}$	$\frac{d^2y}{dx^2} = \frac{d\theta}{dx} = -\frac{M}{EI} = \frac{1}{R}$	$y_{(x)}$ The vertical deflection in bending	$\frac{1}{R^2} = \frac{2}{EI} \times \frac{U}{L}$
Twisting torque $T_{(x)}$	$T = GI_t \frac{d\theta}{dx} = \mu I_t \frac{d\theta}{dx}$	$\theta_{(x)}$ Twist rotation	$\frac{1}{R_T^2} = \frac{2}{GI_t} \times \frac{U}{L}$
Normal effort $N_{(x)}$	$N = ES \frac{du}{dx} = ES\varepsilon = \sigma S$	$u_{(x)}$ Longitudinal displacement (lengthening/shortening)	$\left(\frac{du}{dx}\right)^2 = \frac{2}{ES} \times \frac{U}{L}$
Shear effort $V_{(x)}$	$V = GS_r \frac{dy}{dx} = GS_r \gamma_{(x)} = \mu S_r \gamma_{(x)}$	$\gamma_{(x)}$ Shear angle distortion	$\left(\frac{dy}{dx}\right)^2 = \frac{2}{GS_r} \times \frac{U}{L}$
General relativity	To be determined	$g_{\mu\nu}$	$R_{\mu\nu} - \frac{1}{2} g_{\mu\nu} R + g_{\mu\nu} = -\frac{8\pi G}{c^4} T_{\mu\nu}$

TAB. 3.3 – Dimensional equation for different formulation space curvature = $K \times$ energy density.

Solicitations	Curvature/energy formula	Dimensional equations
Bending moment of $M_{(x)}$	$\frac{1}{R^2} = \frac{2}{EI} \times \frac{U}{L} = \frac{2}{YI} \times \frac{U}{L}$	$\frac{1}{\text{m}^2} = \frac{\text{s}^2}{\text{kgm}^3} \times \frac{U}{\text{m}}$
Twisting torque $T_{(x)}$	$\frac{1}{R_T^2} = \frac{2}{GI_t} \times \frac{U}{L} = \frac{2}{\mu I_t} \times \frac{U}{L}$	$\frac{1}{\text{m}^2} = \frac{\text{s}^2}{\text{kgm}^3} \times \frac{U}{\text{m}}$
Normal effort $N_{(x)}$	$\left(\frac{du}{dx}\right)^2 = \frac{2}{ES} \times \frac{U}{L} = \frac{2}{YS} \times \frac{U}{L}$	$1 = \frac{\text{s}^2}{\text{kgm}} \times \frac{U}{\text{m}}$
Shear effort $V_{(x)}$	$\left(\frac{dy}{dx}\right)^2 = \frac{2}{GS_r} \times \frac{U}{L} = \frac{2}{\mu S_r} \times \frac{U}{L}$	$1 = \frac{\text{s}^2}{\text{kgm}} \times \frac{U}{\text{m}}$
General relativity	$R_{\mu\nu} - \frac{1}{2} g_{\mu\nu} R + g_{\mu\nu} = -\frac{8\pi G}{c^4} T_{\mu\nu}$	$\frac{1}{\text{m}^2} = \frac{\text{s}^2}{\text{kgm}} \times \frac{U}{\text{m}^3}$

With Hooke's law:

$$\sigma = \varepsilon E$$

By introducing this equation in the above equation of the curvature, we get:

$$\left(\frac{d^2y}{dx^2}\right)^2 = \frac{\varepsilon^2 E^2}{E^2 v^2} = \frac{1}{R^2}$$

$$\left(\frac{d^2y}{dx^2}\right)^2 = \frac{\varepsilon^2}{v^2} = \frac{1}{R^2}$$

So we finally have:

$$\frac{1}{R^2} = \left(\frac{d^2y}{dx^2}\right)^2 = \left(\frac{M}{EI}\right)^2 = \frac{\sigma^2}{E^2 v^2} = \frac{\varepsilon^2}{v^2} = \frac{2}{EI}\frac{U}{L}$$

The curvature of the beam can therefore be expressed as a function of the deformations of the elastic medium.

We will see later that the linearized form of the curvature in Einstein's equation (partial derivative functions of $h_{\mu\nu}$) can indeed be related to the deformations ($\varepsilon_{\mu\nu}$) of space (see chapter 5).

3.6 Analysis of Einstein's Equation Applied to the Entire Universe (Case of Cosmology)

Finally, if we apply Einstein's equation to the entire universe, we end up with Friedmann–Lemaitre's equation (see appendix E for their detailed demonstration) whose dimensions effectively intersect those announced by T. Damour (curvature $= 1/\text{m}^2$).

Einstein's equation is written:

$$R_{\mu\nu} - \frac{1}{2} g_{\mu\nu} R + \Lambda g_{\mu\nu} = -\frac{8\pi G}{c^4} T_{\mu\nu}$$

In this expression we have:

$$\kappa = 8\pi \frac{\frac{\mathrm{m}^3}{\mathrm{kgs}^2}}{\left(\frac{\mathrm{m}}{\mathrm{s}}\right)^4} = \frac{\mathrm{s}^2}{\mathrm{kgm}} = \mathrm{N}^{-1}$$

It is important to note also that κ has the same dimension (N^{-1}) as the coefficient of proportionality between curvature and energy that in the case of normal and shear forces (see table 3.3) in link with compression/traction strains ε (measured by the Ligo/Virgo interferometers) and shear strains γ that should be measured by the interferometers (see chapter 9).

Or in its expanded form by removing the cosmological constant:

$$\boldsymbol{R}^{\mu\nu} - \frac{\mathbf{1}}{\mathbf{2}} \boldsymbol{g}^{\mu\nu} \boldsymbol{R} = -\frac{\mathbf{8}\boldsymbol{\pi G}}{\boldsymbol{c}^4} \boldsymbol{T}^{\mu\nu} = -\frac{\mathbf{8}\boldsymbol{\pi G}}{\boldsymbol{c}^4} \{-\boldsymbol{p g}^{\mu\nu} + (\boldsymbol{\rho} + \boldsymbol{p}) \boldsymbol{u}^{\mu} \boldsymbol{u}^{\nu}\}$$

The application of Einstein's equation to the space-time metric of the universe leads to Friedmann's equations. The system of equations that leads to Friedmann's equations is given explicitly below. The following development comes from the "cosmology" lecture of Richard Taillet teacher-researcher at the University of Savoie Mont Blanc.

The interval can still be written as follows:

$$ds^2 = c^2 dt^2 - a^2(t) \left\{ \frac{dr^2}{1 - kr^2} + r^2 \left(d\theta^2 + \sin^2\theta \, d\varphi^2 \right) \right\}$$

The dimensional equation is checked:

$$\mathrm{m}^2 = \left(\frac{\mathrm{m}}{\mathrm{s}}\right)^2 \mathrm{s}^2 - 1 \left\{ \frac{\mathrm{m}^2}{1 - \frac{1}{\mathrm{m}^2}\mathrm{m}^2} + \mathrm{m}^2 \right\}$$

With k curvature:

$$S_{k\left(\frac{r}{R}\right)} = \begin{pmatrix} R\sin\left(\frac{r}{R}\right) \rightarrow \text{spheric} & k = 1 \\ r \rightarrow \text{flat} & k = 0 \\ R\sinh\left(\frac{r}{R}\right) \rightarrow \text{hyperbolic} & k = -1 \end{pmatrix}$$

Note: Astrophysical measurements give a zero curvature ($k \approx 0$), which seems to say that the universe has a gigantic radius of curvature, giving us the impression of a local flatness. The inflation hypothesis explains this very large radius (very large increase in volume of the universe in a very short time).

When the universe expands (Hubble Law), that is, when $a(t)$ increases, this physical distance there ℓ, the distance between the moving spheres increases: $\ell = a(t)\, r$.

The scalar factor, $a(t)$ replaces $R(t)$ we note:

$$a(t) = \frac{R(t)}{R_0}$$

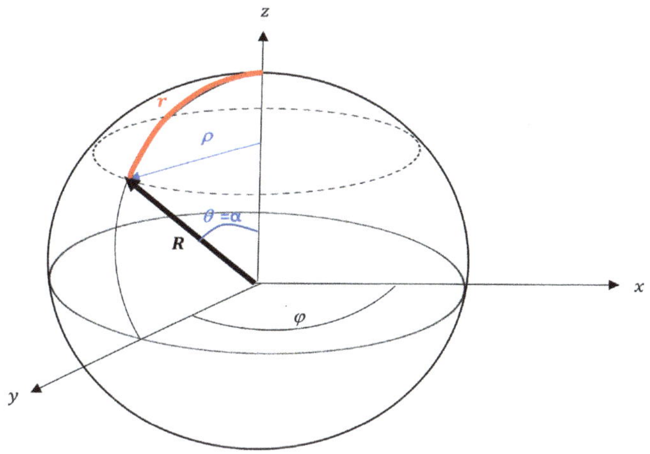

FIG. 3.5 – Spherical comoving coordinates.

where R_0 is a reference radius (universe today for example):

$$r = R(t)\alpha \,\text{et}\, r = R_0\alpha$$

The comoving coordinates are given in figure 3.5.
We choose the following parameters:

$$dt^2;\, dr^2;\, d\theta^2;\, d\varphi^2$$

We define the metric as follows:

$$g_{\mu\nu} = \begin{bmatrix} c^2 & 0 & 0 & 0 \\ 0 & -\frac{a^2_{(t)}}{1-kr^2} & 0 & 0 \\ 0 & 0 & -a^2_{(t)}r^2 & 0 \\ 0 & 0 & 0 & -a^2_{(t)}r^2\sin^2\theta \end{bmatrix}$$

Note: If we choose for time (indexes 0) the parameter c^2dt^2 rather than dt^2 the dimensional equation for the first parameter becomes 1 rather than $\mathrm{m}^2/\mathrm{s}^2$ compatible with an elasticity deformation without dimension.

Note that components 3 and 4 have the size of the square meter.

$$g_{\mu\nu} = \begin{bmatrix} 1 & 0 & 0 & 0 \\ 0 & -\frac{a^2_{(t)}}{1-kr^2} & 0 & 0 \\ 0 & 0 & -a^2_{(t)}r^2 & 0 \\ 0 & 0 & 0 & -a^2_{(t)}r^2\sin^2\theta \end{bmatrix}$$

For the calculations which follow we take for the term dt^2 for the first term of the matrix.

The Ricci tensor that derives from the curvature tensor is written (see appendix E for the proof):

$$R_{\mu\nu} = \begin{bmatrix} 3\left(\frac{\ddot{a}}{a}\right) & 0 & 0 & 0 \\ 0 & -\frac{2k}{1-kr^2} - \frac{\ddot{a}a}{c^2(1-kr^2)} - \frac{2\dot{a}^2}{c^2(1-kr^2)} & 0 & 0 \\ 0 & 0 & -\frac{2r^2\dot{a}^2}{c^2} - 2kr^2 - \frac{r^2a\ddot{a}}{c^2} & 0 \\ 0 & 0 & 0 & -2\left(\frac{\dot{a}^2r^2\sin^2\theta}{c^2}\right) - 2kr^2\sin^2\theta - \frac{\ddot{a}ar^2\sin^2\theta}{c^2} \end{bmatrix}$$

The contraction of Ricci's tensor $R_{\mu\nu}$ gives us the scalar curvature:

$$R = \left\{\frac{6}{c^2}\left(\frac{\ddot{a}}{a}\right) + \frac{6}{c^2}\left(\frac{\dot{a}}{a}\right)^2 + \frac{6k}{a^2}\right\}$$

The stress energy tensor takes the following form in the case of a perfect fluid consisting of stars and galaxies distributed uniformly and isotropically throughout the universe.

$$T_{\mu\nu} = \begin{bmatrix} \rho c^2 & 0 & 0 & 0 \\ 0 & -p & 0 & 0 \\ 0 & 0 & -p & 0 \\ 0 & 0 & 0 & -p \end{bmatrix} \begin{bmatrix} c^2 & 0 & 0 & 0 \\ 0 & -\frac{a^2}{1-kr^2} & 0 & 0 \\ 0 & 0 & -a^2r^2 & 0 \\ 0 & 0 & 0 & -a^2r^2\sin^2\theta \end{bmatrix}$$

Applying Einstein's equations to the different matrices defined above, we obtain the explicit form of the 10 differential equations:

$$\begin{bmatrix} 3\left(\frac{\ddot{a}}{a}\right) & 0 & 0 & 0 \\ 0 & -\frac{2k}{1-kr^2} - \frac{\ddot{a}a}{c^2(1-kr^2)} - \frac{2\dot{a}^2}{c^2(1-kr^2)} & 0 & 0 \\ 0 & 0 & -\frac{2r^2\dot{a}^2}{c^2} - 2kr^2 - \frac{r^2a\ddot{a}}{c^2} & 0 \\ 0 & 0 & 0 & -2\left(\frac{\dot{a}^2r^2\sin^2\theta}{c^2}\right) - 2kr^2\sin^2\theta - \frac{\ddot{a}ar^2\sin^2\theta}{c^2} \end{bmatrix}$$

$$-\frac{1}{2}\begin{bmatrix} c^2 & 0 & 0 & 0 \\ 0 & -\frac{a^2}{1-kr^2} & 0 & 0 \\ 0 & 0 & -a^2r^2 & 0 \\ 0 & 0 & 0 & -a^2r^2\sin^2\theta \end{bmatrix}\left\{\frac{6}{c^2}\left(\frac{\ddot{a}}{a}\right) + \frac{6}{c^2}\left(\frac{\dot{a}}{a}\right)^2 + \frac{6k}{a^2}\right\}$$

$$+\Lambda\begin{bmatrix} c^2 & 0 & 0 & 0 \\ 0 & -\frac{a^2}{1-kr^2} & 0 & 0 \\ 0 & 0 & -a^2r^2 & 0 \\ 0 & 0 & 0 & -a^2r^2\sin^2\theta \end{bmatrix}$$

$$= -\frac{8\pi G}{c^4}\begin{bmatrix} \rho c^2 & 0 & 0 & 0 \\ 0 & -p & 0 & 0 \\ 0 & 0 & -p & 0 \\ 0 & 0 & 0 & -p \end{bmatrix} \begin{bmatrix} c^2 & 0 & 0 & 0 \\ 0 & -\frac{a^2}{1-kr^2} & 0 & 0 \\ 0 & 0 & -a^2r^2 & 0 \\ 0 & 0 & 0 & -a^2r^2\sin^2\theta \end{bmatrix}$$

The dimensional equation of the first equation is $1/\mathrm{s}^2$:

$$\frac{1}{\mathrm{s}^2}+\frac{1}{2}\left(\frac{\mathrm{m}^2}{\mathrm{s}^2}\right)\times\left\{\frac{\mathrm{s}^2}{\mathrm{m}^2}\times\frac{1}{\mathrm{s}^2}+\frac{\mathrm{s}^2}{\mathrm{m}^2}\times\frac{1}{\mathrm{s}^2}+\frac{1}{\mathrm{m}^2}\right\}+\frac{1}{\mathrm{m}^2}\left[\frac{\mathrm{m}^2}{\mathrm{s}^2}\right]=\frac{\left(\frac{\mathrm{m}^3}{\mathrm{kgs}^2}\right)}{\left(\frac{\mathrm{m}}{\mathrm{s}}\right)^4}\times\frac{\mathrm{kg}}{\mathrm{m}^3}\times\frac{\mathrm{m}^4}{\mathrm{s}^4}=\frac{1}{\mathrm{s}^2}$$

The dimensional equation of the second equation is $1/\mathrm{m}^2$:

$$-\frac{1}{\mathrm{m}^2}-\frac{1}{\frac{\mathrm{m}^2}{\mathrm{s}^2}}\left(\frac{1}{\mathrm{s}^2}\right)-\frac{1}{\frac{\mathrm{m}^2}{\mathrm{s}^2}}\left(\frac{1}{\mathrm{s}^2}\right)-\frac{1}{2}(1)\left\{\frac{\mathrm{s}^2}{\mathrm{m}^2}\times\frac{1}{\mathrm{s}^2}+\frac{\mathrm{s}^2}{\mathrm{m}^2}\times\frac{1}{\mathrm{s}^2}+\frac{1}{\mathrm{m}^2}\right\}+\frac{1}{\mathrm{m}^2}[1]$$
$$=\frac{\left(\frac{\mathrm{m}^3}{\mathrm{kgs}^2}\right)}{\left(\frac{\mathrm{m}}{\mathrm{s}}\right)^4}\times\frac{\mathrm{kgm}}{\mathrm{s}^2\mathrm{m}^2}\times\frac{1}{1}=\frac{\mathrm{m}^3\mathrm{s}^4}{\mathrm{kgs}^2\mathrm{m}^4}\times\frac{\mathrm{kgm}}{\mathrm{s}^2\mathrm{m}^2}\times\frac{1}{1}=\frac{1}{\mathrm{m}^2}$$

The dimensional equation of the 3rd and 4th lines are the unit:

$$-\frac{\mathrm{m}^2}{\frac{\mathrm{m}^2}{\mathrm{s}^2}}\left(\frac{1}{\mathrm{s}^2}\right)-\frac{\mathrm{m}^2}{\mathrm{m}^2}-\frac{\mathrm{m}^2}{\frac{\mathrm{m}^2}{\mathrm{s}^2}}\left(\frac{1}{\mathrm{s}^2}\right)+\frac{1\mathrm{m}^2}{2}\left\{\frac{\mathrm{s}^2}{\mathrm{m}^2}\times\frac{1}{\mathrm{s}^2}+\frac{\mathrm{s}^2}{\mathrm{m}^2}\times\frac{1}{\mathrm{s}^2}+\frac{1}{\mathrm{m}^2}\right\}+\frac{1}{\mathrm{m}^2}\left[\mathrm{m}^2\right]$$
$$=\frac{\left(\frac{\mathrm{m}^3}{\mathrm{kgs}^2}\right)}{\left(\frac{\mathrm{m}}{\mathrm{s}}\right)^4}\times\frac{\mathrm{kgm}}{\mathrm{s}^2\mathrm{m}^2}\times\frac{\mathrm{m}^2}{1}=\frac{\mathrm{m}^3\mathrm{s}^4}{\mathrm{kgs}^2\mathrm{m}^4}\times\frac{\mathrm{kgm}}{\mathrm{s}^2\mathrm{m}^2}\times\frac{\mathrm{m}^2}{1}=1$$

If we multiply each member by $1/\mathrm{m}^2$ we get well $1/\mathrm{m}^2$.

3.7 Conclusions

At the conclusion of this chapter, we have obtained a first solid piece of the puzzle. Einstein's equation seems to be a 4-dimensional generalization of the theory of continuum mechanics which itself leads to the elasticity theory, the simplification of which is the strength of materials that we have used here. This approach consisting in studying general relativity *via* the elasticity theory has been carried out by many authors, see [31–55]. However, we need to rise to a higher level and go into tensor mode to have expressions independent of the frame of reference in which we place ourselves, this is what makes tensors strong. We will therefore see if there are bridges between the stress energy tensor $T_{\mu\nu}$ and the curvature tensor $G_{\mu\nu}$ on the one hand, with classical tensors in Continuum mechanics respectively, the stresses σ_{ij} and strains ε_{ij} tensors on the other hand. Conclusion Here's the first piece of our puzzle:

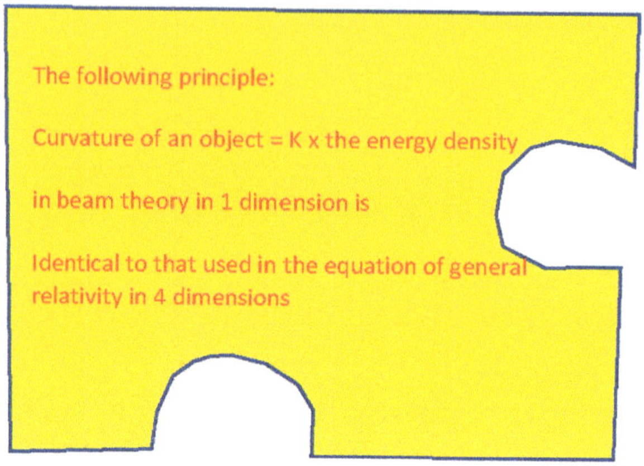

1st piece of the puzzle

Chapter 4

The Stress Energy Tensor in Theory of General Relativity and the Stress Tensor in Elasticity Theory are Similar

Chapter Summary

$$T_{\mu\nu} = \rho u_\mu u_\nu \leftrightarrow \sigma_{ij} = \rho v_i v_j$$

4.1 Definition of the Stress Energy Tensor in General Relativity

In the case of the homogeneous and isotropic universe, the right part (mass-energy) of the stress energy tensor is written [57]:

$$T_{\mu\nu} = \begin{bmatrix} \rho c^2 & 0 & 0 & 0 \\ 0 & -p & 0 & 0 \\ 0 & 0 & -p & 0 \\ 0 & 0 & 0 & -p \end{bmatrix} g_{\mu\nu}$$

Each of the terms to the following dimension:

$$\rho c^2 = \frac{\text{kg}}{\text{m}^3} \times \left(\frac{\text{m}}{\text{s}}\right)^2 = \frac{\frac{\text{kgm}}{\text{s}^2}}{\text{m}^2} = \frac{\text{kg}}{\text{m.s}^2}$$

$$p = \frac{\frac{\text{kgm}}{\text{s}^2}}{\text{m}^2} = \frac{\text{N}}{\text{m}^2} = \frac{\text{kg}}{\text{m.s}^2}$$

That is, a force divided by a surface in other words a pressure.

DOI: 10.1051/978-2-7598-2573-8.c004

4.2 Definition of Stress Tensor in Elasticity Theory

It turns out that the 3-dimensional stress tensor in mechanics also has this dimension. It is usually expressed in MPa (N/mm^2) or Pa (N/m^2).

$$\sigma_{ij} = \begin{bmatrix} \sigma_{xx} & \tau_{xy} & \tau_{xz} \\ \tau_{yx} & \sigma_{yy} & \tau_{yz} \\ \tau_{zx} & \tau_{zy} & \sigma_{zz} \end{bmatrix}$$

Like the stress energy tensor, the stress tensor is symmetrical. The terms on either side of the diagonal passing through σ_{xx} and σ_{zz} are equal.

It should be noted, however, that the stress energy tensor in general relativity corresponds to masses-energy that distorts space-time, much like external forces acting on the beam, and not as internal stresses to the beam as the stress tensor represents. However, it is well known that in mechanics the work of external forces corresponds to the work of the internal forces and thus in a way the stress tensor would be the internal image of the deformations of space-time and the stress energy tensor the external actions applied to the fabric of space. This equivalence we demonstrated in chapter 3 or we showed on this simple example of pure bending beam that the work of external forces is always equal to the work of inner forces in elasticity.

Is it therefore possible to demonstrate mathematically that both are similar?

This demonstration was done in free online physics courses [56]. We demonstrate it again below.

4.3 Demonstration of the Correlation between the Stress Tensor and the Stress Energy Tensor

According to the elasticity theory that derives from the continuum mechanics we consider the stress vector $\overrightarrow{T}$ applied on a facet of surface S and normal $\overrightarrow{n}$ charged by force $\overrightarrow{Q}$:

$$\overrightarrow{T}_{(x,y,z,t,\vec{n})} = \overrightarrow{T}_{(\vec{x},t,\vec{n})} = \lim_{S\to 0} \frac{\overrightarrow{Q}}{S}$$

The relationship between the stress vector $\overrightarrow{T}$, the stress tensor σ_{ij} and the normal to the facet $\overrightarrow{n}$ is written for the component T_i of the stress vector (see figure 4.1):

$$T_i = \sigma_{ij} n_j$$

In a P-point, in a three-dimensional $\overrightarrow{e}_i(\overrightarrow{e}_1; \overrightarrow{e}_2; \overrightarrow{e}_3)$ base, each component of T_i the stress vector $\overrightarrow{T}$ is the sum of the 3 parts of stresses applied to the facet:

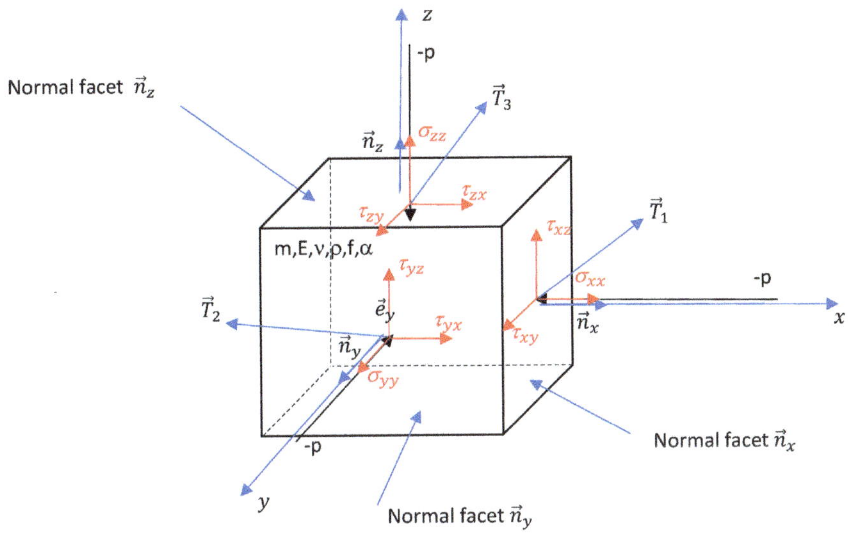

FIG. 4.1 – Visualization of the stress tensor.

$$\begin{cases} T_1 = \sigma_{11}n_1 + \sigma_{12}n_2 + \sigma_{13}n_3 \\ T_2 = \sigma_{21}n_1 + \sigma_{22}n_2 + \sigma_{23}n_3 \\ T_3 = \sigma_{31}n_1 + \sigma_{32}n_2 + \sigma_{33}n_3 \end{cases} \rightarrow \begin{cases} T_1 = \sigma_{xx}n_x + \tau_{xy}n_y + \tau_{xz}n_z \\ T_2 = \tau_{yx}n_x + \sigma_{yy}n_y + \tau_{yz}n_z \\ T_3 = \tau_{zx}n_x + \tau_{zy}n_y + \sigma_{zz}n_z \end{cases}$$

What can still be written with Einstein's sum convention on index j here:

$$T_i = \sigma_{ij}n^j$$

The relationship between force and stress is written:

$$Q_i = \sigma_{ij}S_j$$

Considering a variational approach, the stress tensor is worth:

$$\sigma_{ij} = \frac{\Delta Q_i}{\Delta S_j} \quad \text{with} \quad \Delta S_j \rightarrow 0$$

ΔS_i is an area:

So, with a mass m, a density ρ, a volume V and acceleration a_i, we have:

$$\sigma_{ij} = \frac{\Delta Q_i}{\Delta S_j} = \frac{\Delta(m \times a_i)}{\Delta S_j} = \frac{\Delta(\rho \times V \times a_i)}{\Delta S_j}$$

Considering that the force variation is due only to a change in volume V a time function of t, we get with, $a_i = \frac{v_i}{\Delta t}$,

$$\sigma_{ij} = \frac{\Delta(\rho \times V \times a_i)}{\Delta S_j} = \rho \frac{1}{\Delta S_j}\left(\frac{\Delta V}{\Delta t}\right)v_i$$

In addition, we introduce the expression of the volume $V = \Delta x_i \times \Delta x_j \times \Delta x_k$ in the above expression:

$$\sigma_{ij} = \rho \frac{1}{\Delta S_j} \left(\frac{\Delta x_i \times \Delta x_j \times \Delta x_k}{\Delta t} \right) v_i$$

We can replace the surface S_j with this value:

$$\Delta S_j = \Delta x_i \times \Delta x_k$$

So, the new expression of the tensor is written:

$$\sigma_{ij} = \rho \frac{v_i}{\Delta t} \left(\frac{\Delta x_i \times \Delta x_j \times \Delta x_k}{\Delta x_i \times \Delta x_k} \right)$$

After simplification we get:

$$\sigma_{ij} = \rho v_i \left(\frac{\Delta x_j}{\Delta t} \right)$$

By definition of speed, we have: v_j

$$v_j = \left(\frac{\Delta x_j}{\Delta t} \right)$$

We finally get the expression of the low-speed stress tensor based on density ρ and speeds v_i and v_j:

$$\boldsymbol{\sigma_{ij} = \rho v_i v_j}$$

In General Relativity, the stress energy tensor results from the product of density ρ by the product of four-speeds u_μ and u_ν following the 4 dimensions of space-time:

$$\boldsymbol{T_{\mu\nu} = \rho u_\mu u_\nu}$$

And for a perfect fluid considering a P pressure

$$T_{\mu\nu} = P g_{\mu\nu} - (\rho + P) u_\mu u_\nu$$

With for the four-vector speed u:

$$u_\mu \left| \begin{array}{l} \gamma \times c \\ \gamma \times v_x \\ \gamma \times v_y \\ \gamma \times v_z \end{array} \right.$$

The Lorentz factor γ resulting from the transformation of Lorentz (see special relativity) is written:

$$\gamma = \frac{1}{\sqrt{1 - \frac{v^2}{c^2}}}$$

The 4-dimensional stress energy tensor is therefore a matrix whose components are given below:

$$T_{\mu\nu} = \begin{bmatrix} \frac{m\gamma^2 c^2}{V} & \rho\gamma^2 cv_x & \rho\gamma^2 cv_y & \rho\gamma^2 cv_z \\ \rho\gamma^2 cv_x & \rho\gamma^2 v_x v_x & \rho\gamma^2 v_x v_y & \rho\gamma^2 v_x v_z \\ \rho\gamma^2 cv_y & \rho\gamma^2 v_y v_x & \rho\gamma^2 v_y v_y & \rho\gamma^2 v_y v_z \\ \rho\gamma^2 cv_z & \rho\gamma^2 v_z v_x & \rho\gamma^2 v_z v_y & \rho\gamma^2 v_z v_z \end{bmatrix}$$

where γ is the Lorentz factor (see special relativity), energy density $\rho = \frac{m}{V}$ with mass m, unit volume V, the speed of light c, a speed in direction i, v_i.

At low speed $\gamma = 1$ and the stress energy tensor is written:

$$T_{\mu\nu} = \begin{bmatrix} \frac{mc^2}{V} & \rho cv_x & \rho cv_y & \rho cv_z \\ \rho cv_x & \rho v_x v_x & \rho v_x v_y & \rho v_x v_z \\ \rho cv_y & \rho v_y v_x & \rho v_y v_y & \rho v_y v_z \\ \rho cv_z & \rho v_z v_x & \rho v_z v_y & \rho v_z v_z \end{bmatrix} = \begin{bmatrix} \rho c^2 & \rho cv_x & \rho cv_y & \rho cv_z \\ \rho cv_x & \rho v_x v_x & \rho v_x v_y & \rho v_x v_z \\ \rho cv_y & \rho v_y v_x & \rho v_y v_y & \rho v_y v_z \\ \rho cv_z & \rho v_z v_x & \rho v_z v_y & \rho v_z v_z \end{bmatrix}$$

Given the definition of the stress tensor, the low-speed stress energy tensor can be written:

$$T_{\mu\nu} = \begin{bmatrix} \frac{mc^2}{V} & \rho cv_x & \rho cv_y & \rho cv_z \\ \rho cv_x & \sigma_{xx} & \tau_{xy} & \tau_{xz} \\ \rho cv_y & \tau_{yx} & \sigma_{yy} & \tau_{yz} \\ \rho cv_z & \tau_{zx} & \tau_{zy} & \sigma_{zz} \end{bmatrix}$$

By comparing the expressions of the 4-dimensional stress energy tensor and the 3 dimensions stress tensors, we thus highlight the similarities between its 2 tensors.

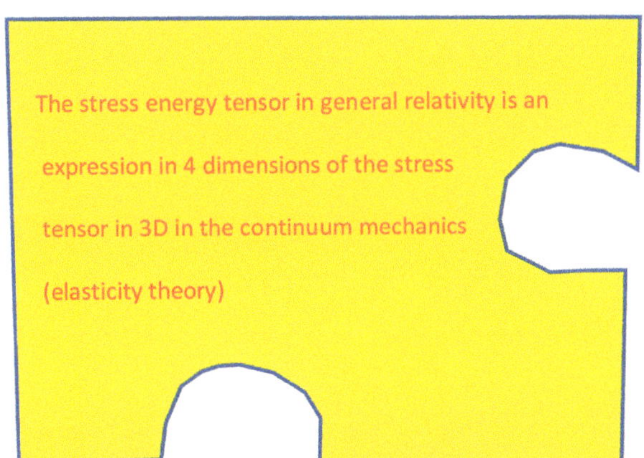

2nd piece of the puzzle

Chapter 5

Relationship between the Metric Tensor and the Strain Tensor in Low Gravitational Field

Chapter Summary

$$g_{ij} = \eta_{ij} + h_{ij} = \eta_{ij} + 2\varepsilon_{ij}$$

5.1 Introduction

The Ligo and Virgo interferometers are made up of two 90° arms in which a vacuum has been created and where a laser goes back and forth between mirrors. Both arms undergo space deformations during the passage of a gravitational wave, that is to say variations in length divided by the initial length before deformation. The question we ask ourselves is therefore the following: To what extent can we relate these deformations to the metric defined in general relativity and thus make a link with the strain tensor?

5.2 Definition of Strain Tensor

We defined the stress tensor σ_{ij} in the previous chapter 4, let's now define the strain tensor ε_{ij}. The general structure of the strain tensors ε_{ij} in the theory of 3-dimensional elasticity is written {4}:

$$\varepsilon_{ij} = \begin{bmatrix} \varepsilon_{xx} & \varepsilon_{xy} & \varepsilon_{xz} \\ \varepsilon_{yx} & \varepsilon_{yy} & \varepsilon_{yz} \\ \varepsilon_{zx} & \varepsilon_{zy} & \varepsilon_{zz} \end{bmatrix}$$

DOI: 10.1051/978-2-7598-2573-8.c005

With the displacements u_i and u_j (see figure 5.1) we have:

$$\varepsilon_{ij} = \frac{1}{2}\left(\frac{\partial u_i}{\partial x_j} + \frac{\partial u_j}{\partial x_i}\right)$$

For lengthening and shortening:

$$\varepsilon_{xx} = \varepsilon_x = \varepsilon = \frac{L_{x,\text{final}} - L_{x,\text{initial}}}{L_{x,\text{initial}}} = \frac{1}{2}\left(\frac{\partial u_x}{\partial x} + \frac{\partial u_x}{\partial x}\right) = \frac{\partial u_x}{\partial x}$$

$$\varepsilon_{yy} = \varepsilon_y = \frac{L_{y,\text{final}} - L_{y,\text{initial}}}{L_{y,\text{initial}}} = \frac{1}{2}\left(\frac{\partial u_y}{\partial y} + \frac{\partial u_y}{\partial y}\right) = \frac{\partial u_y}{\partial y}$$

$$\sigma = \varepsilon E$$

For the shear strain (distortion angle):

$$\varepsilon_{xy} = \gamma' = \frac{\gamma_{xy}}{2} = \frac{1}{2}\left(\frac{\partial u_x}{\partial \text{y}} + \frac{\partial u_y}{\partial x}\right)$$

$$\tau = \gamma_{xy} G = \gamma G$$

The proof is given in figure 5.1.

$$\text{tang}(2\gamma') \approx 2\gamma' = \frac{\frac{\partial u_x}{\partial y}\, dy}{dy} + \frac{\frac{\partial u_y}{\partial x}\, dx}{dx} = \frac{\partial u_x}{\partial y} + \frac{\partial u_y}{\partial x}$$

$$\gamma' = \varepsilon_{xy} = \frac{1}{2}\left(\frac{\partial u_x}{\partial y} + \frac{\partial u_y}{\partial x}\right)$$

The angle OAB initial is reduced to the angle $2\gamma' = \gamma_{xy}$; $O'A'B' = OAB - 2\gamma'$

$$\gamma_{xy} = 2\gamma' = 2\varepsilon_{xy}$$

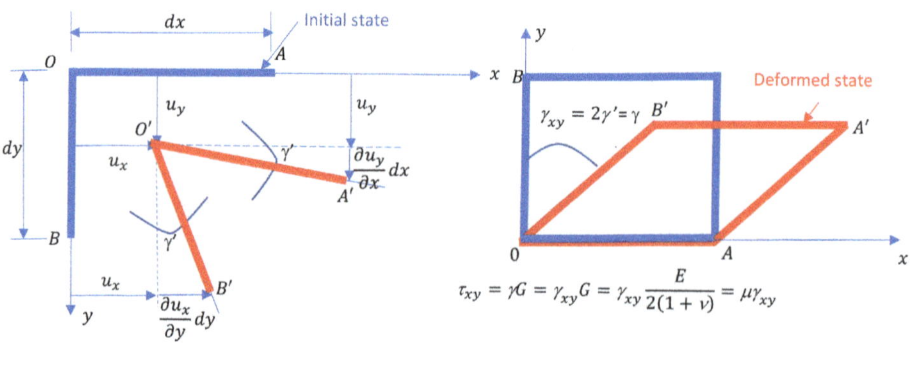

FIG. 5.1 – Visualization of the shear strains and displacements after deformations.

$$\gamma_{xy} = \left(\frac{\partial u_x}{\partial y} + \frac{\partial u_y}{\partial x}\right)$$

$$\varepsilon_{xy} = \gamma' = \frac{\gamma_{xy}}{2} = \frac{\gamma}{2}$$

5.3 Determination of the Link between the Metric and the Strain

The metric is defined as the sum of a flat metric $\eta_{\mu\nu}$ and a small perturbation $h_{\mu\nu}$ that will be determined:

$$g_{\mu\nu} = \eta_{\mu\nu} + h_{\mu\nu}$$

Minkowski's flat metric is written:

$$\eta_{\mu\nu} = \begin{bmatrix} 1 & 0 & 0 & 0 \\ 0 & -1 & 0 & 0 \\ 0 & 0 & -1 & 0 \\ 0 & 0 & 0 & -1 \end{bmatrix}$$

That is, the terms on the measurement of the interval (development of Pythagorean theorem in 4 dimensions) in general relativity:

$$ds^2 = 1\,dt^2 - 1\,dx^2 - 1\,dy^2 - 1\,dz^2$$

Einstein's linearized equation in weak gravitational field allowing him to calculate this metric perturbation is written:

$$\partial^{\lambda}\partial_{\lambda}\overline{h}_{\mu\nu} = \Box\overline{h}_{\mu\nu} = -\frac{16\pi G}{c^4}T_{\mu\nu}$$

In the vacuum the stress energy tensor is zero and Einstein's equation becomes equivalent to the equation of a wave:

$$\partial^{\lambda}\partial_{\lambda}\overline{h}_{\mu\nu} = \Box\overline{h}_{\mu\nu} = 0$$

Expression in which we have:
$\Box$ The D'alembertian
The disturbance of the metric is written:

$$h_{\mu\nu} = \overline{h}_{\mu\nu} - \frac{1}{2}\eta_{\mu\nu}\overline{h}$$

Or:

$$\overline{h}_{\mu v} = h_{\mu v} + \frac{1}{2}\eta_{\mu v}\overline{h}$$

Since we have a wave equation, the solution to this equation is:

$$\overline{h}_{\mu v} = A_{\mu v}\cos(k_\sigma x^\sigma)$$

$k^\sigma = \left(\frac{\omega}{c}; \overrightarrow{k}\right)$ four vector wave of the plane wave.

The coordinates in space-time:

$$x^\sigma = (ct, x, y, z)$$

$$\|\overrightarrow{k}\|^2 = \frac{\omega^2}{c^2}$$

ω is the circular frequency of the wave.

The amplitude of the wave shows that there are two polarizations of the wave rotated 45° to each other $(A_+; A_\times)$:

$$A_{\mu v} = A_+\begin{bmatrix} 0 & 0 & 0 & 0 \\ 0 & +1 & 0 & 0 \\ 0 & 0 & -1 & 0 \\ 0 & 0 & 0 & 0 \end{bmatrix} + A_\times\begin{bmatrix} 0 & 0 & 0 & 0 \\ 0 & 0 & +1 & 0 \\ 0 & +1 & 0 & 0 \\ 0 & 0 & 0 & 0 \end{bmatrix}$$

h is the trace (sum of terms following the diagonal = 0) of $h_{\mu v}$. The geometric translation (particle displacements in the plane xy initially on a circle) during the passage of a gravitational plane wave propagating in the z direction is given below:

For polarization A_+ (see figure 5.2).

The final position of the particle squared l^2 of the center of the circle is equal to the initial position (radius R squared) R^2 minus the variation in length squared related to the oscillating perturbation of the metric $R^2A_+(\cos(2\theta))\cos(\frac{\omega}{c}(ct - z))$ caused by the gravitational wave:

$$l^2 = R^2 - R^2A_+(\cos(2\theta))\cos\left(\frac{\omega}{c}(ct - z)\right)$$

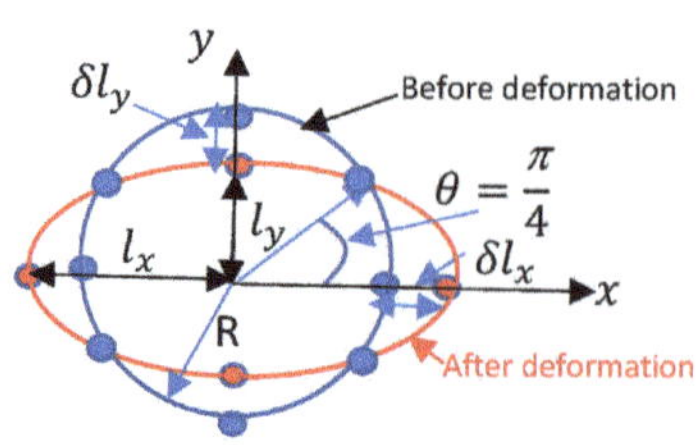

Fig. 5.2 – Example of particle coordinates being modified in the plane xy, for a subsequent polarized gravitational wave A_+ propagating in the z direction.

The proof is given below. The perturbation of the metric $h_{\mu\nu}$ is written:
For a polarized wave A_+:

$$h_{\mu\nu} = A_+ \cos\left(\frac{\omega}{c}(ct - z)\right) \begin{bmatrix} 0 & 0 & 0 & 0 \\ 0 & \mathbf{+1} & \mathbf{0} & \mathbf{0} \\ 0 & \mathbf{0} & \mathbf{-1} & \mathbf{0} \\ 0 & \mathbf{0} & \mathbf{0} & \mathbf{0} \end{bmatrix}$$

For a polarized wave $A_\times$:

$$h_{\mu\nu} = A_\times \cos\left(\frac{\omega}{c}(ct - z)\right) \begin{bmatrix} 0 & 0 & 0 & 0 \\ 0 & \mathbf{0} & \mathbf{+1} & \mathbf{0} \\ 0 & \mathbf{+1} & \mathbf{0} & \mathbf{0} \\ 0 & \mathbf{0} & \mathbf{0} & \mathbf{0} \end{bmatrix}$$

The two polarizations are similar but offset by an angle of 45°. The spatial part of the perturbation of the $h_{\mu\nu}$ metric is in bold. The interval is written (see lectures of R. Taillet [57]):

$$ds^2 = g_{\mu\nu}\, dx^\mu\, dx^\nu$$

We are looking for the displacement of particles positioned on a circle of radius R (see figure 5.2) in the xy plane perpendicular to the direction of propagation z of the gravitational wave. The infinitely small interval becomes the final length l of the particles position from the center of the circle. Moreover, we pass in space coordinates ij:

$$l^2 = -g_{ij}\Delta x^i \Delta x^j$$

We can replace the metric tensor g_{ij} by its expression function of the perturbation h_{ij} and of the flat metric η_{ij}:

$$l^2 = -(\eta_{ij} + h_{ij})\Delta x^i x^j$$

$$l^2 = -\eta_{ij}\Delta x^i x^j - h_{ij}\Delta x^i \Delta x^j$$

The first term corresponds to a distance in flat geometry. Only the terms on the diagonal are different from 0:

$$\eta_{11}\Delta x^1 \Delta x^1 + \eta_{22}\Delta x^2 \Delta x^2 = -x^2 - y^2 = -R^2 \cos\theta^2 - R^2 \sin\theta^2 = -R^2$$

So:

$$l^2 = R^2 - h_{ij}\Delta x^i \Delta x^j$$

With the definition of $h_{\mu\nu}$ for the polarization A_+:

$$l^2 = R^2 - x^2 A_+ \cos\left(\frac{\omega}{c}(ct - z)\right) + y^2 A_+ \cos\left(\frac{\omega}{c}(ct - z)\right)$$

So:

$$l^2 = R^2 - \left(x^2 - y^2\right) A_+ \cos\left(\frac{\omega}{c}(ct - z)\right)$$

With the definition of $x = R\cos\theta$ and $y = R\sin\theta$:

$$l^2 = R^2 - R^2 A_+ \left(\cos\theta^2 - \sin\theta^2\right) \cos\left(\frac{\omega}{c}(ct - z)\right)$$

So:

$$l^2 = R^2 \left[1 - A_+ (\cos(2\theta)) \cos\left(\frac{\omega}{c}(ct - z)\right)\right]$$

$$l = R\sqrt{\left[1 - A_+ (\cos(2\theta)) \cos\left(\frac{\omega}{c}(ct - z)\right)\right]}$$

$$l \approx R\left[1 - \frac{A_+}{2} (\cos(2\theta)) \cos\left(\frac{\omega}{c}(ct - z)\right)\right]$$

For the other polarization we have:

$$l^2 = R^2 \left[1 - A_\times (\sin(2\theta)) \cos\left(\frac{\omega}{c}(ct - z)\right)\right]$$

$$l \approx R\left[1 - \frac{A_\times}{2} (\sin(2\theta)) \cos\left(\frac{\omega}{c}(ct - z)\right)\right]$$

By showing a strain:

$$\frac{l - R}{R} = -\varepsilon \approx -\frac{A_+}{2} (\cos(2\theta)) \cos\left(\frac{\omega}{c}(ct - z)\right) \approx -\frac{h_{\mu\nu}}{2}$$

Indeed, l^2 is the final squared length and R^2 is the initial squared length.

We can therefore calculate the length variation due to the movement of particles on the circle:

$$\frac{l^2 - R^2}{R^2} = \frac{(\text{Final length})^2 - (\text{Initial length})^2}{(\text{Initial length})^2} = -A_+ (\cos(2\theta)) \cos\left(\frac{\omega}{c}(ct - z)\right)$$

where $L_F = L_i + \delta_i$ the final length, L_i the initial length and δ_i is the variation in length.

With δ_i small front L_i:

$$\frac{l^2 - R^2}{R^2} = \frac{L_F^2 - L_i^2}{L_i^2} = \frac{L_i^2 + 2L_i\delta_i + \delta_i^2 - L_i^2}{L_i^2}$$

Neglecting the terms δ_i^2 in front of the terms δ_i, in it comes:

$$\frac{l^2 - R^2}{R^2} \cong \frac{2\delta_i}{L_i}$$

And if we do not neglect the squared terms:

$$\frac{l^2 - R^2}{R^2} = \frac{L_F^2 - L_i^2}{L_i^2} = \frac{2\delta_i}{L_i} + \frac{\delta_i^2}{L_i^2}$$

And with the definition of a strain:

$$\varepsilon = \frac{L_F - L_i}{L_i} = \frac{L_i + \delta_i - L_i}{L_i} = \frac{\delta_i}{L_i}$$

We finally get:

$$\frac{l^2 - R^2}{R^2} = \frac{L_F^2 - L_i^2}{L_i^2} \cong \frac{2\delta_i}{L_i} = 2\varepsilon$$

And if we do not neglect the squared terms:

$$\frac{l^2 - R^2}{R^2} = \frac{L_F^2 - L_i^2}{L_i^2} = 2\varepsilon + \varepsilon^2$$

To correlate with Hooke's law $\sigma = \varepsilon E$.

$$\frac{l^2 - R^2}{R^?} \cong -A_+ (\cos(2\theta)) \cos\left(\frac{\omega}{c}(ct - z)\right) \cong 2\varepsilon$$

So, in low field the perturbation of the metric $h_{\mu\nu}$ is well related to the metric $g_{\mu\nu}$.

$$\boldsymbol{g_{\mu\nu} = \eta_{\mu\nu} + h_{\mu\nu} = \eta_{\mu\nu} + 2\varepsilon_{\mu\nu}}$$

$$\boldsymbol{g_{\mu\nu} - \eta_{\mu\nu} = h_{\mu\nu} = 2\varepsilon_{\mu\nu}}$$

$$\boldsymbol{\frac{g_{\mu\nu} - \eta_{\mu\nu}}{2} = \frac{h_{\mu\nu}}{2} = \varepsilon_{\mu\nu}}$$

And if we do not neglect the squared terms:

$$g_{\mu\nu} = \eta_{\mu\nu} + h_{\mu\nu} = \eta_{\mu\nu} + 2\varepsilon_{\mu\nu} + \varepsilon_{\mu\nu}^2$$

$$\boldsymbol{h_{\mu\nu} = 2\varepsilon_{\mu\nu} + \varepsilon_{\mu\nu}^2}$$

For memory Ligo measured $\varepsilon = 10^{-21}$ that is why ε^2 can be neglected.

Remarque

It is interesting to compare the last expression with the beam curvature expression obtained in strength of material (see chapter 3) we well see a relation between a curvature and a strain square: $\frac{1}{R^2} = \left(\frac{d^2y}{dx^2}\right)^2 = \left(\frac{M}{EI}\right)^2 = \frac{\sigma^2}{E^2 v^2} = \frac{\varepsilon^2}{v^2} = \frac{2}{EI}\frac{U}{L}$.

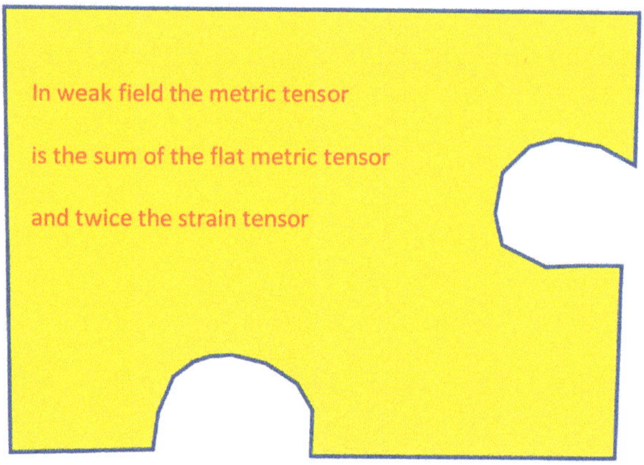

3rd piece of the puzzle

Chapter 6

Relationship between the Stress Tensor and the Strain Tensor in Elasticity (*K*) and between the Curvature and the Stress Energy Tensor (*κ*) in General Relativity in Weak Gravitational Fields

Chapter Summary

General relationship between the strain tensor and elastic strain energy

$$\left[\left(\varepsilon_{ij} + \frac{\nu}{1-2\nu}\varepsilon_{kk}\delta_{ij}\right)\right]\varepsilon^{ij} - \frac{2(1+\nu)}{E}U_{ij} - \frac{(1+\nu)}{E}\sigma_{ij}\varepsilon^{ij} \rightarrow f_{(\varepsilon_{ij}{}^{2})} = K_{\left(\frac{1+\nu}{E}\right)}U$$

Relationship between strains and elastic strain energy in plate theory:

$$\frac{1}{z^2}\left[(\varepsilon_{xx})^2 + (\varepsilon_{yy})^2 + 2(1-\nu)\frac{1}{4}\{\varepsilon_{xy}\}^2 + 2\nu\{\varepsilon_{xx}\varepsilon_{yy}\}\right] = \frac{24(1-\nu^2)}{Eh^3} \times \frac{\mathrm{d}U}{\mathrm{d}x\mathrm{d}y}$$

Relationship between strains and curvature in plate theory:

$$\varepsilon_{xx} = -\frac{z}{R_x}, \varepsilon_{yy} = -\frac{z}{R_y}, \gamma' = \varepsilon_{xy} = -2z\frac{\partial^2 w_{(x,y)}}{\partial x \partial y} = -2z\frac{1}{R_{xy}}.$$

Relationship between the curvature and the stress energy tensor in general relativity:

$$G^{\mu\nu} = \left(R^{\mu\nu} - \frac{1}{2}g^{\mu\nu}R\right) = -\frac{8\pi G}{c^4}T^{\mu\nu} = -\kappa T^{\mu\nu}$$

Relationship between the metric tensor and the strain tensor:

$$g_{\mu\nu} = \eta_{\mu\nu} + h_{\mu\nu} = \eta_{\mu\nu} + 2\varepsilon_{\mu\nu}$$

DOI: 10.1051/978-2-7598-2573-8.c006

6.1 Reminder of Results

We showed in chapter 3 that the strength of the materials had similarities to general relativity in the case of the bending of a beam as well as for all other types of stresses (principle: curvature space = $K \times$ energy density, scalar curvature R of a sphere of radius r in general relativity = $2/r^2$ see appendix D).

We showed in chapter 4 the relationship between the stress energy tensor T_{ij} and the stress tensor σ_{ij} ($\sigma_{ij} = \rho v_i v_j$; $T_{ij} = \rho u_i u_j$) on the one hand, and chapter 5 between the metric g_{ij}, h_{ij} and the strain tensor $\varepsilon_{ij}(h_{\mu v} = 2\varepsilon_{\mu v})$ on the other hand.

We will now investigate the proportionality constant between these terms (stresses, strains, strain energy, stress energy) by reasoning from tensors that have the advantage of not depending on the frame of reference in which we find ourselves.

6.2 Some Reminders about the Elasticity Theory

First of all, let us recall the expression of Hooke's law expressed in a tensorial way between the stress tensor σ_{ij} and the strain tensor ε_{ij}:

$$\sigma_{ij} = \frac{E}{(1+v)}\left(\varepsilon_{ij} + \frac{v}{(1-2v)}\varepsilon_{kk}\delta_{ij}\right)$$

Or by showing Lamé's coefficients,

$$\sigma_{ij} = 2\mu\varepsilon_{ij} + \varepsilon_{kk}\delta_{ij}$$

With the definition of Lamé's coefficients:

$$\mu = \frac{E}{2(1+v)} = G$$

$$\lambda = \frac{Ev}{(1+v)(1-2v)}$$

In this expression ε_{kk} is the trace of the strain tensor, δ_{ij} is the Kronecker's symbol and v Poisson's ratio.

The elasticity strain energy density that is equal to the work of external forces is written:

$$U_{ij} = \frac{1}{2}\sigma_{ij}\varepsilon^{ij}$$

$$\frac{\text{N.m}}{\text{m}^3} = \frac{\text{N}}{\text{m}^2}$$

By introducing the expression of the stress tensor in the expression of the strain energy density U_{ij}, we obtain for an elastic medium characterized by its Young's modulus $E = Y$ and Poisson's ratio ν:

$$U_{ij} = \frac{E}{2(1+\nu)}\left[\left(\varepsilon_{ij} + \frac{\nu}{1-2\nu}\varepsilon_{kk}\delta_{ij}\right)\right]\varepsilon^{ij}$$

Or:

$$\left[\left(\varepsilon_{ij} + \frac{\nu}{1-2\nu}\varepsilon_{kk}\delta_{ij}\right)\right]\varepsilon^{ij} = \frac{2(1+\nu)}{E}U_{ij} = \frac{(1+\nu)}{E}\sigma_{ij}\varepsilon^{ij}$$

6.3 Highlighting the Parallelism between Elasticity Theory and General Relativity

We can now highlight the parallelism between elasticity theory and general relativity theory (see figure 6.1).

Thus, to build a bridge between the elasticity theory and the formalism of general relativity, we will highlight at first, the parallelism between the strength of material and general relativity on the basis of the principle: curvature $= K \times$ energy density on the one hand (see Introduction) and then, secondly, the transversality of physical terms (curvature, metric, strain energy, external work, stress, deformations, proportionality coefficient K and κ) between general relativity and the elasticity theory on the other hand.

Curvature	= K	Energy density
$\left(\frac{du}{dx}\right)^2$	$= \frac{2}{ES}$	$\frac{U}{L}$
$\left(\frac{dy}{dx}\right)^2$	$= \frac{2}{GS_r}$	$\frac{U}{L}$
$\frac{1}{R^2}$	$= \frac{2}{EI}$	$\frac{U}{L}$
$\frac{1}{z^2}\left[(\varepsilon_{xx})^2 + (\varepsilon_{yy})^2 + 2(1-\nu)\frac{1}{4}\{\varepsilon_{xy}\}^2 + 2\nu\{\varepsilon_{xx}\varepsilon_{yy}\}\right]$	$= \frac{24(1-\nu^2)}{Eh^2}$	$\frac{dU}{dxdyh}$
$G_{\mu\nu} + \Lambda g_{\mu\nu} = R_{\mu\nu} - \frac{1}{2}g_{\mu\nu}R + \Lambda g_{\mu\nu}$	$= 2.0766\ 10^{-43}$	$T_{\mu\nu}$

FIG. 6.1 – Parallelism between elasticity theory, plate theory and general relativity.

The parallelism between general relativity and strength of material on the principle of curvature = $K \times$ energy density is given in figure 6.1 for different solicitations.

The first line corresponds to the normal effort case, the second to the shear effort case, the third to the bending of a beam [3], the fourth to the bending of a plate [58, 59] and the last to general relativity.

We see that in all 5 cases, we have a curvature in the left term and an energy density in the right term. In the first 4 cases, the constant of proportionality connecting each side of the equation depends on the materials (Young 's modulus) and on the geometric characteristics of the physical object (thickness h, section S, reduced shear section S_r, Poisson ratio ν). The dimension N^{-1} is respected for the 1st, 2nd, 4th and 5th case. On the other hand in the 5th case of general relativity ($\kappa = \frac{8\pi G}{c^4} = 2.0766 \times 10^{-43}\mathrm{N}^{-1}$), the proportionality coefficient κ has a well-established value without showing any characteristic of the physical medium [2]. This is due to the fact that Einstein calibrated his equation, not on considerations of mechanical characteristics of an elastic space-time medium undergoing deformations, but by a passage to the limits with the gravitation according to Newton using Poisson formula (use of time components of its equations). However, the dimension N^{-1} of κ is exactly the same as for the first two lines and 4th (see from the third if we decompose the inertia). So, given this parallelism, and taking into account of the elastic behavior of the spatial medium, it seems reasonable to think that constant κ includes hidden mechanical variables of the materials constituting the space, a sort of dark elastic material. This is what we are going to study. This is precisely the purpose of this book. For memory the correspondence between external and internal work in the case of a beam in strength of material and general relativity would be defined if Einstein had established a space deformation energy tensor $M^{\mu\nu}$ (see figure 6.2).

The same correspondence can be written in terms of tensors. The transversality between the general relativity and the elasticity theory on key parameters is as follows (see chapters 4 and 5 and figure 6.3): see also [51–53] where parallelism is studied. The reader keeps in mind the links between (ε_{ij}, h_{ij} and g_{ij}) and between ($T_{\mu\nu}$ and σ_{ij}).

In figure 6.3, the first and second lines are Einstein's expressions for general relativity (the second being a linearization of the first in weak fields). The third line is the expression of the strain energy density from the elasticity theories. As $T^{\mu\nu}$ is of similar construction to σ_{ij} (see chapter 4) on the one hand, and as $\bar{h}_{\mu\nu}$ depends on $h_{\mu\nu}$ which can be attached to $\varepsilon_{\mu\nu}$ (see chapter 5) on the other hand, we can therefore look

$$\frac{1}{R^2} = \frac{2}{EI}\left(\frac{W_{ext(total)}}{L}\right) = K\left(\frac{W_{ext(total)}}{L}\right) = \frac{2}{EI}\left(\frac{U}{L}\right)$$

$$G^{\mu\nu} = -\frac{8\pi G}{c^4}T^{\mu\nu} \quad = -\kappa\ T^{\mu\nu} \quad = -\kappa\, M^{\mu\nu}$$

FIG. 6.2 – Parallelism between elasticity theory and general relativity theory.

$$G^{\mu\nu} = R^{\mu\nu} - \frac{1}{2} g^{\mu\nu} R = -\frac{8\pi G}{c^4} \quad T^{\mu\nu}$$

$$\partial^{\lambda}\partial_{\lambda}\bar{h}_{\mu\nu} = \Box\, \bar{h}_{\mu\nu} = -\frac{16\pi G}{c^4} \quad T_{\mu\nu}$$

$$\left[\left(\varepsilon_{ij} + \frac{\nu}{1-2\nu}\varepsilon_{kk}\delta_{ij}\right)\right]\varepsilon^{ij} = \frac{2(1+\nu)}{E} \quad U_{ij} = \frac{(1+\nu)}{E}\sigma_{ij}\varepsilon^{ij}$$

FIG. 6.3 – The transversalism between elasticity theory and general relativity theory in weak field.

for it again *via* the transversalism of the equations in figure 6.3, a link between the constant of proportionality appearing in the theory of elasticity and the same (κ) appearing in general relativity. From this analysis, it is clear that constant κ in which case, $\frac{8\pi G}{c^4}$ approaches the mechanical constant of the spatial medium as a function of Young's modulus E and Poisson's ratio ν. In other words, κ should be proportional to $\frac{(1+\nu)}{E}$ (corrected to find the dimension N^{-1}) if the space is a deformable elastic medium. T. Damour in his book "If Einstein was told to me, chapter 3 elastic space-time" also highlights the parallelism between Einstein's formulation and the theory of elasticity (Hooke's law) by modeling Einstein's equation by the expression $D_{(g)} = \kappa T$ where D is a strain tensor as a function of the metric g and T a tension tensor. κ represents the flexibility of space. In this case the space is not flexible or in other words it is extremely stiff ($1/\kappa$), it is not easily deformed because the coefficient κ is 2.043×10^{-43} N^{-1}

Note: According to table 3.3 that this expression of Hooke's law can be understood in two ways: in elongation and compression strains of the elastic space fabric but also in shear angle ($\varepsilon = \frac{1}{E}\sigma$) or ($\gamma = \frac{1}{G}\tau$).

6.4 Consequence of Parallelism and Transversalism between the Elasticity Theory and General Relativity

To finalize the construction of the parallelism between general relativity and the theory of elasticity and to find a mechanical transposition of Einstein's constant κ, we must now look for to find the expression of κ in terms of mechanical characteristics of the elastic spatial medium. The comparison of K and κ member to member makes it possible to identify the mechanical parameters associated with G and c. If the spatial medium is elastic, then the constant of proportionality between the curvature and the energy density can be written as according to the elasticity theory as shown below. $f_{(\varepsilon_{ij}{}^2)} = K_{\left(\frac{1+\nu}{E}\right)} U$.

So we still have à to verify that physically the space is indeed a continuous elastic medium, a kind of "elastic jelly" as described by T. Damour [4, 5], condition sine qua none to be able to apply the classical continuum mechanic theories and elasticity theories and thus try to define mechanical characteristics $(\mathrm{E}, v, \mathrm{F}, \rho \rightarrow \frac{\frac{E_{\text{vacuum}}}{c^2}}{V})$ of the dark material that constitutes space.

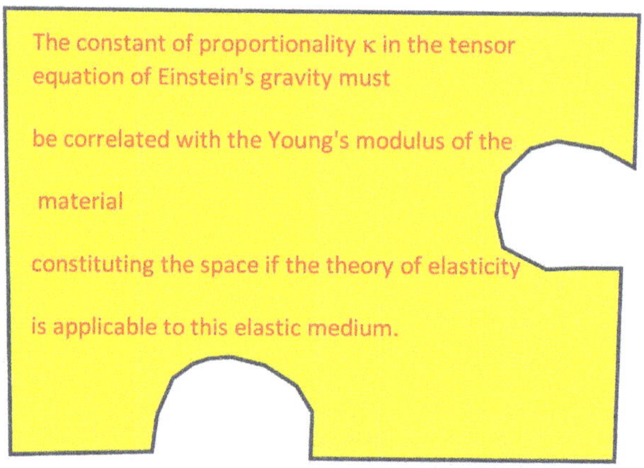

4th piece of the puzzle

Chapter 7

Can Space-be Considered as an Elastic Medium? New Ether?

7.1 The Conclusions of Michelson and Morley's Experiment

We showed in the preamble that the shape of galaxies was similar to vortexes observed at sea and cyclones observed in the atmosphere. The last two having a medium in which they develop, we may wonder whether if there is not an equivalent medium for the space in which the galaxies are immersed, a new kind of ether [28]. Let us recall that this new ether has nothing to do with the luminiferous ether which does not exist (experiment of Michelson and Morley [63]) since the light evolves in the vacuum by self-simulation of the electric field and the field magnetic as shown by Maxwell and without material support.

We first recall Michelson and Morley's conclusions, which are:

> Of all the above, it seems reasonably certain that if there is a relative movement between the Earth and the luminous ether, it must be small enough to refute the explanation of Fresnel's aberration…

Some comments:

- It is important not to confuse a vibration of a medium supposed to be a luminescent ether that does not exist (light propagates alone without the need for a medium) with a fabric of curved space whose light follows curvatures (*e.g.* star lights deflected during the passage around the Sun and observed during an eclipse).
- Contrary to what is often said, Michelson and Morley did not write that there was no ether but that it had to be small enough not to be detected.
- It is well known that the Higgs field does not interfere with photon or light (light having no mass, photons move at speed c and the objective of the Higgs boson is precisely to give mass to particles by interacting with them, and therefore to impose a speed lower than c).

DOI: 10.1051/978-2-7598-2573-8.c007

- It is then obvious that it is impossible to detect this Higgs field with light. It is therefore possible that Michelson and Morley's experience with light is not the right experiment to detect the material that makes up the elastic space fabric.
- We can see the torsion deformation (the frame dragging effect) of the space created by the earth's rotation in this medium and measured by the satellite probe B by onboard gyroscopes, as the expression of the twisting of space around the earth similar to a certain Ether characterized by $g_{\mu\nu}$. This experiment can be considered the right one to measure the twisting of space by the Earth rotation and the dynamic state of this new relativistic ether.
- We can see the gravitational wave as a vibration of an elastic medium (see appendix G).

Light follows the curvature of space but does not need a luminescent medium to propagate.

7.2 Einstein's View of the Ether

Einstein's letter to Lorentz of June 17, 1916 [27], reads:

> I agree with you that the general theory of relativity is closer to the Ether hypothesis than special relativity. This new theory of ether, however, would not violate the principle of relativity, for the state of this g_{mn} ether would not be that of the rigid body in a state of independent movement, but each state of motion would depend on the position determined by the material processes.

The important words here are that rigid space does not exist. It is according to general relativity malleable by the matter found there. Einstein in 1905 with special relativity had suppressed the Luminiferous Ether, but he recognizes the existence of a certain ether at a conference in Leiden in 1920.

If the luminescent ether is dead the relativistic ether seems to be alive...

7.3 Observations Made Demonstrate the Elastic Behaviour of Space-Time

Let us return to what we have learned from general relativity.

Einstein's gravitational field equation has been well verified experimentally. Indeed, all the tests and gravitational measurements carried out over 100 years constitute evidence that confirms that space is an elastic deformable physical object, which is not made of "nothing" but that is filled with "something", a field, an elastic substance, indeed:

(a) The acceleration of the expansion of the Universe has been observed: galaxies are moving away all the more so faster that they are far from us (Hubble Law [21]). In other words, galaxies are "immobile" and it is the space between them that expands like dots on a balloon, which is inflated (see figures 7.1a and 7.1b).

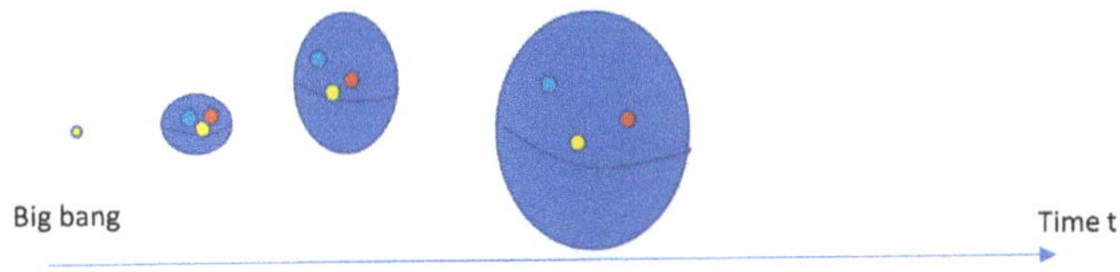

FIG. 7.1a – Expanding space-time like an inflating balloon.

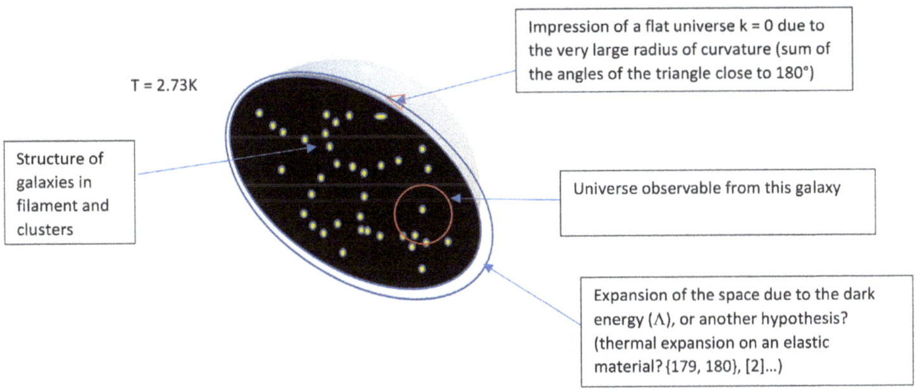

FIG. 7.1b – Flat universe in expansion.

Thus, the space fabric is an elastic body that stretches $v = H_0 d$, where v is the rate of recession of the galaxy, H_0 is the Hubble constant and d is the distance of the observer galaxy.

But what swells? What is the structure of this fabric? What material is it made of? Does it have a thermal expansion coefficient?

(b) The tangential beam of light in the Sun is bent by the deformation of the space around it (see figure 7.2). The light rays from the stars follow the curvature of an elastic body deformed by the mass (Eddington 1919 [11] measured a deviation angle of stars around the Sun during an eclipse of 1.75″).

But the star follows the curvature of which fabric? Made in what material?

(c) Star S2 revolves around the central black hole of the Milky Way (see figure 7.3).

(d) The space is elastic. As the Sun moves, the curvature of space in its wake disappears and becomes flat. The beam of light becomes straight again; the stars resume their original positions [11].

(e) The rotation of the Earth twists the space fabric around it, indeed the gyroscopes placed in orbit 400 km around the Earth are deflected by this rotation/twisting/curvature of space (see the experiment Probe B. Thus, a horizontal angle θ_{x2} of 0.000010833°/year was measured from vacuum-placed gyroscopes (see table 7.1) in accordance with the prediction of general relativity) (see figures 7.4 and 7.5).

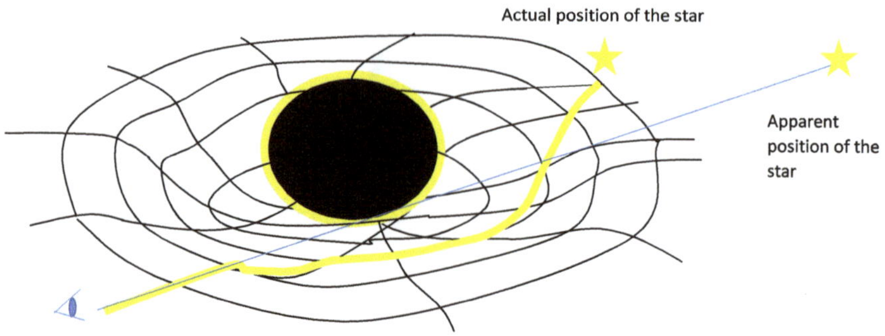

FIG. 7.2 – In the plane of the equator light follows the curvatures of space-time.

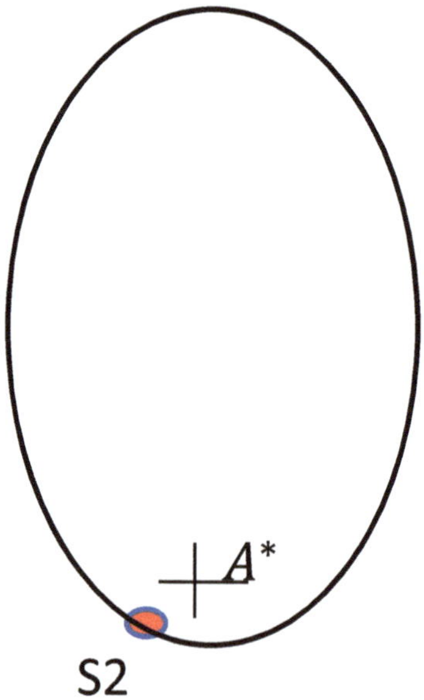

FIG. 7.3 – Movement of the star S2 around the black hole Sagittarius A* in the center of the Milky Way.

The calculation of the effects of the rotation of the Earth on vacuum space according to general relativity (see figure 7.4):

$$\Omega = \frac{3GM}{2c^2R^3}(R \times v) + \frac{GI}{c^2R^3}\left[\frac{3R}{R^2}(\omega \cdot R) - \omega\right]$$

Precession effect Geodetic effect Lense Thirring (Source Wikipedia Gravity Probe B).

TAB. 7.1 – Comparison of results obtained according to general relativity with those measured by the Gravity Probe B satellite (Source Wikipedia Gravity Probe B).

Considered effect	Predicting by general relativity	Measurements made by Gravity Probe B in milli arc second	Error %
Geodetic	−6606.1	−6601.8/−18.3	0.28
Frame dragging	−39.2	−37.2/−7.2	19

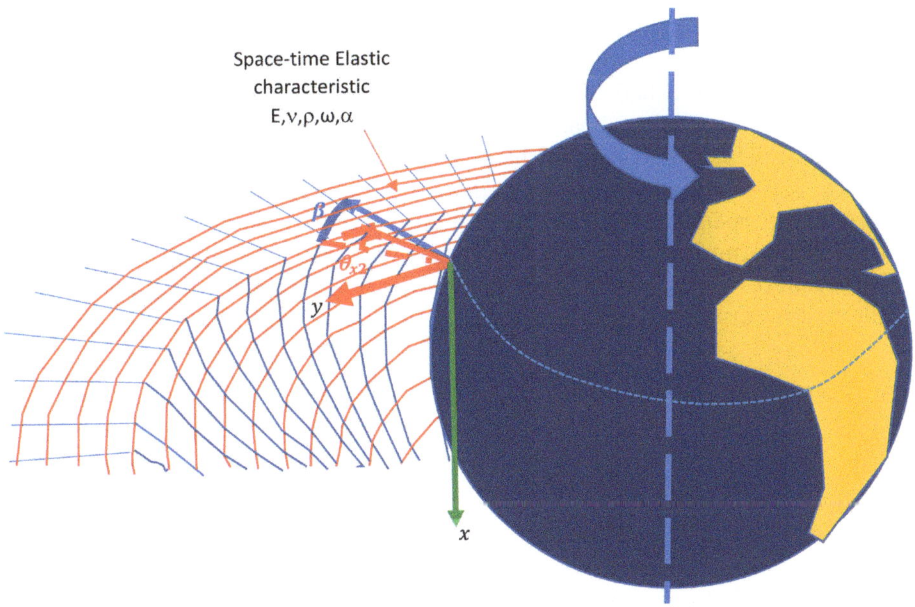

FIG. 7.4 – Space twisted by the rotation of the Earth.

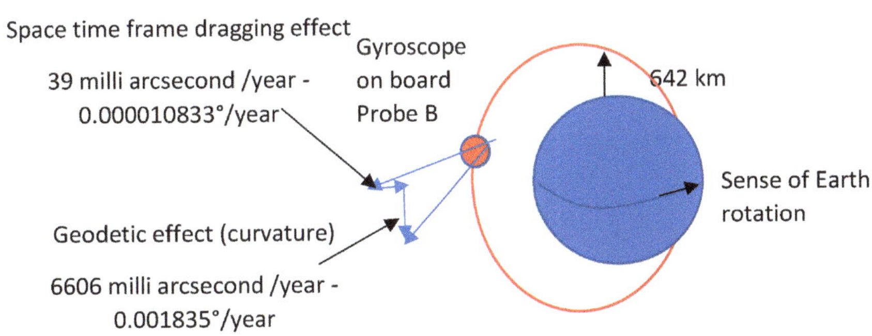

FIG. 7.5 – Distortion of space around the earth (Source Wikipedia Gravity Probe B).

But what is physically distorted?

(f) The spatial deformations produced by the coalescence, for example, of two black holes in the form of gravitational waves, were first measured by Ligo interferometers in 2015 (official announcement of GW150914 detected by Ligo on February 11, 2016) [23]. Other observations followed [24], including a fusion of neutron stars (pulsars see GW170817) also observed in electromagnetic radiation. Interferometers have therefore effectively measured signals that move at the speed of light c within the structure of the space medium. So, if there are deformations, then there is an equivalent elastic physical body that deforms in connection with active stresses (see appendix G). Therefore, according to the elasticity theory, the material constituting this physical body can be characterized by Young's modulus $E(=Y)$ and Poisson's ratio ν. As strain tensors ε_{ij} can thus be established from the $\delta L/L$ deformations in correlation with these deformations of 10^{-21} measured in each arm. Mechanically speaking, these measured deformations are in the main direction. They are associated with the elastic transverse waves of the medium (see appendix G) with a propagation direction perpendicular to the plane built by the two arms of the interferometers. The formulas below describe the metric space g_{ij} consisting of h_{ij} a perturbation of the flat metric η_{ij} in low gravitational field and in connection with the strain tensors ε_{ij} (see figure 7.6).

$$g_{ij} = \eta_{ij} + h_{ij} = \eta_{ij} + 2\varepsilon_{ij}$$

$$h_{ij} = 2\varepsilon_{ij}$$

$$\frac{\delta L}{L} = \frac{\delta_i}{L} = 10^{-21} = \varepsilon_{ij} = \frac{1}{2} h_{ij}$$

But these waves propagate in which medium, what vibrates by transmitting these elastic oscillations from the space and then after their passage resume a position at rest?

(g) And finally, space can be considered like any elastic material according to the study conducted by Ringermacher and Mead.

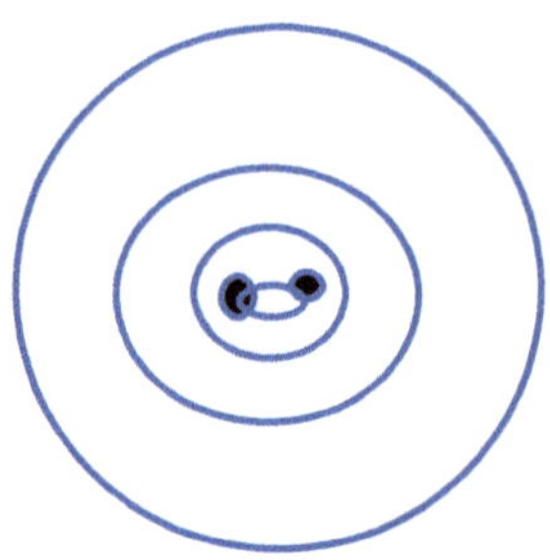

FIG. 7.6 – Gravitational waves created by two rotating black holes.

7.4 Consequence of Measurements

On the basis of all these points, it seems logical to consider that space is made of a strange elastic material, a new type of ether as explained by Einstein himself, a kind of "elastic jelly" as proposed by T. Damour [4, 5, 15, 16].

From the above observations, it appears that:

(a) The space fabric seems to be physically made up of "something" since galaxies are driven by this "something" that expands in an accelerated way.
(b) Space bends (analogy with the notion of curvature in the theory of elasticity see in the presence of energy density, mass). The sum of the angles of a triangle in the presence of mass is no more than 180 degrees [4, 5, 15, 16].
(c) The space fabric is distorted and transmits gravitational waves through lengthening and shortenings. Thus, small deformations (interferometers) or angles (Gravity Probe B) are measured when we place them far enough away from said mass/energy (weak field principle).
(d) The space is elastic, the curvature disappears when the mass that created it disappears.
(e) A black hole is a type of space yielded charged to the extreme (see appendix F).
(f) Space is homogeneous in all the directions.

We will therefore be able to study what it means to unify the elasticity theory with general relativity, focusing our investigations on the constant of proportionality κ.

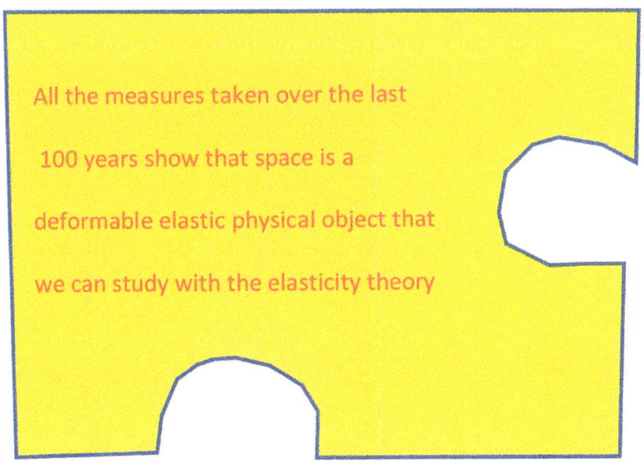

5th piece of the puzzle

Chapter 8

And if We Reconstructed the Formula of Einstein's Gravitational Field by No Longer Considering the Temporal Components of the Tensors, but the Spatial Components

Chapter Summary

$$\begin{bmatrix} G^{00} & G^{01} & G^{02} & G^{03} \\ G^{10} & G^{11} & G^{12} & G^{13} \\ G^{20} & G^{21} & G^{22} & G^{23} \\ G^{30} & G^{31} & G^{32} & G^{33} \end{bmatrix} = \kappa \begin{bmatrix} T^{00} & T^{01} & T^{02} & T^{03} \\ T^{10} & T^{11} & T^{12} & T^{13} \\ T^{20} & T^{21} & T^{22} & T^{23} \\ T^{30} & T^{31} & T^{32} & T^{33} \end{bmatrix}$$

8.1 Let Us Step Back from Gravitation According to Newton

In fact, Newton's formula is the basis of the definition of gravitational constant since the Newtonian gravitational force F is proportional to the gravitational constant G, the product of the M and m masses, and is inversely proportional to the square of the r radius that separates the gravity centroid of these two masses.

$$F = G \times \frac{M \times m}{r^2}$$

DOI: 10.1051/978-2-7598-2573-8.c008

Considering that this mathematical expression of Newton's gravitation is merely a weak field simplification of general gravitation, that an illusion of force, should not we also consider that constant G is also an illusion disappearing with Newton's formula that created it?

Just as we must abandon Newton's formulation in a highly gravitational field, should we not also abandon G as an indivisible universal constant because it is the basis of this Newtonian formula which turns out not to be the true formulation characterizing gravitation, but only a simplification? However this cannot be done simply, because it turns out that G reappeared in Einstein's constant κ, since Einstein calibrated his gravitational field equation by imposing that in weak fields it gives back the results of Newton's formulation which has been fully verified!

But in this case, is it possible to reconstruct Einstein's constant κ without going through G but going through a different theory? Can we separate G into more fundamental parameters?

To answer these questions, it is interesting to compare the strengths and weaknesses of gravitation between Newton and Einstein's approaches.

8.2 The Strengths and Weaknesses of Newton's Gravitational Approach

The strengths of Newton's gravitational approach are:

(a) It enabled the mathematical discovery of the planet Neptune by Urban Le Verrier. Therefore, it works very well in a weak gravitational field obviously.
(b) It explains all the effects of gravity on Earth and in the solar system, with the exception of the Mercury perihelion delay (the effect of the strong gravitational field near the Sun).

The weak points of Newton's gravitation are:

(a) It only works if objects have masses. It is therefore not possible to predict the curvature of a ray of light tangent to the Sun by the effect of gravitation, although predicted by general relativity and verified by Arthur Eddington on May 29, 1919 [11].
(b) The forces are instantaneous (thus applied faster than the speed of light) and their mode of transmission is still unexplained while special relativity shows that no phenomenon can be faster than the speed of light.
(c) Its results are imprecise for the action of gravity in a strong field. Indeed, the delay in the perihelion of the Mercury planet is defined exactly by the general relativity at 43 arc-second while it is much weaker using Newton's gravitational approach.
(d) The formulation depends on the r parameter, and if the M and m mass objects rotate in relation to each other at speeds towards the speed of light, the effects of special relativity change the notion of distance for each observer. What r value should be used in the calculations in this case?

(e) In this formulation, space-time is a rigid non-deformable object, whereas in general relativity it is precisely the distortion of space-time that generates gravitation giving us the illusion that forces act and attract objects between them.
(f) According to Newton an object falls to earth in a straight line, or in gravitation according to Einstein it is also deflected (no more straight line).

In view of all these points, it is therefore clear that Newton's concept of force makes no sense in several situations (strong gravitational field). It is an illusion. This formulation is only a simplification of a broader theory, general relativity. This formulation works in weak gravity fields but is wrong in strong fields.

8.3 *G* a Gravitational Constant of Strange Dimensions as a Combination of Underlying Parameters

We may also wonder why the universe would depend on a G constant of strange dimensions: inverse of a density by a square frequency. Some authors suggest, on the basis of this dimensional equation, that G actually depends on a density and a frequency [13].

$$G = 1/\left(\mathrm{kg/m^3}\right) \times 1/\ \mathrm{s}^2$$

$$G(\rho, f^2) = cte \times \frac{1}{\rho} f^2$$

In general relativity [8, 9], G is introduced because it comes from the use of Poisson's equation to calibrate constant κ (analysis at component 00, time t of the different tensors of Einstein's gravitational formula):

Indeed, the Laplacian of the gravitational field, $\Delta\phi$, follows Poisson's equation:

$$\Delta\phi = 4\pi G\rho$$

And the 00 component of the metric $g_{\mu v}$ is then:

$$g_{00} \approx 1 + \frac{2\phi}{c^2}$$

It is also surprising, in hindsight, to see that Einstein precisely calibrates his equations on the temporal component of his tensors, whereas precisely the formulas of Newton and Poisson that follow are independent of time!

8.4 How to Re-parameterize κ in Einstein's Gravitational Field Equation

κ connects Ricci's tensor $R_{\mu v}$ (from the contraction of Riemann's curvature tensor, a function of the metric $g_{\mu v}$ and in particular of its partial derivatives, see glossary and definitions), to the stress energy tensor $T_{\mu v}$ (mass/external energy applied to space-time) in Einstein's gravitational field equation.

$$\kappa = \frac{8\pi G}{c^4}$$

$$\kappa = \frac{\frac{\mathrm{m}^3}{\mathrm{kgs}^2}}{\frac{\mathrm{m}^4}{\mathrm{s}^4}} = \frac{\mathrm{s}^2}{\mathrm{kgm}} = \frac{1}{\mathrm{Newton}} = \mathrm{N}^{-1}$$

Einstein's gravitational field equation is written:

$$G^{\mu v} = \left(R^{\mu v} - \frac{1}{2} g^{\mu v} R\right) = -\frac{8\pi G}{c^4} T^{\mu v} = -\kappa\, T^{\mu v}$$

$$\frac{1}{\mathrm{m}^2} = \frac{1}{\mathrm{N}} \times \frac{\mathrm{N.m}}{\mathrm{m}^3}$$

We do not show here the cosmological constant Λ as a possible source of dark energy. In addition, the scalar curvature R is a tensorial contraction of $R_{\mu v}$.

Moreover, as Newton's expression is false for the concept of force (it is the distortion of space-time that gives the illusion of a force of attraction between two objects of mass M and m and therefore there is no force of attraction), and as the gravitational constant is directly related to this concept of force (see Newton's gravitational formulation), is it not necessary to rethink the G proportionality factor associated with this force in general relativity?

Therefore, the question that needs to be asked is: if Newton's gravitation did not exist, could Einstein have calibrated κ without going through Newton's limit, without using Poisson's equation and the temporal component (00) of his tensor, but directly using the spatial components (1, 2, 3) of his tensors indicated in bold and red below?

$$\begin{bmatrix} G^{00} & G^{01} & G^{02} & G^{03} \\ G^{10} & \mathbf{G^{11}} & \mathbf{G^{12}} & \mathbf{G^{13}} \\ G^{20} & \mathbf{G^{21}} & \mathbf{G^{22}} & \mathbf{G^{23}} \\ G^{30} & \mathbf{G^{31}} & \mathbf{G^{32}} & \mathbf{G^{33}} \end{bmatrix} = -\kappa \begin{bmatrix} T^{00} & T^{01} & T^{02} & T^{03} \\ T^{10} & \mathbf{T^{11}} & \mathbf{T^{12}} & \mathbf{T^{13}} \\ T^{20} & \mathbf{T^{21}} & \mathbf{T^{22}} & \mathbf{T^{23}} \\ T^{30} & \mathbf{T^{31}} & \mathbf{T^{32}} & \mathbf{T^{33}} \end{bmatrix}$$

$$\begin{bmatrix} G^{00} & G^{01} & G^{02} & G^{03} \\ G^{10} & G^{11} & G^{12} & G^{13} \\ G^{20} & G^{21} & G^{22} & G^{23} \\ G^{30} & G^{31} & G^{32} & G^{33} \end{bmatrix} = -\kappa \begin{bmatrix} \sigma^{tt} & \tau^{tx} & \tau^{ty} & \tau^{tz} \\ \tau^{xt} & \sigma^{xx} & \tau^{xy} & \tau^{xz} \\ \tau^{yt} & \tau^{yx} & \sigma^{yy} & \tau^{yz} \\ \tau^{zt} & \tau^{zx} & \tau^{zy} & \sigma^{zz} \end{bmatrix}$$

Or is it possible to find Einstein's constant κ by going through a different theory using the spatial components of gravitational field tensors?

To define the theory to be used, we must now explore the strengths and weaknesses of general relativity.

8.5 The Strengths and Weaknesses of Gravitation According to Einstein

The strengths of general relativity are:

(a) It fills all the gaps in Newton's formulation with extraordinary precision.
(b) It predicts the curvature of a ray of light passing near the Sun during an eclipse. By the principle of equivalence, an elevator on Earth is equivalent to an accelerated elevator at g in empty space. A ray of light crossing horizontally through a hole in one of the vertical walls of this elevator in space, appears curved to an observer in this elevator. By the principle of equivalence, on Earth, gravitation therefore bends the light.
(c) It introduces time into four-dimensional tensorial writing. Phenomena are no longer instantaneous and respect special relativity.
(d) It accurately predicts the delay of the perihelion of the planet Mercury, in a strong gravitational field.
(e) It no longer considers space-time as a rigid physical object but as a deformable object, consistent with observations made more than a century ago.

The weak points of general relativity are:

(a) Einstein's field equation is a mathematical description of the distortion of space-time due to the energy density present in that space. Gravitation is therefore mathematically described by the metric of space-time. This metric is a four-dimensional mathematical description of the deformation $g_{\mu v}$ of space-time with respect to the Minkowski's flat metric.

But this mathematical (geometric) description of space says nothing about the physical nature and possible mechanical properties of this deformed elastic medium constituting the space fabric.

(b) There is a fracture between quantum field theory that describes the standard model, vacuum energy, Casimir's force and general relativity. General relativity perfectly explains the mechanics of the infinitely large and the quantum mechanics that of the infinitely small. The problem is that these two theories do not overlap. Gravitation has not yet been quantified (h independent) and is deterministic (not probabilistic as in quantum mechanics)! Graviton, the boson vector of the force of gravity and keystone of quantum gravity in relation to quantum field theory has not yet been measured.

8.6 Approach to Reconstructing General Relativity from the Elasticity Theory

So, in the next chapters we will reconstruct the equation of Einstein's gravitational field, and in particular Einstein's constant κ without going through the temporal components of the tensors of the gravity field (which use Newton's G gravity constant), but through the spatial components of the tensors ($G_{\mu v}$; $T_{\mu v}$) in the elasticity theory and the results of measurements carried out for more than 100 years.

The measurements in elasticity theory making by strain gauges deposited on the solids, we will consider the interferometers of Ligo and Virgo as gauges of deformation of space positioned on earth since they measure strains of this structure of space during the passage of gravitational waves (see chapters 9–12).

We will then study with the results obtained, the temporal components of the tensors of the gravitational field (see chapter 17).

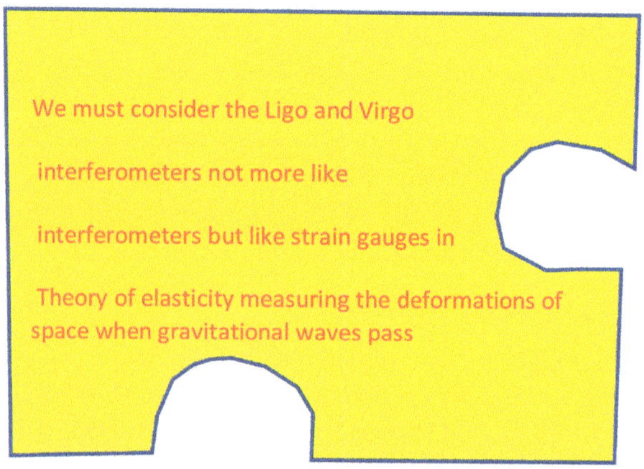

6th piece of the puzzle

Chapter 9

Re-interpretation of the Results of the Theoretical Calculation of General Relativity on Gravitational Waves in Weak Field from the Windows of Elasticity Theory

Chapter Summary

Relationship between the metric tensor $g_{\mu\nu}$ and the strain tensor $\varepsilon_{\mu\nu}$:

$$g_{\mu\nu} = \eta_{\mu\nu} + h_{\mu\nu} = \eta_{\mu\nu} + 2\varepsilon_{\mu\nu}$$

Relationship between metric perturbation $h_{\mu\nu}$ and strain tensor $\varepsilon_{\mu\nu}$ applied to the deformations to spatial elastic material ε_{ij}:

DOI: 10.1051/978-2-7598-2573-8.c009

$$h_{\mu\nu} = 2\varepsilon_{\mu\nu} \rightarrow \begin{cases} h_{\mu\nu} = A_{+}\cos\left(\frac{\omega}{c}(ct - z)\right)\begin{bmatrix} 0 & 0 & 0 & 0 \\ 0 & \mathbf{+1} & \mathbf{0} & \mathbf{0} \\ 0 & \mathbf{0} & \mathbf{-1} & \mathbf{0} \\ 0 & \mathbf{0} & \mathbf{0} & \mathbf{0} \end{bmatrix} \rightarrow \varepsilon_{xy(A_{+})} \\ = \frac{1}{2}A_{+}\cos\left(\frac{\omega}{c}(ct - z)\right)\begin{bmatrix} \varepsilon_{xx} & 0 & 0 \\ 0 & -\varepsilon_{yy} & 0 \\ 0 & 0 & 0 \end{bmatrix} \\ h_{\mu\nu} = A_{\times}\cos\left(\frac{\omega}{c}(ct - z)\right)\begin{bmatrix} 0 & 0 & 0 & 0 \\ 0 & \mathbf{0} & \mathbf{+1} & \mathbf{0} \\ 0 & \mathbf{+1} & \mathbf{0} & \mathbf{0} \\ 0 & \mathbf{0} & \mathbf{0} & \mathbf{0} \end{bmatrix} \rightarrow \varepsilon_{xy(A_{\times})} \\ = \frac{1}{2}A_{\times}\cos\left(\frac{\omega}{c}(ct - z)\right)\begin{bmatrix} 0 & \varepsilon_{xy} & 0 \\ \varepsilon_{yx} & 0 & 0 \\ 0 & 0 & 0 \end{bmatrix} \end{cases}$$

Relationship between the stress tensor σ_{ij} applied to the spatial elastic material and the spatial strain tensor ε_{ij}:

$$\sigma_{ij} = \frac{E}{(1+\nu)}\left(\varepsilon_{ij} + \frac{\nu}{(1-2\nu)}\varepsilon_{kk}\delta_{ij}\right)$$

$$\rightarrow \begin{cases} \varepsilon_{xy(A_{+})} = \frac{1}{2}A_{+}\cos\left(\frac{\omega}{c}(ct - z)\right)\begin{bmatrix} \varepsilon_{xx} & 0 & 0 \\ 0 & -\varepsilon_{yy} & 0 \\ 0 & 0 & 0 \end{bmatrix} \rightarrow \sigma_{xy(A_{+})} \\ = \begin{bmatrix} \sigma_{xx} & 0 & 0 \\ 0 & \sigma_{yy} & 0 \\ 0 & 0 & 0 \end{bmatrix} \\ \varepsilon_{xy(A_{\times})} = \frac{1}{2}A_{\times}\cos\left(\frac{\omega}{c}(ct - z)\right)\begin{bmatrix} 0 & \varepsilon_{xy} & 0 \\ \varepsilon_{yx} & 0 & 0 \\ 0 & 0 & 0 \end{bmatrix} \rightarrow \sigma_{xy(A_{\times})} \\ = \begin{bmatrix} 0 & \tau_{xy} = \sigma_{xx} & 0 \\ \tau_{yx} = \sigma_{yy} & 0 & 0 \\ 0 & 0 & 0 \end{bmatrix} \end{cases}$$

9.1 Introduction

Ligo's observations in 2015 announced on February 11, 2016, as well as all those that followed with the Virgo interferometer, have now scientifically demonstrated the existence of black holes, their fusions and the resulting gravitational waves [23, 24]. Einstein was therefore still right in predicting in 1916 his waves of elastic vibrations of the space structure.

We therefore consider this situation from an elasticity theory point of view, namely that the rotation of a binary system (like two black holes, for example) creates a kind of twist wave in the elastic spatial fabric (see figures 9.1a and 9.1b).

Indeed, this point was presented and demonstrated by Professor Kip Thorne (Nobel Prize 2017) during his lecture on March 6, 2018 at UCI USA [19] "exploring the universe with gravitational waves ranging from big bang to black holes". At this conference accessible on the net, he explains that after numerical simulations of general relativity of a coalescence of two black holes, there are different types of vortexes (turning right, turning left) emerging from black holes rotating, causing, twisting the space environment around them. These swirls merge to create a ring that consists of these swirling mixtures and the result is compression and traction tendencies measured by Ligo and Virgo. The consequences are that gravitational waves are not a conventional shear wave but a mixture of vortexes whose results are compression/traction tendencies measured in interferometers.

9.2 Re-interpretation of the 2 Gravitational Wave Polarizations in Terms of Space Deformation Tensors in the Sense of Elasticity Theory

If we compare the results of general relativity in weak field ($h_{\mu\nu}$) with the elastics strain tensors (ε_{ij}) (see chapter 5) we can deduce information about the deformation states of the elastic medium in the xy plane of the arms of the interforemeter during the passage of a gravitational wave coming from the z direction. Kip Thorne also explained this at a conference on the geometry of space-time in warped space [20, 65]. R. Feynman also touches on this interpretation in his lectures on gravitation

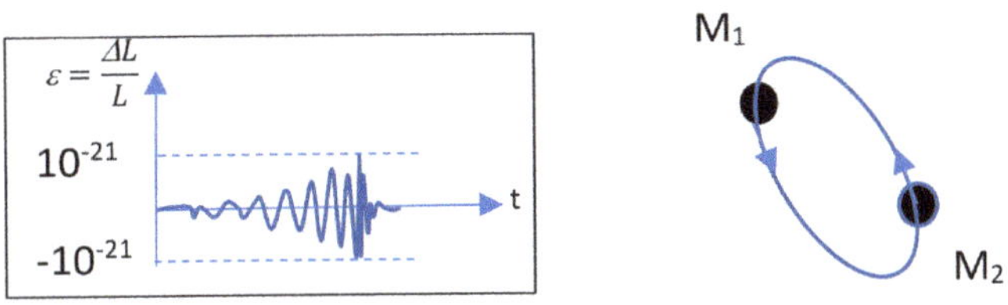

Fig. 9.1a – Example of measured gravitational wave.

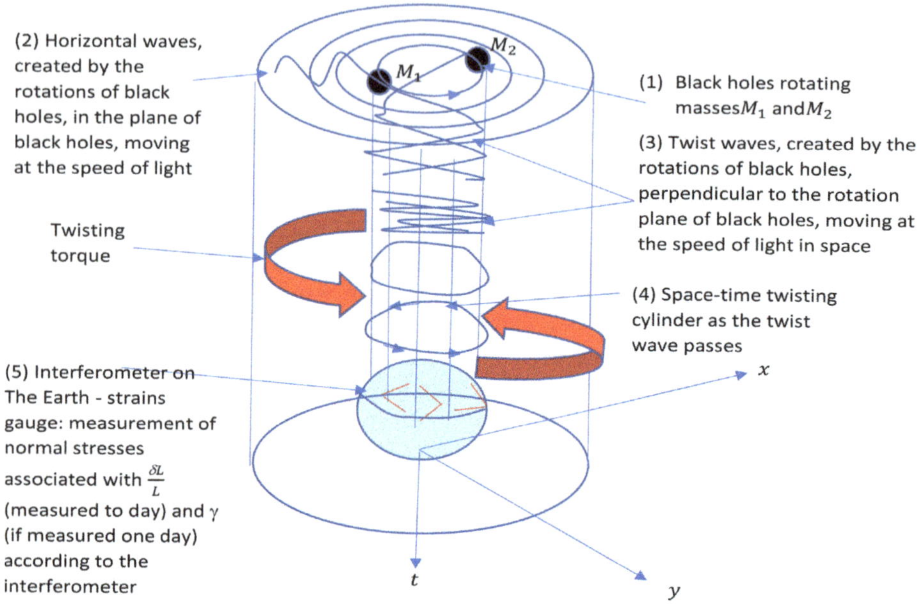

FIG. 9.1b – Space twist waves created by a binary system of rotating black holes.

[17]. In fact, we found it on our side and then reconfirmed with these different presentations of K. Thorne and R. Feynman. See also [51] at [53].

Here is the interpretation according to the elasticity theory of the physical calculations of gravitational wave polarizations:

First point: It is known that there are two types of polarization clearly separated (at 45° from each other) from the gravitational wave produced by the coalescence of two massive objects turning in relation to each other called: A_+ and $A_\times$.

$$\begin{cases} h_{\mu\nu} = A_+ \cos\left(\frac{\omega}{c}(ct - z)\right) \begin{bmatrix} 0 & 0 & 0 & 0 \\ 0 & \mathbf{+1} & \mathbf{0} & \mathbf{0} \\ 0 & \mathbf{0} & \mathbf{-1} & \mathbf{0} \\ 0 & \mathbf{0} & \mathbf{0} & \mathbf{0} \end{bmatrix} \\ h_{\mu\nu} = A_\times \cos\left(\frac{\omega}{c}(ct - z)\right) \begin{bmatrix} 0 & 0 & 0 & 0 \\ 0 & \mathbf{0} & \mathbf{+1} & \mathbf{0} \\ 0 & \mathbf{+1} & \mathbf{0} & \mathbf{0} \\ 0 & \mathbf{0} & \mathbf{0} & \mathbf{0} \end{bmatrix} \end{cases}$$

Second point: By the relationship between $h_{\mu\nu} = 2\varepsilon_{\mu\nu}$, there exists for each polarization $h_{\mu\nu}$ the equivalent of an associated strain tensor $\varepsilon_{\mu\nu}$.

Based on the expressions connecting the perturbation $h_{\mu\nu}$ of the metric to the strain tensors $\varepsilon_{\mu\nu}$ we obtain for the polarization A_+ the following strain tensor:

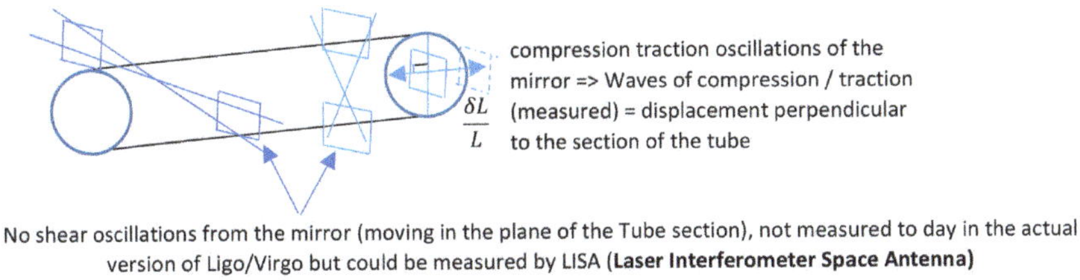

FIG. 9.2 – Movements of the laser mirror inside the interferometric tubes created by the transverse gravitational wave.

$$\varepsilon_{xy(A_+)} = \begin{bmatrix} \varepsilon_{xx} & 0 & 0 \\ 0 & -\varepsilon_{yy} & 0 \\ 0 & 0 & 0 \end{bmatrix}$$

This state of deformation is obtained in the tubes of the interferometer by analyzing the front and rear movements δL or strains $(\frac{\delta L}{L})$ of the laser mirrors in the two tube sections (see figure 9.2).

Based on the expressions connecting the perturbation of the metric $h_{\mu\nu}$ to the strain tensor $\varepsilon_{\mu\nu}$ we also obtain for polarization $A_\times$ the following strain tensor:

$$\varepsilon_{xy(A_\times)} = \begin{bmatrix} 0 & \varepsilon_{xy} & 0 \\ \varepsilon_{yx} & 0 & 0 \\ 0 & 0 & 0 \end{bmatrix}$$

This state of deformations must be obtained in the tubes of the interferometer by analyzing the lateral movements of the laser mirrors in the two tube sections (see figures 9.3a and 9.3b).

These two states of deformation certainly prove that $h_{\mu\nu}$ corresponds to a state of shear strain for spatial layers (multi sandwich see [51] at [53] 2018 figure 1.1 the cosmic fabric is as a stack of three-dimensional hypersurfaces) perpendicular to direction z of the gravitational wave (see figures 9.1b and 12.2).

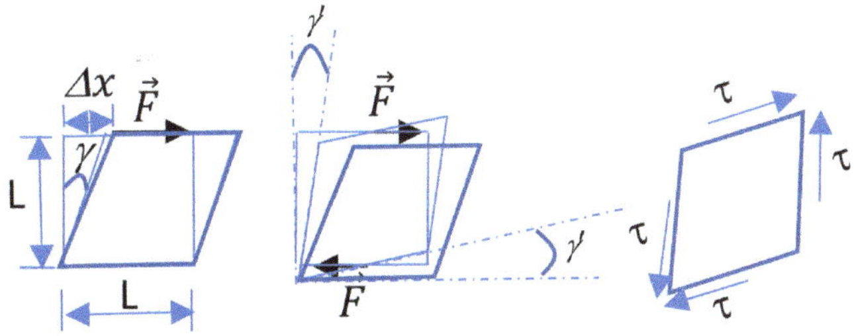

FIG. 9.3a – Shear deformation of the interferometer arms.

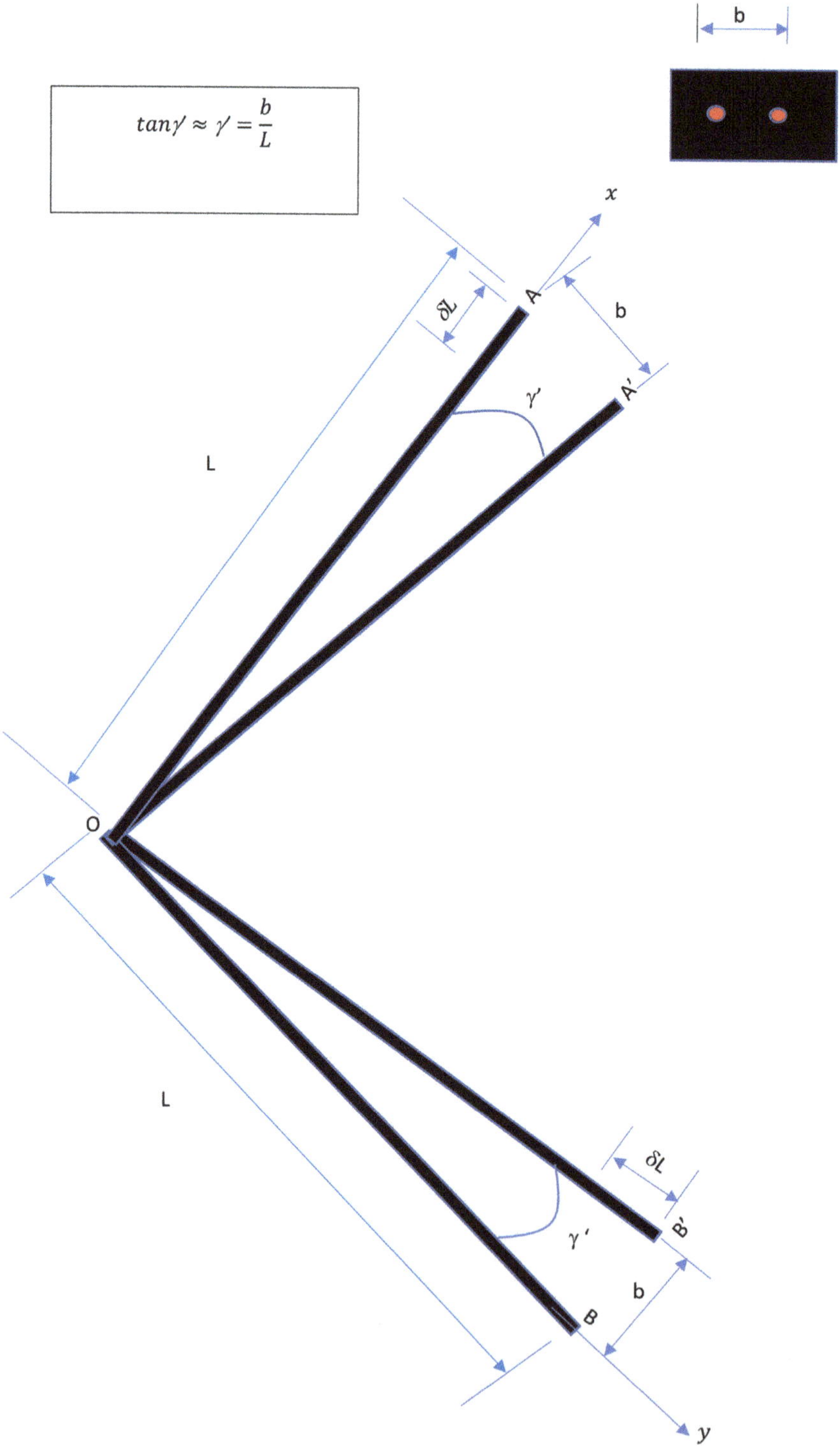

Fig. 9.3b – Measured and possible movements of interferometer arms.

It also proves, in connection with the fact that gravitational waves are elastic waves of space (space medium see appendix G and Ligo/Virgo measurements [23, 24]), that the elastic behavior of space is in perfect correlation with the elasticity theory and therefore the existence of an equivalent elastic material in constituting the space vacuum (see appendix G). This fundamental research also proves that it is possible to unify the elasticity theory to general relativity and thus that it is possible to define an equivalent elastic material constituting space (see [51] at [53] and [13]). This elastic medium is so a new sort of ether deformable and elastic (not rigid as it was considered in the past). This approach can be considered as an interesting addition to looking at the dynamics of general relativity through the elasticity theory. Therefore, we completely agree with the Pr Thibault Damour when he said: "space-time is an elastic structure which is deformed by the presence within it of mass-energy" and "the speed of sound for the elastic deformation of space in Einstein's theory is the speed of light; gravitational waves are waves of elastic deformations of space" see [5].

Third point: By the elasticity theory, there is therefore, for each deformation tensors, a stress tensor connecting by Hooke's law the components of the strain tensor ε_{ij} to the stress tensor σ_{ij} [66] with $G = E/(2(1 + v))$:

$$\sigma_{ij} = \frac{E}{(1+v)}\left(\varepsilon_{ij} + \frac{v}{(1-2v)}\varepsilon_{kk}\delta_{ij}\right)\begin{cases} \sigma_{11} = 2G\left[\varepsilon_{11} + \frac{v}{(1-2v)}(\varepsilon_{11}+\varepsilon_{22}+\varepsilon_{33})\right]; (i=1, j=1) \\ \sigma_{22} = 2G\left[\varepsilon_{22} + \frac{v}{(1-2v)}(\varepsilon_{11}+\varepsilon_{22}+\varepsilon_{33})\right]; (i=2, j=2) \\ \sigma_{33} = 2G\left[\varepsilon_{22} + \frac{v}{(1-2v)}(\varepsilon_{11}+\varepsilon_{22}+\varepsilon_{33})\right]; (i=3, j=3) \\ \sigma_{13} = 2G[\varepsilon_{13}]; \quad (i=2, j=3) \\ \sigma_{23} = 2G[\varepsilon_{23}]; \quad (i=1, j=2) \\ \sigma_{12} = 2G[\varepsilon_{12}]; \quad (i=3, j=1) \end{cases}$$

And the reciprocal in strains:
$\sigma_{kk} = \text{trace of } \sigma_{ij} = \sigma_{11} + \sigma_{22} + \sigma_{33};\ \delta_{ij} = 1 \text{ if } i = j;\ 0 \text{ otherwise}$

$$\varepsilon_{ij} = \frac{1}{E}\left((1+v)\sigma_{ij} - v\sigma_{kk}\delta_{ij}\right)\begin{cases} \varepsilon_{11} = \frac{1}{E}[\sigma_{11} - v(\sigma_{22}+\sigma_{33})]; (i=1, j=1) \\ \varepsilon_{22} = \frac{1}{E}[\sigma_{22} - v(\sigma_{11}+\sigma_{33})]; (i=2, j=2) \\ \varepsilon_{33} = \frac{1}{E}[\sigma_{33} - v(\sigma_{11}+\sigma_{22})]; (i=3, j=3) \\ \varepsilon_{12} = \frac{1+v}{E}[\sigma_{12}]; \quad (i=2, j=3) \\ \varepsilon_{13} = \frac{1+v}{E}[\sigma_{13}]; \quad (i=1, j=2) \\ \varepsilon_{23} = \frac{1+v}{E}[\sigma_{23}]; \quad (i=3, j=1) \end{cases}$$

In association with the strain tensors above, in the particular case of gravitational waves, we obtain the following two stress tensors:

(a) Associated with polarization A_+, the stress tensor corresponding to the pure compression/traction of the space medium in the plane xy is:

$$\sigma_{xy(A_+)} = \begin{bmatrix} \sigma_{xx} & 0 & 0 \\ 0 & \sigma_{yy} & 0 \\ 0 & 0 & 0 \end{bmatrix}$$

(b) Associated with polarization $A_\times$, the stress tensor corresponding to pure shear of the space medium in the plane xy is:

$$\sigma_{xy(A_\times)} = \begin{bmatrix} 0 & \tau_{xy} = \sigma_{xx} & 0 \\ \tau_{yx} = \sigma_{yy} & 0 & 0 \\ 0 & 0 & 0 \end{bmatrix}$$

The two stress tensors above, depending on the orientation of the facet considered, are characteristic of a pure shear associated with a pure twist of space medium following the rotation of two massive objects that merge. Figure 9.4 shows the flat deformation of the space medium created perpendicular to the direction of the spread gravitational wave as a transverse wave.

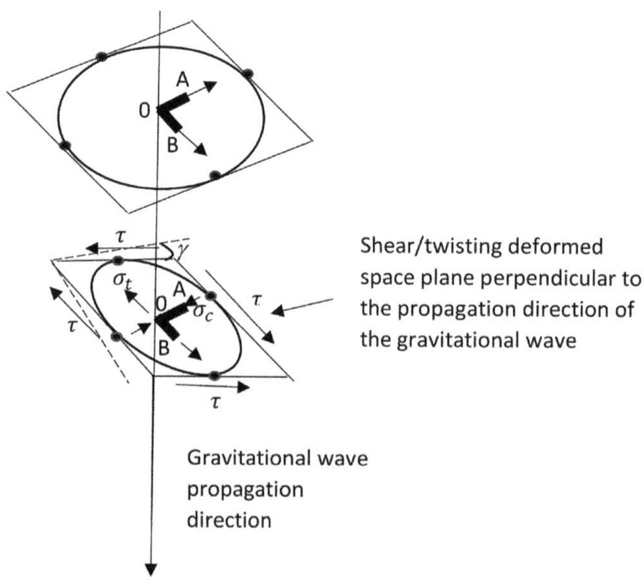

FIG. 9.4 – Deformations in plane of space elastic medium created by gravitational transverse waves.

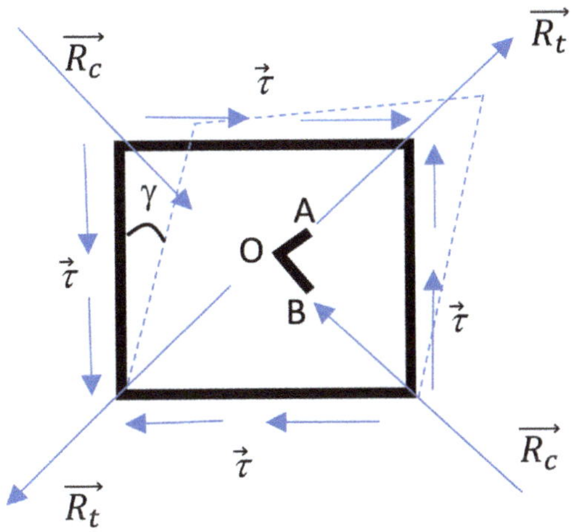

FIG. 9.5 – Creating normal stresses from the combination of shear stresses.

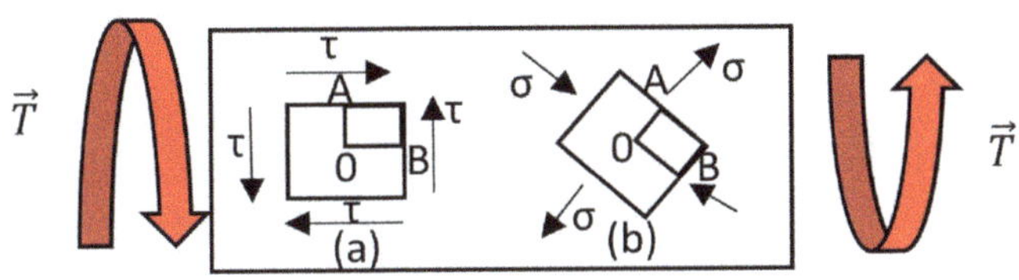

FIG. 9.6 – Normal stresses and shear stresses measured on interferometers based on their orientations on a twisted space cylinder.

Figure 9.5 shows how a combination of shear stress τ create normal stresses σ in the directions $\overline{0A}$ and $\overline{0B}$.

Depending on the orientation of the interferometers, they do not measure anything (case a) or deformations, $\frac{\delta L}{L}$ (case b) as shown in the figure 9.6.

9.3 Consequence in Terms of Oscillating Waves in the Arms of Interferometers

This will be the fourth point: Each of these stress states corresponds to a wave velocity characteristic of the elastic medium measured in the interferometric plane by the longitudinal oscillation of the laser mirror.

(a) Combined with polarization A_+, a pure longitudinal-traction compression wave speed in each direction x or y is measured by the laser mirror of the Ligo and

Virgo interferometers connected to a stress tensor (traction/compression) according to the following expression:

$$\frac{\partial^2 u_{(x,t)}}{\partial x^2} - \frac{1}{c^2}\frac{\partial^2 u_{(x,t)}}{\partial t^2} = 0$$

D'Alembert's equation is written:

$$\frac{\partial^2 u(x,t)}{\partial x^2} - \frac{\rho}{E} \times \frac{\partial^2 u(x,t)}{\partial t^2} = 0$$

By comparing the two equations below we obtain the well-known equation linking density ρ of the medium, Young's modulus E (=Y) characteristic of the elasticity of the material and the velocity c of the elastic wave [60]:

$$\frac{1}{c^2} = \frac{\rho}{E}$$

$$c = \sqrt{\frac{E}{\rho}} = \sqrt{\frac{Y}{\rho}}$$

(b) Associated with polarization $A_\times$, a pure shear wave speed, not yet measured by the laser mirror of the Ligo and Virgo interferometers (lateral movements of the mirrors (see figures 9.2, 9.3a and 9.3b), connected to a tensor and a twisting torque.

Indeed, if we consider that the elastic fabric that constitutes space has a behavior similar to that we have on Earth in case of an earthquake but with only transverse waves, in this case we only have a shear wave S. Indeed, according to what we measure on Ligo and Virgo, only deformations in the plane xy perpendicular to the direction of the gravitational wave are observed (see figures 9.4 and 9.5).

Thus, the following expression from the elasticity theory of applies (see appendix G), with u_i the component i of the displacement vector $\vec{u}$, time t, the stress tensors σ_{ij} and density ρ (see [51, 53]). The fundamental law of dynamics is written:

$$\rho\frac{\partial^2 u_i}{\partial t^2} = \frac{\partial \sigma_{ij}}{\partial j}$$

The dimensional equation is the following:

$$\frac{\mathrm{kg}}{\mathrm{m}^3} \times \frac{\frac{\mathrm{m}}{\mathrm{s}}}{\mathrm{s}} = \frac{\frac{\mathrm{N}}{\mathrm{m}^2}}{\mathrm{m}}$$

With for the strain tensors ε_{ij}:

$$\varepsilon_{ij} = \frac{1}{2}\left(\partial_i u_j + \partial_j u_i\right)$$

The classic motion of the elastic wave equation is written:

$$\rho\frac{\partial^2\vec{u}}{\partial t^2} = (\lambda+2\mu)\vec{\nabla}\left(\mathrm{div}(\vec{u})\right) - \mu\overrightarrow{\mathrm{rot}}\left(\overrightarrow{\mathrm{rot}}\left(\vec{u}\right)\right) + f_{\mathrm{external}}$$
$$\rho\frac{\partial^2\vec{u}}{\partial t^2} = (\lambda+2\mu)\vec{\nabla}\left(\mathrm{div}(\vec{u})\right) - \mu\vec{\nabla}\left(\mathrm{div}(\vec{u})\right) + \mu\vec{\Delta}\left(\vec{u}\right) + f_{\mathrm{external}}$$

After calculations (see appendix G):

$$\rho\frac{\partial^2\vec{u}}{\partial t^2} = (\lambda+\mu)\vec{\nabla}\left(\mathrm{div}(\vec{u})\right) + \mu\vec{\Delta}\left(\vec{u}\right) + f_{\mathrm{external}}$$

Or:

$$\rho\frac{\partial^2\vec{u}}{\partial t^2} = (\lambda+\mu)\vec{\nabla}\left(\vec{\nabla}\cdot\vec{u}\right) + \mu\vec{\nabla}^2\vec{u} + f_{\mathrm{external}}$$

where

$$\begin{aligned}\vec{\nabla}\vec{u} &= \overrightarrow{\mathrm{grad}}\left(\vec{u}\right)\\ \vec{\nabla}^2\vec{u} &= \vec{\Delta}\left(\vec{u}\right)\end{aligned}$$

When, the solution of the equation follows the decomposition of Helmoltz which gives two waves that propagate in the elastic medium: $f_{\mathrm{external}} = 0$.

$$\vec{u} = \vec{u}_{\mathrm{longitudinal,\,pressure}} + \vec{u}_{\mathrm{transversal,\,shear}}$$

A velocity pressure wave c_{pressure}:

$$c_{\mathrm{pressure}} = \sqrt{\frac{\lambda+2\mu}{\rho}}$$

This hypothesis of compression of the waves in the z direction is not acceptable because in this case the deformations are in the same direction as the propagation of the waves. Of course, this is not the case for gravitational waves where deformations are in a plane perpendicular to the direction of wave propagation (z). We will use this to determine Poisson's ration in the next chapter.

A shear wave of velocity c_{shear}:

$$c_{\mathrm{shear}} = \sqrt{\frac{\mu}{\rho}} = \sqrt{\frac{E}{2(1+\nu)\rho}}$$

$$c^2_{\mathrm{shear}} = \frac{\mu}{\rho} = \frac{E}{2(1+\nu)\rho}$$

T. Tenev and M.F. Horstemeyer, arrive at the same expression in their publication (see formula 3.9, at 3.14 of Mechanics of space-time – A Solid Mechanics perspective on the theory of General Relativity version March 2018).

This shear wave hypothesis is also not strictly acceptable, as the deformations measured on the interferometer are not shear deformations (angles γ) associated with shear waves but $\frac{\delta L}{L}$, as, strains that are always related to possible compression waves; but we will still consider it because it is possible that shear strain exists and is not already measured.

In strict terms, gravitational waves cannot be equated with conventional elastic waves in an elastic medium. They have the particularity of creating deformations perpendicular to the direction of propagation of waves (characteristic of transverse shear waves), but with lengthening and shortening (characteristics of compression waves/traction).

As we have deformations and stresses in planes, the proposal of a spatial medium consisting of a multi sandwich of thin sheared sheets (or stacking of hypersurfaces following [51] at [53]) perpendicular to the direction of propagation of the gravitational wave therefore seems a reasonable hypothesis. This thousand-sheet structure is the only way to explain why under shear stresses we measure deformations and stresses in planes without correlation with deformations or stresses perpendicular to the plane.

9.4 Expression of Einstein's Linearized Gravitational Equation in the Form of Strains

The linearized form of Einstein's equation in weak gravitational fields is:

$$\partial^{\lambda}\partial_{\lambda}\overline{h}_{\mu\nu} = \Box\overline{h}_{\mu\nu} = -\frac{16\pi G}{c^4}T_{\mu\nu}$$

In the vacuum (in the case of gravitational waves), we have:

$$\partial^{\lambda}\partial_{\lambda}\overline{h}_{\mu\nu} = \Box\overline{h}_{\mu\nu} = 0$$

$\Box$: D'Alembertian's wave operator,

$$\overline{h}_{\mu\nu} = h_{\mu\nu} + \frac{1}{2}\eta_{\mu\nu}\overline{h}$$

$$\overline{h}_{\mu\nu} = A_{\mu\nu}\cos(k_{\sigma}x^{\sigma})$$

Four-vector wave of the plane wave:

$$k^{\sigma} = \left(\frac{\omega}{c}; \overrightarrow{k}\right)$$

$$x^{\sigma} = (ct, x, y, z)$$

$$\|\overrightarrow{k}\|^{2} = \frac{\omega^{2}}{c^{2}}$$

ω the circular frequency of the wave:

$$A_{\mu\nu} = A_{+}\begin{bmatrix} 0 & 0 & 0 & 0 \\ 0 & +1 & 0 & 0 \\ 0 & 0 & -1 & 0 \\ 0 & 0 & 0 & 0 \end{bmatrix} + A_{\times}\begin{bmatrix} 0 & 0 & 0 & 0 \\ 0 & 0 & +1 & 0 \\ 0 & +1 & 0 & 0 \\ 0 & 0 & 0 & 0 \end{bmatrix}$$

$$\eta_{\mu\nu} = \begin{bmatrix} 1 & 0 & 0 & 0 \\ 0 & -1 & 0 & 0 \\ 0 & 0 & -1 & 0 \\ 0 & 0 & 0 & -1 \end{bmatrix}$$

h the trace of $h_{\mu\nu}$:

$$\overline{h} = -h$$

From the above formulas we get a new equation of Einstein's gravitational field linearized function of the strain tensor ε_{ij}:

$$\partial^{\lambda}\partial_{\lambda}\left(h_{\mu\nu} + \frac{1}{2}\eta_{\mu\nu}\overline{h}\right) = \Box\left(h_{\mu\nu} + \frac{1}{2}\eta_{\mu\nu}\overline{h}\right) = -\frac{16\pi G}{c^{4}}T_{\mu\nu}$$

By replacing $h_{\mu\nu}$ with $2\varepsilon_{\mu\nu}$ we get:

$$\partial^{\lambda}\partial_{\lambda}\left(2\varepsilon_{\mu\nu} + \frac{1}{2}\eta_{\mu\nu}2\overline{\varepsilon}\right) = \Box\left(2\varepsilon_{\mu\nu} + \frac{1}{2}\eta_{\mu\nu}2\overline{\varepsilon}\right) = -\frac{16\pi G}{c^{4}}T_{\mu\nu}$$

After simplification by 2, we get the same formula of the weak-field constant κ:

$$\boldsymbol{\partial^{\lambda}\partial_{\lambda}\left(\varepsilon_{\mu\nu} + \frac{1}{2}\eta_{\mu\nu}\overline{\varepsilon}\right) = \Box\left(\varepsilon_{\mu\nu} + \frac{1}{2}\eta_{\mu\nu}\overline{\varepsilon}\right) = -\frac{8\pi G}{c^{4}}T_{\mu\nu}}$$

This formula is similar to T. Damour's formula in his book "if Einstein had told me"

$$D_{(g)} = \kappa T$$

The deformations D, function of the metric g are equal to κ multiplied by the tensions T which act.

That is, Hooke's law. $\frac{\Delta L}{L} = \frac{1}{E} \times \frac{F}{S}$ or $\varepsilon = \frac{1}{E} \times \sigma$.

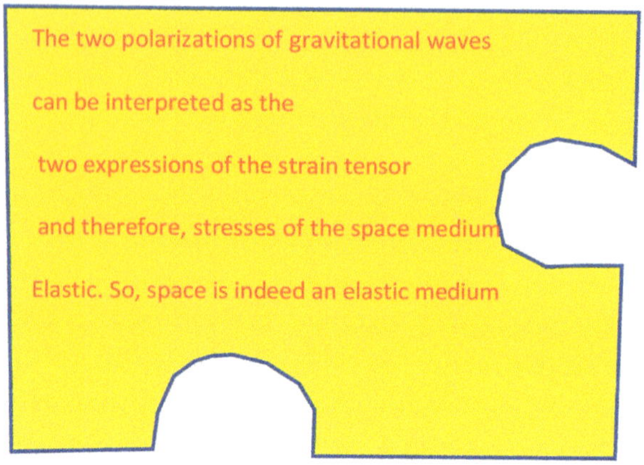

7th piece of the puzzle

Chapter 10

Determination of Poisson's Ratio of the elastic Space Material

Chapter Summary

Poisson's ratio ν is worth 1.

10.1 Introduction

As we have an elastic medium constituting the space (see chapter 9), we must define the mechanical characteristics of this medium, that is to say it's Young's modulus and Poisson's ratio. We can conclude from the wave celerity equations seen in the previous chapter and appendix G, a potential value of Poisson's ratio ν.

3 approaches allow us to establish this value and in particular the results of measurement of the strains made on the interferometers and resulting from the general relativity calculations as shown in figure 10.1.

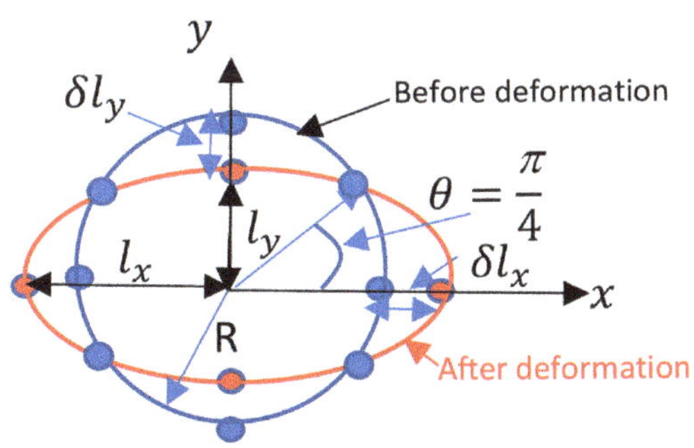

FIG. 10.1 – Example of particle coordinates subjected to a polarized gravitational wave A_+ propagating through the xy plane.

DOI: 10.1051/978-2-7598-2573-8.c010

A first approach linked to the movement of particles on a circle during the passage of a gravitational wave following the z direction.

A second approach by considering that the gravitational wave is a transverse wave and is not a compression wave.

A third approach to the state of the art available.

We will study these 3 approaches in the following paragraphs.

10.2 First Approach: Analysis of the Movements of Particles Positioned in Space on a Circle Undergoing the Passage of a Gravitational Wave

The analysis of the figure 10.1 from the calculation of general relativity shows that an object on the xy plane positioned perpendicular to the z direction of the propagation of the transverse gravitational waves, is simultaneously compressed in one direction and stretched in the direction perpendicular. The deformations are equal but of opposite sign $\varepsilon_{xx} = -\nu\varepsilon_{yy}$:

With the definition of Poisson's ratio we have: $v = \frac{\text{relative transverse shrinkage}}{\text{relative longitudinal elongation}} = 1.$

10.3 Second Approach: In the z Direction, the Gravitational Wave is a Transverse Wave and is Not a Compression Wave

There is no compression wave perpendicular to the plane of the interferometer. So, we can write:

$$c_{\text{pressure}} = \sqrt{\frac{\lambda + 2\mu}{\rho}} = 0$$

where:

$$\mu = \frac{E}{2(1+\nu)}$$

$$\lambda = \frac{E\nu}{(1+\nu)(1-2\nu)}$$

The equation above should be equal to 0 which gives $\nu = 1$ again.

$$c_{\text{pression}} = \sqrt{\frac{\frac{E\nu}{(1+\nu)(1-2\nu)} + \frac{E}{(1+\nu)}}{\rho}} = 0$$

$$c_{\text{pression}} = \sqrt{\frac{E(1-\nu)}{\rho(1+\nu)(1-2\nu)}} = 0 \rightarrow \nu = 1$$

10.4 Third Approach: Based on Available Datas

Based on the results of T.G. Tenev and M.F. Horstemeyer [51] version 2017, we have for Young's modulus $E = Y = 4.4 \times 10^{113}$ Pa (see 3.4, formula 3.13) and a density of 1.30×10^{96} kg/m^3 (see 3.4, formula 3.14).

$$\rho = \frac{\mu}{c^2} \rightarrow \nu = \frac{E}{2c^2\rho} - 1$$

With the equation above and the results of T.G. Tenev and M.F. Horstemeyer [51], we get:

$\nu_z = 0.8829$. T.G. Tenev and M.F. Horstemeyer in "mechanic of space time chapter 4" obtain $\nu = 1$.

Acceptable value given the high energy intensity of the vacuum being discussed.

Conclusion: One retains for Poisson's ratio: $\nu = 1$.

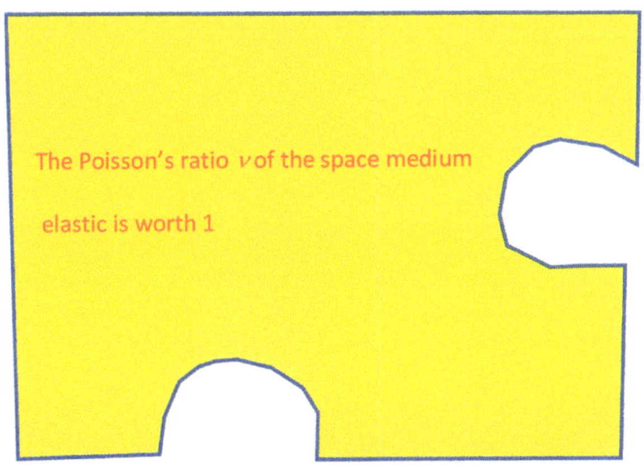

8th piece of the puzzle

Chapter 11

Dynamic Study of the elastic Space Strains in an Arm of an Interferometer

Chapter Summary

$$\frac{1}{L^2}(\varepsilon)^2 = 8\pi\left(\frac{\pi f^2}{\rho}\right)\frac{1}{c^4} \times \frac{U}{V}$$

$$\frac{1}{L^2}(\varepsilon)^2 = \frac{8\pi G}{c^4} \times \frac{U}{V}$$

$$G = \frac{\pi f^2}{\rho}$$

11.1 Study of an Interferometric Arm Subjected to Gravitational Waves Causing Compressions and Tractions of the Volume of Empty Space within It

11.1.1 Assumptions

The objective of this chapter is to propose a tensorial equation between curvature and strain energy based on space strains measured by the Ligo and Virgo interferometers.

In this first part we will study a horizontal space cylinder in a direction solicited by a gravitational wave (without taking into account the effect of Poisson's ratio ν), using the fact that oscillating compression/traction deformations are associated with compression/traction waves vacuum in the interferometric tube (see chapter 9).

DOI: 10.1051/978-2-7598-2573-8.c011

We assume in this chapter that the coalescence of two black holes, for example, creates by their rotations, a twist of space as shown in figure 9.1b. This twisting of the space sheets creates, in successive planes xy, tensions and compressions (and thus lengthenings and shortenings) of the sheets of space materials (multi sandwich) that finally arrive on Earth in each arm xy of the interferometers (see figures 9.3a,b–9.5). In weak field, for space only, the variation of the metric from the flat metric is given by $h_{\mu\nu}$. The result is for a polarized wave A_+, the stress tensor associated with it is as we have seen in chapter 9:

$$h_{\mu\nu} = 2\varepsilon_{\mu\nu} = A_+ \cos\left(\frac{\omega}{c}(ct - z)\right) \begin{bmatrix} 0 & 0 & 0 & 0 \\ 0 & \mathbf{+1} & \mathbf{0} & \mathbf{0} \\ 0 & \mathbf{0} & \mathbf{-1} & \mathbf{0} \\ 0 & \mathbf{0} & \mathbf{0} & \mathbf{0} \end{bmatrix} \rightarrow \varepsilon_{xy(A_+)}$$

$$= \frac{1}{2} A_+ \cos\left(\frac{\omega}{c}(ct - z)\right) \begin{bmatrix} \varepsilon_{xx} & 0 & 0 \\ 0 & -\varepsilon_{yy} & 0 \\ 0 & 0 & 0 \end{bmatrix} \rightarrow \sigma_{xy(A_+)} = \begin{bmatrix} \sigma_{xx} & 0 & 0 \\ 0 & \sigma_{yy} & 0 \\ 0 & 0 & 0 \end{bmatrix}$$

Note: We choose this polarization because it corresponds to the displacements measured in the interferometric arms (compression/traction volume that create the advances and retreats of laser mirrors see figure 9.2).

In this section, we therefore consider a tube of interferometer of a length L and section S (associated with the strains ε_x measured by Hooke's Law) as defined in figure 11.1. Inside the tube, the vacuum is considered to consist of a spatial elastic substance (see chapter 7) made from very small particles to form a quantum-sized r granulometry.

Note: In this simplified approach, we deliberately disconnect the correlation between directions x and y interferometric tubes. This is done in order to see already if the basic principle space curvature $= K \times$ energy density is respected. The consequence is that Poisson's ratio ν is not taken into account here in this first approach in order to be compatible with the information of the compression/traction tensor. It will be taken into account in chapter 12.

Since there are movements of the laser mirror in the direction of the tube corresponding to the compression/traction of the space medium (see figure 9.2), there are associated strains and stresses in this elastic medium and therefore a normal dynamic force N and a compression/traction wave inside the tube.

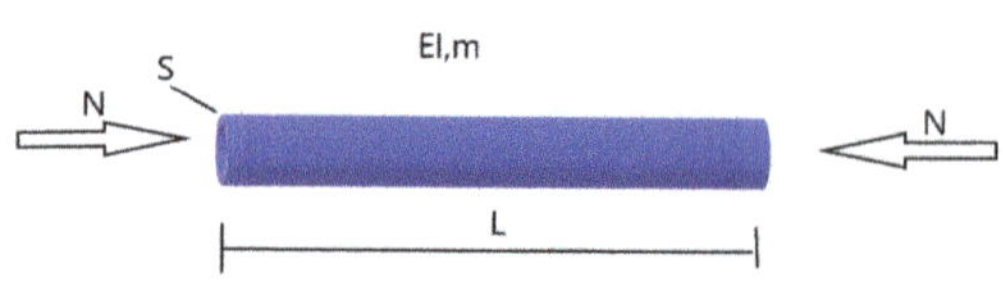

FIG. 11.1 – Tube charged with normal effort N.

11.1.2 Determination of Tensorial Equations Associated with Each Arm of the Interferometer

Hooke's law can be written according to the displacement $u_{(x)}$:

$$\sigma_{xx} = \varepsilon E = \frac{N}{S} = \left(\frac{u_{(x+dx)} - u_{(x)}}{dx}\right) E = \frac{\delta L}{L} E$$

where σ is the normal stress in N/m^2, ε is deformation in % ($\varepsilon = \frac{N}{ES}$) is measured by Ligo and Virgo, N is the normal force in Newton, S is the section of the tube in m^2, L is the length of the tube in m, $V = S \times L$ is the volume of the tube in m^3, $E = Y$ is Young's modulus of the material constituting the tube in N/m^2, δL is the variation of length in m under normal force N, and $u_{(x)}$ is the longitudinal displacement in m along the longitudinal axis x (see figure 11.1).

The relationship effort-displacement according to the rigidity $K = ES/L$ is written as follows:

$$N = \frac{ES}{L}\delta L = K\delta L$$

The strain energy U (N.m) of the static space tube, when N is constant is worth:

$$U = \frac{1}{2}\int_0^L \frac{N^2}{ES}dx = \frac{1}{2}\frac{N^2}{K}$$

The strain energy of the rigidity spring K (N/m) can be written by substituting N by its expression function of rigidity K:

$$U = \frac{1}{2}K(\delta L)^2$$

In addition, according to the definition of strain, the variation in displacement δL is:

$$\varepsilon \times L = \delta L$$

By deferring this expression in the expression of the strain energy we obtain:

$$(\varepsilon)^2 = \frac{2}{K}\frac{U}{L^2}$$

By starting from the rigidity K according to the length L, of Young's modulus $E(= Y)$, of the tube section S and by substituting the expression of L by K we obtain with $S = V/L$:

$$\frac{1}{L^2}(\varepsilon)^2 = 2\frac{K}{V}\frac{1}{E^2} \times \frac{U}{V}$$

So:

$$\frac{U}{V} = \frac{\frac{1}{L^2}(\varepsilon)^2}{2\frac{K}{V}\frac{1}{E^2}}$$

We now consider the tube under a dynamic behavior and with pure compression/traction longitudinal waves following x in the arm of the interferometer as a consequence of polarized gravitational waves A_+. Density ρ is noted as a function of mass m and volume V:

$$\rho = \frac{m}{V}$$

The fundamental dynamic equation allows us to find the tube's Eigen circular frequency (harmonic oscillator) in space material:

$$U_c + U = \frac{1}{2} m\dot{x}_{(t)}{}^2 + \frac{1}{2} Kx_{(t)}{}^2 = E_0 = 0$$

where U_c is kinetic energy.

By derivative in respect to t the previous formula we get:

$$\ddot{x}_{(t)} + \frac{K}{m} x_{(t)} = 0$$

And according to Newton's definition of force:

$$\text{Force} = m\ddot{x} = -Kx$$

This allows to express the Eigen circular frequency ω depending on the rigidity K of the tube and its mass m:

$$\omega^2 = \tfrac{K}{m} = (2\pi f)^2$$

$$\omega = 2\pi f$$

We can write a new expression of volume V:

$$V = \frac{m}{\rho} = \frac{K}{\omega^2 \rho}$$

The total energy density U_T of the spring-mass system depends on the density of kinetic energy U_c and the strain energy density U:

$$\frac{U_c}{V} + \frac{U}{V} = \frac{U_T}{V}$$

The strain energy density is therefore:

$$\frac{U}{V} = \frac{U_T}{V} - \frac{U_c}{V} = T$$

We can still write:

$$\frac{\frac{1}{L^2}(\varepsilon)^2}{2\frac{K}{V} \times \frac{1}{E^2}} = T$$

Or in an equivalent way:

$$\frac{1}{L^2}(\varepsilon)^2 = 2\frac{K}{V} \times \frac{1}{E^2} T$$

By substituting V by its expression function of circular frequency we get:

$$\frac{1}{L^2}(\varepsilon)^2 = 2\rho\frac{\omega^2}{E^2} \times T$$

By replacing the circular frequency ω with the frequency f, we get:

$$\frac{1}{L^2}(\varepsilon)^2 = 8\pi^2\rho\frac{f^2}{E^2} \times T$$

In the plane x, y of the interferometer, the movements of the mirror are perpendicular to the section of the tube (see figure 9.2). Thus, in the plane of the interferometer, the mirror follows the compression/traction movements of the space material inside the volume of the tube ($\frac{\delta L_x}{L_x}$, in connection with the displacement,$u_{(x)}$). These ossilations in the plane of the interferometer are therefore pure longitudinal compression/traction waves correlated to transverse waves perpendicular to the interferometer plane.

So, in this section, depending on the type of tensor considered, the velocity c of the wave is given in the formula function of Young's modulus ($E = Y = \rho c^2$). The speed is limited because it spreads in an elastic medium of density ρ. A similar experiment would involve shooting a metal ball more or less quickly in the middle of a pile of sand. Depending on the density and compaction of the sand, the ball will move more or less quickly. In our case, the bullets should be photons and the medium would consist of an extremely fine granulometry material $r = 1 \times 10^{-35}$ m (see chapter 16, table 16.1). We now return to the tube under a gravitational wave. We multiplies and divides by ρ the above expression:

$$\frac{1}{L^2}(\varepsilon)^2 = 8\pi^2 f^2\left(\frac{\rho}{E}\right)^2\frac{1}{\rho} \times T$$

Finally, we have the relationship between dynamic curvature and strain ε in the tube:

$$\frac{1}{L^2}(\varepsilon)^2 = 8\pi\left(\frac{\pi f^2}{\rho}\right)\left(\frac{\rho}{E}\right)^2 \times \frac{U}{V}$$

Taking into account:

$$E = Y = \rho c^2$$

We get the expression of energy density U/V:

$$\frac{1}{L^2}(\varepsilon)^2 = 8\pi\left(\frac{\pi f^2}{\rho}\right)\frac{1}{c^4}\times\frac{U}{V}$$

We notice that:

- the term $\frac{\pi f^2}{\rho}$ has the dimension of the gravitational constant G ($m^3/(kg.s^2)$);
- the term $\frac{1}{L^2}(\varepsilon)^2$ has the dimension of a curvature ($1/m^2$). Indeed, L corresponds to an infinitely large curvature of radius R (beam of length L) and it is recalled that the scalar curvature R (see $\frac{1}{2}g_{\mu\nu}R$ in the Einstein field equation) is worth $2/R^2$ (see appendix D) and that $g_{\mu\nu} = \eta_{\mu\nu} + h_{\mu\nu} = \eta_{\mu\nu} + 2\varepsilon_{\mu\nu} + \varepsilon^2_{\mu\nu}$ (see chapter 5);
- the term (U/V) to the dimension of the strain energy density (the internal strain work) ($N.m/m^3$) and that it is equal to the external work applied to the space tube;
- the term $\left(\frac{\pi f^2}{\rho}\right)\frac{1}{c^4}$ has the dimension of the reverse of the load (N^{-1}) to compare with κ.

The above formula therefore satisfies the principle space curvature $= K \times$ energy density. We found Einstein's equation in the case of the oscillations of a space tube stressed by a gravitational wave without going through the temporal components of the tensors, without going through the passage to Newton's limits simply using the continuum mechanic and by applying it to a potential elastic material filling the space.

Therefore, we learn through the elasticity theory that the term $(\frac{\pi f^2}{\rho})$ can actually be identified with Newton's G constant but depending on the mechanical parameters of the space medium:

$$G = \frac{\pi f^2}{\rho} = \frac{\pi f^2 c^2}{E} = \frac{\pi f^2 c^2}{Y}$$

Or if we extract Young's modulus Y of the elastic space-time:

$$Y_{\text{space time}} = \frac{\pi f^2 c^2}{G}$$

Note: This expression is close to the formula (5) of [13] "What is the Stiffness of Spacetime of Kirk T. McDonald May 5, 2018 foot note 5 exchange with Prof R. Weiss Nobel price (email April 15, 2018) $Y_{\text{space time}} = \frac{\pi f^2 c^2}{4G}$".

In these formulations f is the natural frequency of the space material inside the tube and ρ is the density of the space material $\left(\rho = \frac{m}{V} = \frac{E_{\text{vaccum}}/c^2}{V}\right)$. In the next chapter, we will do the same but taking into account Poisson's ratio ν.

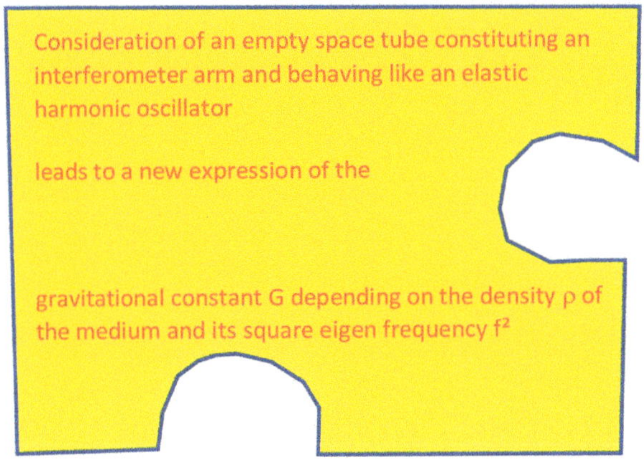

9th piece of the puzzle

Chapter 12

Dynamic Study of Simultaneous Elastic Space Strains in the 2 Arms of an Interferometer

Chapter Summary

$$\frac{1}{L^2}(\varepsilon_{xx})^2 = 4(1+v) \times \pi \times \frac{\pi f^2}{\rho} \times \frac{1}{c^4} \times \frac{U}{V}$$

$$\frac{1}{L^2}(\varepsilon_{yy})^2 = 4(1+v) \times \pi \times \frac{\pi f^2}{\rho} \times \frac{1}{c^4} \times \frac{U}{V}$$

$$G = \frac{\pi f^2}{\rho}$$

$$v = 1$$

$$\begin{bmatrix} \frac{1}{L^2} & 0 \\ 0 & \frac{1}{L^2} \end{bmatrix} \begin{bmatrix} (\varepsilon_{xx})^2 & 0 \\ 0 & (\varepsilon_{yy})^2 \end{bmatrix} = \frac{8\pi G}{c^4} \begin{bmatrix} T_{xx} & 0 \\ 0 & T_{yy} \end{bmatrix}$$

$$R_{ij} = \frac{8\pi G}{c^4} T_{ij}$$

DOI: 10.1051/978-2-7598-2573-8.c012

12.1 Study of Two Interferometric Arms Subjected to Gravitational Waves Resulting in Compression/Traction of the Volume of Space within Them

12.1.1 Assumptions

In this chapter, we will study two horizontal tubes of space, perpendicular to each other, solicited by a gravitational wave compressing them and dilating them successively and simultaneously ($\delta_{xx} = -\delta_{yy}$) as shown in figure 12.1. Poisson's ratio v will therefore be taken into account in this chapter. We will use traction-compression waves as in the previous chapter. These waves are correlated with the strain and stress tensor associated with the perturbation $h_{\mu v}$ of the metric.

We assume the same assumptions about deformation and stress tensors as in the previous chapter. Both arms of the interferometers are defined in figure 12.1.

In this case, both tubes behave like a gigantic stress/strain gauge. Each arm is a main direction in the plane xy. So, we can write a tensor expression associated with the two dimensions and consider a strain tensor and a strain energy tensor of the whole. We also assume that the plan xy is disconnected in the z direction (elasticity theory: problem of flat stress, sandwich frame of the space-time). The consequence is that the space material is considered made of thin sheets multi-sandwiches $2r$ thick twisted successively during the passage of the shear wave (see figure 12.2).

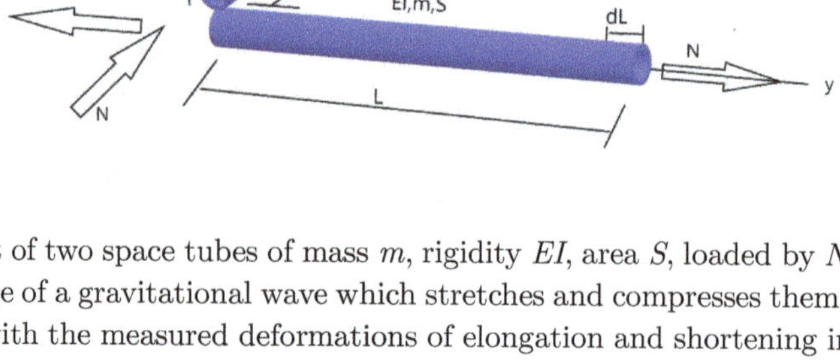

FIG. 12.1 – Set of two space tubes of mass m, rigidity EI, area S, loaded by N-force resulting from the passage of a gravitational wave which stretches and compresses them simultaneously in correlation with the measured deformations of elongation and shortening in application of the Hooke's law.

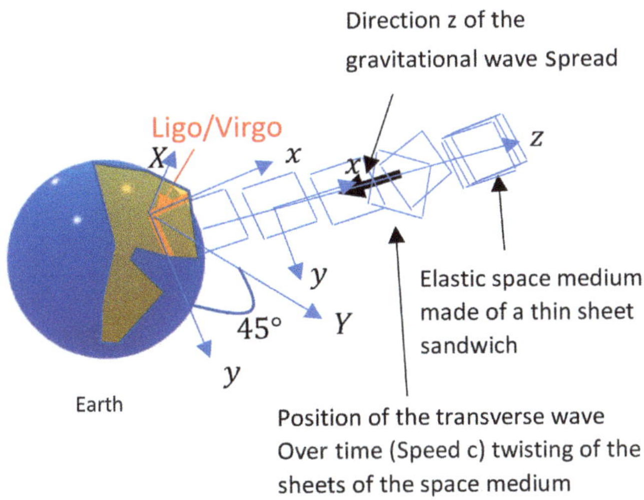

FIG. 12.2 – Sandwich structure of the space medium made of elastic thin sheets made of an elastic dark material twisted during the passage of a gravitational wave.

With the actual measurements made by Ligo and Virgo, we know that when one arm is compressed, the other is in tension simultaneously. We will be interested here in what happens only in the plane of interferometers placed in one of these thin elastic sheets constituting the space medium (x, y planes).

12.1.2 Determination of Tensorial Equation Associated with the Two Arms of the Interferometer

Step One: Determining the strains and stresses of a twisted torsion space sheet

Depending on the axis $\overrightarrow{x}; \overrightarrow{y}$, we have these two strains $\varepsilon_{xx} = -\nu\varepsilon_{yy}$ which are correlated with general relativity by $h_{\mu\nu} = 2\varepsilon_{\mu\nu}$. The strain tensor according to the axis system $\overrightarrow{x}; \overrightarrow{y}$ is given by:

$$h_{\mu\nu} = 2\varepsilon_{\mu\nu} = A_{+}\cos\left(\frac{\omega}{c}(ct - z)\right)\begin{bmatrix} 0 & 0 & 0 & 0 \\ 0 & \mathbf{+1} & \mathbf{0} & \mathbf{0} \\ 0 & \mathbf{0} & \mathbf{-1} & \mathbf{0} \\ 0 & \mathbf{0} & \mathbf{0} & \mathbf{0} \end{bmatrix} \rightarrow \varepsilon_{xy(A_{+})}$$

$$= \frac{1}{2}A_{+}\cos\left(\frac{\omega}{c}(ct - z)\right)\begin{bmatrix} \varepsilon_{xx} & 0 & 0 \\ 0 & -\varepsilon_{yy} & 0 \\ 0 & 0 & 0 \end{bmatrix} \rightarrow \sigma_{xy(A_{+})} = \begin{bmatrix} \sigma_{xx} & 0 & 0 \\ 0 & -\sigma_{yy} & 0 \\ 0 & 0 & 0 \end{bmatrix}$$

So, the relationships between distortions (shear strains) and stresses are written:

$$\varepsilon_{xx} = \frac{1}{E}\left(\sigma_{xx} - \nu\sigma_{yy}\right)$$

$$\varepsilon_{yy} = \frac{1}{E}(\sigma_{yy} - \nu\sigma_{xx})$$

Using the definition of the stress tensor we get:

$$\sigma_{xx} = \frac{E}{(1+\nu)}\left\{\varepsilon_{xx} + \frac{\nu}{(1-2\nu)}(\varepsilon_{xx} + \varepsilon_{yy})\right\}$$

$$\sigma_{yy} = \frac{E}{(1+\nu)}\left\{\varepsilon_{yy} + \frac{\nu}{(1-2\nu)}(\varepsilon_{xx} + \varepsilon_{yy})\right\}$$

Taking into account that $\varepsilon_{xx} = -\nu\varepsilon_{yy}$, as seen from what is measured on the interferometers (simultaneously strains of $+10^{-21}$ in one tube and -10^{-21} in the other see figure 12.2), with $\nu = 1$, the above expressions become (second term null):

$$\sigma_{xx} = \frac{E}{(1+\nu)}\{\varepsilon_{xx}\} \quad \sigma_{yy} = -\frac{E}{(1+\nu)}\{\varepsilon_{yy}\}$$

Note: With $\varepsilon_{xx} = \frac{\gamma_{xy}}{2}$ (see figure 12.4) we obtain well a shear of the space sheet in correlation with the traction/compression (see figure 9.6) [66]:

$$\tau_{xy} = \frac{E}{2(1+\nu)}\{\gamma_{xy}\}$$

Note: If one section of the tube is compressed (direction $\overrightarrow{x}$) the other $\overrightarrow{y}$ is in traction (origin of the sign minus). According to the axis system $\overrightarrow{x}; \overrightarrow{y}$ the stress tensor becomes [66]:

$$\sigma_{xy} = \begin{bmatrix} \sigma_{xx} & 0 & 0 \\ 0 & -\sigma_{yy} & 0 \\ 0 & 0 & 0 \end{bmatrix}$$

Using Mohr's circle (see figure 12.3) we can confirm the global shear along a facet at 45° in real life (90° on Mohr's circle) (System of Axes $\overrightarrow{X}; \overrightarrow{Y}$) (see figure 12.3).

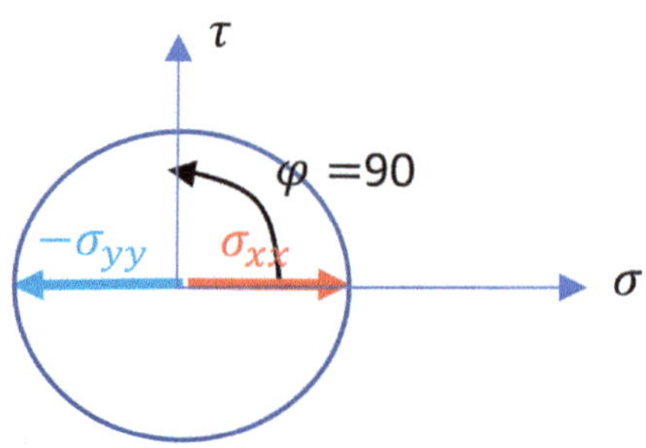

FIG. 12.3 – Mohr's circle in stresses.

When we turn at 90° on Mohr's circle, we turn at 45° on the real side (image of spin 2 of the graviton). So, on a facet at 45° we are pure shear as shown in the figure 12.3. The deformation tensor according to the axis system $\overrightarrow{X}; \overrightarrow{Y}$ is:

$$\varepsilon_{XY} = \begin{bmatrix} 0 & \varepsilon_{XY} & 0 \\ \varepsilon_{YX} & 0 & 0 \\ 0 & 0 & 0 \end{bmatrix}$$

The stress tensor according to the axis system $\overrightarrow{X}; \overrightarrow{Y}$ is:

$$\sigma_{XY} = \begin{bmatrix} 0 & \tau_{XY} = \sigma_{XX} = \sigma & 0 \\ \tau_{YX} = \sigma_{YY} = \sigma & 0 & 0 \\ 0 & 0 & 0 \end{bmatrix}$$

The shear strain is written according to Hooke's law in shear $\tau = \gamma G$:

$$\tau_{XY} = \frac{E}{2(1+v)}\gamma_{XY} = \mu\gamma_{XY}$$

In addition, we have in elasticity theory (see figures 5.1 and 12.4):

$$\varepsilon_{XY} = \frac{1}{2}\gamma_{XY}$$

From the above expressions, we get:

$$\tau_{XY} = \frac{E}{(1+v)}\varepsilon_{XY}$$

On Mohr's circle in strains (figure 12.4) we see that [66]:

$$\frac{\gamma_{XY}}{2} = \varepsilon_{XX}$$

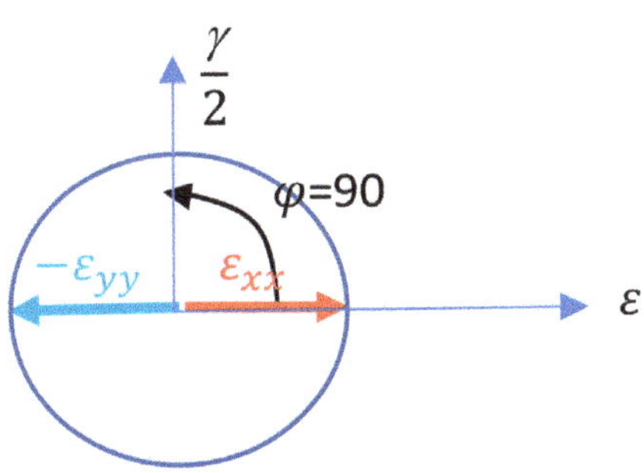

FIG. 12.4 – Mohr's circle in strains.

According to this equation, it seems that depending on the orientation of the interferometer in the plane $\overrightarrow{x}$, $\overrightarrow{y}$ or $\overrightarrow{X}$, $\overrightarrow{Y}$ (see figure 12.2) compared to the direction of propagation of the gravitational wave ($\overrightarrow{z}$), lateral movements of laser mirrors are possible with the same shape and wave intensity as that conventional compression/traction displacements (see figure 9.2). It should be interesting to measure these movements especially when traditional compression/traction movements are not measured because of the position of the interferometer in relation to the direction of propagation of the gravitational wave.

From the above expressions we get:

$$\varepsilon_{XX} = \varepsilon_{XY} \rightarrow \sigma = \tau$$

The deformation of the circle containing the particles is therefore the same (traction in one direction and compression in the other), but the circle rotates from an angle of 45°.

Step Two: Determination of the strain energy of the two connected tubes in the main axis system $\overrightarrow{x}\overrightarrow{y}$

The strain energy of the two tubes connected in traction/compression is now considered according to the axis system $\overrightarrow{x}\,\overrightarrow{y}$:

Note: The results will be the same when we consider the axis system $\overrightarrow{X}$; $\overrightarrow{Y}$ because of the equivalence between shear stresses and normal stresses on the one hand and the equivalence between deformations and angles on the other hand.

Energy density or energy per unit volume is written:

$$U_{ij} = \frac{U}{V} = \frac{1}{2}\sigma_{ij}\varepsilon^{ij}$$

In the axis system $\overrightarrow{x}\,\overrightarrow{y}$, we are in pure compression/traction in the tube, and the total strain energy per unit of volume is:

$$\frac{U}{V} = \frac{1}{2}\sigma_{xx}\varepsilon_{xx} + \frac{1}{2}\sigma_{yy}\varepsilon_{yy}$$

As the S section of the tube is constant and as the stresses and strains are constant on S, for a fixed section S of abscise (x) or (y), we have the following surface density of strain energy:

$$\frac{U}{S} = \frac{1}{2}\int_0^L \sigma_{xx}\varepsilon_{xx}\,dx + \frac{1}{2}\int_0^L \sigma_{yy}\varepsilon_{yy}\,dx$$

When one arm is in compression, the other is in traction and the measured strains are identical (except the sign) and therefore:

$$\varepsilon_{xx} = -\varepsilon_{yy} = \frac{N}{ES}$$

And with Hooke's law:

$$N_x = N_y = N = \frac{ES}{L}\delta L = K\delta L$$

$$\sigma_{xx} = \frac{E}{(1+v)}\{\varepsilon_{xx}\}$$

$$\sigma_{yy} = -\frac{E}{(1+v)}\{\varepsilon_{yy}\}$$

From the stress formula we can write again the surface density of strain energy:

$$\frac{U}{S} = \frac{E}{2(1+v)}\int_0^L \{\varepsilon_{xx}\}^2 dx + \frac{E}{2(1+v)}\int_0^L \{\varepsilon_{yy}\}^2 dx$$

Replacing the strains with their expressions (N/ES) results we obtain:

$$U = \frac{E}{2(1+v)}\int_0^L \left\{\frac{N}{ES}\right\}^2 Sdx + \frac{E}{2(1+v)}\int_0^L \left\{\frac{N}{ES}\right\}^2 Sdx$$

We obtain after simplification the energy in the case in two space dimensions:

$$U = \frac{1}{2(1+v)}\int_0^L \frac{N^2}{ES}dx + \frac{1}{2(1+v)}\int_0^L \frac{N^2}{ES}dx$$

Or after simplification:

$$U = \frac{1}{(1+v)} \times \frac{N^2 L}{ES} = \frac{1}{(1+v)} \times \frac{(K\delta L)^2}{K}$$

Taking into account the strain energy of a rigidity spring K (N/m) $= ES/L$ we get:

$$U = \frac{1}{(1+v)}K(\delta L)^2$$

In addition, according to the definition of strain, the variation in displacement δL is with $v = 1$:

$$\varepsilon_{xx}L = -\varepsilon_{yy}L = \pm\varepsilon \times L = \pm\varepsilon_{ii} \times L = \pm\delta L$$

with $i = x$ or y we get:

$$U = \frac{1}{(1+v)}K(\varepsilon_{ii})^2 L^2$$

Or:

$$(\varepsilon_{ii})^2 = (1+v)\frac{1}{K} \times \frac{U}{L^2}$$

Starting from the rigidity K based on length L, Young's modulus E, the tube section S and substituting the resulting L expression of K ($L = ES/K$) in the above formula, we obtain:

$$(\varepsilon_{ii})^2 = (1+\nu)K\frac{U}{E^2 S^2}$$

Given volume V of the tube, the section S is extracted ($V/L = S$): Substituting the S formula in the above expression results in the U/V strain energy density we obtain:

$$\frac{1}{L^2}(\varepsilon_{ii})^2 = (1+\nu)\frac{K}{V}\times\frac{1}{E^2}\times\frac{U}{V}$$

Or:

$$\frac{U}{V} = \frac{\frac{1}{L^2}(\varepsilon_{ii})^2}{(1+\nu)\frac{K}{V}\times\frac{1}{E^2}} = T$$

We now consider the tubes under dynamic behavior and with a compression/traction wave following x and y in both arms simultaneously because of gravitational transverse waves. By proceeding as in the previous chapter in one dimension we get:

$$V = \frac{m}{\rho} = \frac{K}{\omega^2\rho}$$

The following formula of the strain energy density is obtained:

$$\frac{1}{L^2}(\varepsilon_{ii})^2 = (1+\nu)\rho\frac{\omega^2}{E^2}\times T$$

By substituting the circular frequency $\omega = 2\pi f$ for f-frequency in the equation above, we get:

$$\frac{1}{L^2}(\varepsilon_{ii})^2 = 4\pi^2(1+\nu)\rho\frac{f^2}{E^2}\times T$$

We have pure compression/traction in the interferometer (not in 3D with a shear wave), and therefore the compression/traction wave equation is the same as at the front. We get with plane wave $E = Y = \rho c^2$:

$$\frac{1}{L^2}(\varepsilon_{ii})^2 = 4\pi^2(1+\nu)\rho\frac{f^2}{c^4\rho^2}\times T$$

Based on the above equations, we can write this equation based on each strain following each direction of each arm of the interferometer:

$$\frac{1}{L^2}(\varepsilon_{xx})^2 = 4(1+\nu) \times \pi \times \frac{\pi f^2}{\rho} \times \frac{1}{c^4} \times \frac{U}{V}$$

$$\frac{1}{L^2}\left(\varepsilon_{yy}\right)^2 = 4(1+\nu) \times \pi \times \frac{\pi f^2}{\rho} \times \frac{1}{c^4} \times \frac{U}{V}$$

And it is interesting to see that with "$\nu = 1$" and $G = \frac{\pi f^2}{\rho}$ we get again:

$$\frac{1}{L^2}(\varepsilon_{ii})^2 = 8\pi \frac{G}{c^4} \times \frac{U}{V}$$

Or:

$$\frac{1}{L^2}(\varepsilon_{xx})^2 = \frac{8\pi G}{c^4} \times \frac{U}{V}$$

$$\frac{1}{L^2}\left(\varepsilon_{yy}\right)^2 = \frac{8\pi G}{c^4} \times \frac{U}{V}$$

We can therefore construct the tensorial expression on the basis of the above equations and following the principle space curvature $= K \times$ the strain energy density which is also equal to K times the work of external forces.

$$\begin{bmatrix} \frac{1}{L^2} & 0 \\ 0 & \frac{1}{L^2} \end{bmatrix} \begin{bmatrix} (\varepsilon_{xx})^2 & 0 \\ 0 & \left(\varepsilon_{yy}\right)^2 \end{bmatrix} = \frac{8\pi G}{c^4} \begin{bmatrix} T_{xx} & 0 \\ 0 & T_{yy} \end{bmatrix}$$

Note: If additional measurements are taken on an interferometer and we can confirm that there are also shear deformations (angles γ), the hypothesis of plate behavior is confirmed, the equation above will be constructed with a matrix 3×3 supplemented by Poisson's ratio ν, strains ε_{xy} and ε_{yx}, strain energies T_{xy} and T_{yx}. By opposition, if measurements are made and there are no shear deformations (angles γ), consider the twist wave hypothesis mentioned above. In this case, the mechanical model could not be a plate, but a twisted space cylinder due to the rotation of the binary system of black holes.

But all these elements do not change the aim of this book: to study coefficient κ. We continue on the basis of the actual data, the length variations $\delta L/L$ measured in each direction of the arms of the interferometer. It should be remembered that in weak field there is a relationship between the g_{ij} metric and the strain tensor ε_{ij} and a relationship between the strain energy tensor T^{ij} and the mechanical stress tensor σ^{ij}. The parallelism between the above expressions and the bending plate formula is noted for the diagonal terms but not for the terms concerning Poisson's ratio (see above). What is important here is that the term $\frac{\varepsilon_{ij}}{z^2}$, can be interpreted as a curvature in comparison with the terms, $\frac{\varepsilon_{ij}}{L^2}$, which we have in a plate [29, 30]. Indeed, in the case of the hypothesis of a plate, the strain energy is with z, the thickness of the plate perpendicular to the plane xy (see {57} for the proof):

$$\frac{1}{z^2}\left[(\varepsilon_{xx})^2+(\varepsilon_{yy})^2+2(1-\nu)\frac{1}{4}\{\varepsilon_{xy}\}^2+2\nu\{\varepsilon_{xx}\varepsilon_{yy}\}\right]=\frac{24(1-\nu^2)}{Eh^3}\times\frac{dU}{dxdy}$$

With for the rigidity D of the plate:

$$D=\frac{Eh^3}{12(1-\nu^2)}$$

$$\frac{1}{z^2}\left\{\varepsilon_{xx};\varepsilon_{yy};\frac{\varepsilon_{xy}}{2}\right\}\begin{pmatrix}1 & \nu & 0\\ \nu & 1 & 0\\ 0 & 0 & 2(1-\nu)\end{pmatrix}\begin{pmatrix}\varepsilon_{xx}\\ \varepsilon_{yy}\\ \frac{\varepsilon_{xy}}{2}\end{pmatrix}=\frac{24(1-\nu^2)}{Eh^2}\times\frac{dU}{dxdyh}$$

$$\frac{1}{\mathrm{m}^2}=\frac{1}{\mathrm{N}}\times\frac{\mathrm{Nm}}{\mathrm{m}^3}$$

We note that the right term in N.m/m^3 is like a strain energy density and the term left in m^{-2} is like a curvature. We note $\frac{24(1-\nu^2)}{Eh^2}$ that has the same dimension as $\kappa=\frac{8\pi G}{c^4}(N^{-1})$. The flexibility of the space fabric κ can be considered as an equivalent flexibility and $1/\kappa$ as an equivalent rigidity of this structure. With the relationships between the curvature and the second derivatives of z (w) displacement we have:

$$\frac{1}{R_x}=\frac{\partial^2 w_{(x,y)}}{\partial x^2}\qquad\frac{1}{R_y}=\frac{\partial^2 w_{(x,y)}}{\partial y^2}\qquad\frac{1}{R_{xy}}=\frac{\partial^2 w_{(x,y)}}{\partial x\partial y}$$

And the relationships between strains and curvatures are written:

$$\varepsilon_{xx}=-\frac{z}{R_x},\quad\varepsilon_{yy}=-\frac{z}{R_y},\quad\gamma=\varepsilon_{xy}=-2z\frac{\partial^2 w_{(x,y)}}{\partial x\partial y}=-2z\frac{1}{R_{xy}}$$

The strain tensor ε_{ij} can then be expressed in terms of the curvature of the thin plate R_{ij}:

$$\varepsilon_{ij}=\begin{bmatrix}\varepsilon_{xx} & \varepsilon_{xy}\\ \varepsilon_{yx} & \varepsilon_{yy}\end{bmatrix}=-z\begin{bmatrix}\frac{1}{R_x} & 2\frac{1}{R_{xy}}\\ 2\frac{1}{R_{xy}} & \frac{1}{R_y}\end{bmatrix}=-zR_{ij}$$

$$\left\{\frac{1}{R_x};\frac{1}{R_y};\frac{1}{R_{xy}}\right\}\begin{pmatrix}1 & \nu & 0\\ \nu & 1 & 0\\ 0 & 0 & 2(1-\nu)\end{pmatrix}\begin{pmatrix}\frac{1}{R_x}\\ \frac{1}{R_y}\\ \frac{1}{R_{xy}}\end{pmatrix}=\frac{24(1-\nu^2)}{Eh^2}\times\frac{dU}{dxdyh}$$

Therefore, the above expression can be considered equivalent to a curvature whose curvature radius R in each direction is infinite ($R => L$).

$$\begin{bmatrix} \frac{1}{L^2} & 0 \\ 0 & \frac{1}{L^2} \end{bmatrix} \begin{bmatrix} (\varepsilon_{xx})^2 & 0 \\ 0 & (\varepsilon_{yy})^2 \end{bmatrix}$$

So, we define:

$$R_{ij} = \begin{bmatrix} \frac{1}{L^2} & 0 \\ 0 & \frac{1}{L^2} \end{bmatrix} \begin{bmatrix} (\varepsilon_{xx})^2 & 0 \\ 0 & (\varepsilon_{yy})^2 \end{bmatrix}$$

and we have as a strain energy density tensor:

$$T_{ij} = \begin{bmatrix} T_{xx} & 0 \\ 0 & T_{yy} \end{bmatrix}$$

The above little expression yet to be written in tensorial form:

$$\boldsymbol{R_{ij} = \frac{8\pi G}{c^4} T_{ij}}$$

We consider in this expression an infinite radius of curvature which therefore implies a zero scalar curvature. Indeed, we assume $\varepsilon_{xx} = -\varepsilon_{yy}$ and after increasing the second index of Ricci's tensor we get the diagonal part $(\varepsilon_{xx}; -\varepsilon_{yy})$, for which the trace is zero.

$$\frac{1}{\mathrm{m}^2} = \frac{1}{\mathrm{N}} \times \frac{\mathrm{N.m}}{\mathrm{m}^3}$$

In this expression, R_{ij} is the equivalent curvature tensor of space and T_{ij} is the strain energy tensor of the deformed elastic space.

To achieve parallelism with Einstein's field equation where $G_{\mu\nu}$ covers the curvature tensor of space and $T_{\mu\nu}$ covers the energy density out of space (for example, the Sun has bent space, but it is not space itself), we must consider in our analogy that the work of external forces is equal to the work of internal forces created by strain energy.

$$T_{\text{external}} = T_{\text{internal}}$$

In this case, our 2 dimensions analogy is close to Einstein's four-dimensional field equation.

We were therefore able to find κ, passing through the spatial components of the stress tensor equivalent to the strain energy tensor (and not *via* the temporal component of the tensor as Einstein did to correlate his equation with Newton's approach in a weak gravitational field) corresponding to the inner work of the space fabric that is equal to the external work of the applied masses ($U = W_{\text{int}} = W_{\text{ext}}$). This passage through continuum mechanics and the elasticity theory leads us to express G according to a density ρ of the vacuum and the square of the Eigen frequency f of the structure of the spatial medium supposed made of a dark elastic material.

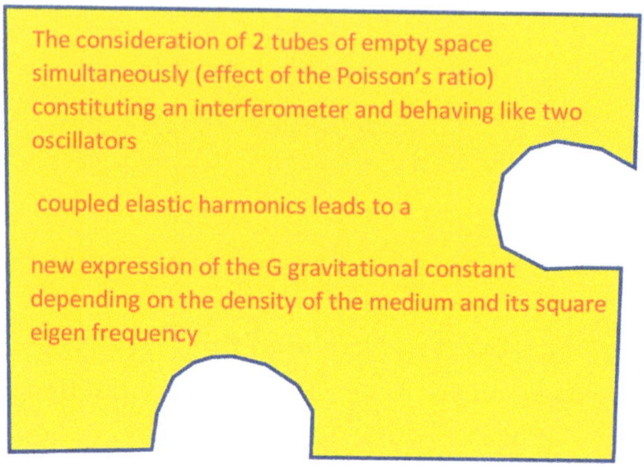

10th piece of the puzzle

Chapter 13

Study of an Elastic Space Cylinder Twisted by the Coalescence of Two Black Holes

Chapter Summary

$$\frac{\gamma^2}{L^2} = 16\pi\frac{\pi f^2}{\rho} \times \frac{1}{c^4} \times \frac{U}{V}$$

$$G = \frac{\pi f^2}{\rho}$$

$$\frac{1}{L^2}\gamma^2 = 16\pi\frac{G}{c^4} \times T$$

Einstein's linearized equation by replacing $h_{\mu\nu}$ with $2\varepsilon_{\mu\nu}$

$$\Box\left(2\varepsilon_{\mu\nu} + \frac{1}{2}\eta_{\mu\nu}2\bar{\varepsilon}\right) = -\frac{16\pi G}{c^4}T_{\mu\nu}$$

DOI: 10.1051/978-2-7598-2573-8.c013

13.1 Study of a Vertical Space Cylinder in Pure Twisting – Use of Shear Speed of the Shear Wave Correlated with the Shear Strains

13.1.1 Assumptions

We assume in this chapter that the coalescence of two black holes, for example, creates by their rotations, a twist of space as shown in figures 9.1b and 13.1. In weak field, for space only, the metric is given by $g_{\mu\nu} = \eta_{\mu\nu} + h_{\mu\nu}$. Einstein's gravitational equation in a weak field for space is:

$$\partial^{\lambda}\partial_{\lambda}\bar{h}_{ij} = \Box\bar{h}_{ij} = -\frac{16\pi G}{c^4}T_{ij}$$

The result for a polarized wave is $A_{\times}$:

$$h_{\mu\nu} = 2\varepsilon_{\mu\nu} = A_{\times}\cos\left(\tfrac{\omega}{c}(ct-z)\right)\begin{bmatrix} 0 & 0 & 0 & 0 \\ 0 & \mathbf{0} & +\mathbf{1} & \mathbf{0} \\ 0 & +\mathbf{1} & \mathbf{0} & \mathbf{0} \\ 0 & \mathbf{0} & \mathbf{0} & \mathbf{0} \end{bmatrix}$$

$$\rightarrow \varepsilon_{XY(A_{\times})} = \tfrac{1}{2}A_{\times}\cos\left(\tfrac{\omega}{c}(ct-z)\right)\begin{bmatrix} 0 & \varepsilon_{XY} & 0 \\ \varepsilon_{YX} & 0 & 0 \\ 0 & 0 & 0 \end{bmatrix}$$

$$\rightarrow \sigma_{XY(A_{\times})} = \begin{bmatrix} 0 & \tau_{XY} = \sigma_{XX} & 0 \\ \tau_{YX} = \sigma_{YY} & 0 & 0 \\ 0 & 0 & 0 \end{bmatrix}$$

With the formula $h_{\mu\nu} = 2\varepsilon_{\mu\nu}$, we have a link between general relativity (perturbation of the spatial part of the metric h_{ij}) and the elasticity theory (the strains of the elastic medium ε_{ij}). Thus, on the basis of this data, the spatial perturbation of the metric $h_{\mu\nu(A_{\times})}$ is in link with the following strain tensor:

$$\varepsilon_{XY} = \begin{bmatrix} 0 & \varepsilon_{XY} = \varepsilon & 0 \\ \varepsilon_{YX} = \varepsilon & 0 & 0 \\ 0 & 0 & 0 \end{bmatrix}$$

Thanks to Hooke's formula in elasticity theory, it corresponds to a strain tensor ε_{XY} a stress tensor σ_{XY}:

$$\sigma_{XY} = \begin{bmatrix} 0 & \tau_{XY} & 0 \\ \tau_{YX} & 0 & 0 \\ 0 & 0 & 0 \end{bmatrix} = \begin{bmatrix} 0 & \tau & 0 \\ \tau & 0 & 0 \\ 0 & 0 & 0 \end{bmatrix}$$

This type of stress tensor is representative of a pure twist cylinder, twisted by a torsion moment M_t (see figure 13.1). This study appears in [61] and in [66]. For this chapter, we consider the cube from normal to the z facet in a Q point (distribution of stresses τ and $\sigma_{ij(A_{\times})}$):

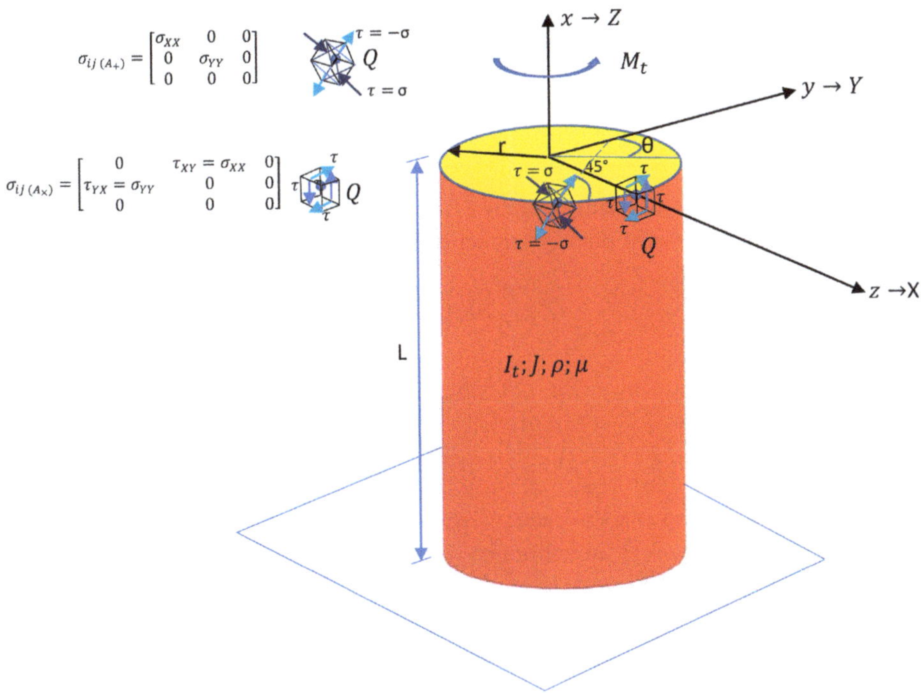

FIG. 13.1 – Space cylinder clamped in foot in pure twisting twisted for example by the rotation of 2 black holes.

Note: The attentive reader will not have failed to notice, however, that the facets subjected to traction and compression are not in the shear plane perpendicular to the direction of the wave (see figure 9.5), as is the case with gravitation, but tilted at 45° (figure 13.1). This approach is therefore only a simplified model to verify the possible value of κ according to the stress tensors related to pure twisting studied according to the elasticity theory.

13.1.2 Determination of Tensorial Equation Associated with Twisting Space Tube

The strain twist energy U of a beam (tube) of length L and area S is written with I_t the torsion inertia (noted J sometimes) and μ the shear modulus (noted G sometimes), considering a homogeneous and isotropic space and a constant twisting moment M_t:

$$U = \frac{1}{2}\int_0^L \frac{{M_t}^2}{\mu I_t}\,dx = \frac{1}{2} \times \frac{{M_t}^2 L}{\mu I_t}$$

With θ the rotating displacement of the point Q on the outer surface of the cylinder:

$$\left(\frac{d\theta}{dx}\right) = \frac{M_t}{\mu I_t}$$

We assume that the twisting torque M_t is constant. Introducing the above formula into the torsion strain energy equation we obtain:

$$U = \frac{1}{2}\mu I_t \left(\frac{d\theta}{dx}\right)^2 L$$

From the expression above, the twisting equivalent curvature $\frac{d\theta}{dx}$ is extracted:

$$\left(\frac{d\theta}{dx}\right)^2 = \frac{2}{\mu I_t} \times \frac{U}{L}$$

The relationship between the twist torque M_t and the twisting rotation θ brings out the twisting stiffness of the cylinder:

$$M_t = k\theta$$

By integrating in respect to x, the formula below:

$$\left(\frac{d\theta}{dx}\right) = \frac{M_t}{\mu I_t}$$

We get:

$$\theta = \frac{M_t L}{\mu I_t}$$

Or by modification of the above formulation:

$$M_t = \frac{\mu I_t}{L}\theta$$

By comparing the two expressions of M_t we obtain:

$$M_t = k\theta$$

$$M_t = \frac{\mu I_t}{L}\theta$$

We get the spring stiffness k in torsion:

$$k = \frac{\mu I_t}{L}$$

By introducing this result into the expression of twisting strain energy, we obtain:

$$U = \frac{1}{2} \times \frac{{M_t}^2 L}{\mu I_t} = \frac{1}{2} \times \frac{(M_t)^2}{k}$$

Taking into account that:

$$M_t = k\theta$$

We get a new expression of the twist strain energy expressed according to the twisting stiffness of the cylinder k:

$$U = \frac{1}{2}k\theta^2$$

By Hooke's law, we have a relationship between the shear stress τ and shear strain (distortion γ). Using figure 13.2, we determine the relationship between shear strain γ and the angular displacement θ.

We have geometrically:

$$\text{tang}\gamma \approx \gamma = \frac{r\theta}{L}$$

And the relationship between tangential stress τ and strain γ allows us to write:

$$\tau = \mu\gamma = G\gamma = G\frac{r\theta}{L}$$

By introducing θ into the strain twisting energy equation of the cylinder we obtain:

$$U = \frac{1}{2}k\theta^2$$

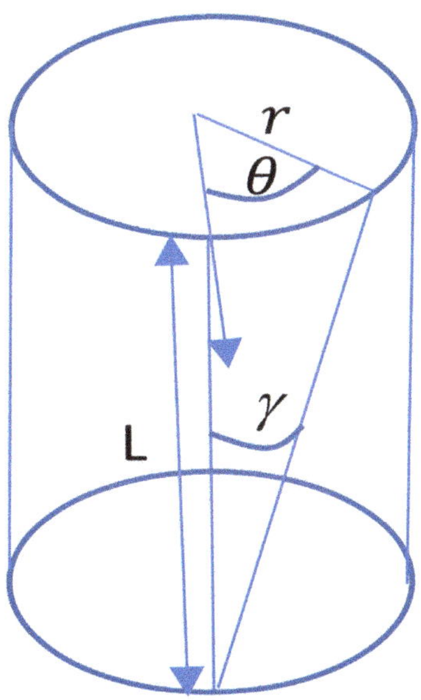

FIG. 13.2 – Definition of the shear strain (distortion γ).

$$U = \frac{1}{2} k \frac{\gamma^2 L^2}{r^2}$$

Which we can reformulate:

$$\frac{\gamma^2}{r^2} = \frac{2U}{kL^2}$$

We can now replace the stiffness in twisting k by its value. We extract the length L of the cylinder and we get:

$$L = \frac{\mu I_t}{k}$$

Introducing this expression of L into the expression above, we obtain:

$$\frac{\gamma^2}{r^2} = \frac{2U}{kL^2}$$

We get so:

$$\frac{\gamma^2}{r^2} = \frac{2kU}{\mu^2 I_t^2}$$

The twisting inertia I_t of a cylinder expressed according to S, the surface of the cylinder, is written:

$$I_t = \frac{\pi d^4}{32} = \frac{\pi r^4}{2} = S\frac{r^2}{2}$$

By introducing the twist inertia I_t in the formula below:

$$\frac{\gamma^2}{r^2} = \frac{2kU}{\mu^2 I_t^2}$$

We get:

$$\gamma^2 r^2 = \frac{8kU}{\mu^2 S^2}$$

Taking into account the definition of volume V:

$$S = \frac{V}{L}$$

If you refer to the above expression, we get:

$$\frac{1}{L^2}\gamma^2 = 8\frac{k}{r^2 \mu^2 V} \times \frac{U}{V}$$

We now consider the dynamic behavior of the cylinder consisting of an elastic substance of density ρ, Young's modulus E, Poisson's ratio v, obtained from the vacuum energy. The fundamental equation of dynamics is derived from the sum of kinetic energy U_C and twisting strain energy U:

$$U_C + U = \frac{1}{2} J\omega^2 + \frac{1}{2} k\theta^2$$

With for the mechanical characteristics of the cylinder, the moment of inertia J of a rotating cylinder is:

$$J = \frac{1}{2}\rho\pi r^4 L$$

The Eigen circular frequency ω is written:

$$\omega = \frac{d\theta}{dt} = 2\pi f$$

By introducing in:

$$U_C + U = \frac{1}{2} J\omega^2 + \frac{1}{2} k\theta^2$$

With the formula of the Eigen frequency, we obtain:

$$U_C + U = \frac{1}{2} J\left(\frac{d\theta_{(t)}}{dt}\right)^2 + \frac{1}{2} k\theta_{(t)}^2$$

By deriving in respect to t we obtain the formula below to search for the cylinder's Eigen frequencies:

$$J\frac{d^2\theta_{(t)}}{dt^2} + k\theta_{(t)} = 0$$

By introducing the expression k and J obtained above, we get:

$$\frac{1}{2}\rho\pi r^4 L\frac{d^2\theta}{dt^2} + \frac{\mu I_t}{L}\theta = 0$$

By introducing the formula of I_t above we get:

$$\frac{d^2\theta}{dt^2} + \frac{2\mu\frac{\pi r^4}{2}}{L\rho\pi r^4 L}\theta = 0$$

That is, after simplification:

$$\frac{d^2\theta}{dt^2} + \frac{\mu}{\rho L^2}\theta = 0$$

We set as a definition of ω^2 which satisfies the dimensional equation:

$$\omega^2 = \frac{k}{J} = \frac{\mu I_t}{JL} = \frac{\mu \frac{\pi r^4}{2}}{\frac{1}{2}\rho\pi r^4 L} = \frac{\mu}{\rho L^2} = \frac{\frac{\text{kgm}}{\text{s}^2\text{m}^2}}{\frac{\text{kg}}{\text{m}^3} \times \text{m}^2} = \frac{1}{\text{s}^2}$$

Replacing the shear modulus μ based on k in the above expression:

$$L = \frac{\mu I_t}{k}$$

We get:

$$\omega^2 = \frac{k}{\rho L I_t}$$

In the formula above, L is extracted, and each side is multiplied by the surface $S = \pi r^2$. A new formula of the cylinder's volume V is obtained:

$$L\pi r^2 = V = \frac{k}{\omega^2 \rho I_t}\pi r^2$$

By introducing this formula into:

$$\frac{1}{L^2}\gamma^2 = 8\frac{k}{r^2\mu^2 V} \times \frac{U}{V}$$

We get:

$$\frac{1}{L^2}\gamma^2 = 8\frac{k}{r^2\mu^2\left(\frac{k}{\omega^2\rho I_t}\pi r^2\right)} \times \frac{U}{V}$$

Or after simplification:

$$\frac{1}{L^2}\gamma^2 = 8\frac{\omega^2\rho I_t}{\pi r^4 \mu^2} \times \frac{U}{V}$$

Taking into account the expression of twist inertia I_t:

$$I_t = \frac{\pi d^4}{32} = \frac{\pi r^4}{2} = S\frac{r^2}{2}$$

We get:

$$\boldsymbol{\frac{1}{L^2}\gamma^2 = 4\rho\left(\frac{\omega}{\mu}\right)^2 \times \frac{U}{V} = 4\rho\left(\frac{\omega}{\mu}\right)^2 \times T}$$

By replacing ω and μ by their values, we get:

$$\frac{1}{L^2}\gamma^2 = 4\frac{4\pi^2 f^2 \rho}{\left(\frac{E}{2(1+\nu)}\right)^2} \times \frac{U}{V}$$

Or after simplification:

$$\frac{1}{L^2}\gamma^2 = 4(1+\nu)^2 \frac{16\pi^2 f^2 \rho}{E^2} \times \frac{U}{V}$$

By multiplying and dividing by ρ the right part of the equation above we get:

$$\frac{1}{L^2}\gamma^2 = 4(1+\nu)^2 \frac{16\pi^2 f^2 \rho^2}{E^2 \rho} \times \frac{U}{V}$$

Or:

$$\frac{1}{L^2}\gamma^2 = 64\pi(1+\nu)^2 \frac{\pi f^2}{\rho} \left(\frac{\rho}{E}\right)^2 \times \frac{U}{V}$$

This time, we use the definition of the velocity of a transversal wave (twist wave) acting by shear:

$$c_{\text{shear}} = \sqrt{\frac{\mu}{\rho}} = \sqrt{\frac{E}{2(1+\nu)\rho}}$$

$$c^2 = \frac{E}{2(1+\nu)\rho}$$

and we get:

$$\left(\frac{\rho}{E}\right)^2 = \frac{1}{4c^4(1+\nu)^2}$$

By introducing this formula into the expression above we obtain:

$$\frac{1}{L^2}\gamma^2 = 64\pi(1+\nu)^2 \frac{\pi f^2}{\rho} \left(\frac{\rho}{E}\right)^2 \times \frac{U}{V}$$

We get:

$$\frac{1}{L^2}\gamma^2 = 64\pi(1+\nu)^2 \times \frac{\pi f^2}{\rho} \times \frac{1}{4c^4(1+\nu)^2} \times \frac{U}{V}$$

And we get the final result after simplification:

$$\frac{\gamma^2}{L^2} = 16\pi \times \frac{\pi f^2}{\rho} \times \frac{1}{c^4} \times \frac{U}{V}$$

Assuming that the work of external forces T is equal to the work of internal forces U:

We confirm the new expression of G already obtained in the previous chapters:

$$G = \frac{\pi f^2}{\rho}$$

$$\frac{1}{L^2}\gamma^2 = 16\pi\frac{G}{c^4} \times T$$

Which we have to compare with:

$$\Box\left(2\varepsilon_{\mu\nu} + \frac{1}{2}\eta_{\mu\nu}2\overline{\varepsilon}\right) = -\frac{16\pi G}{c^4}T_{\mu\nu}$$

So, we have a factor 2 on the left term of (dimension $1/\mathrm{m}^2$) that allows us to find the usual value of κ with a factor 8 (factor related to the fact that $h_{\mu\nu} = 2\varepsilon_{\mu\nu}$).

$$\frac{1}{L^2}\gamma^2 => \Box\left(2\varepsilon_{\mu\nu} + \frac{1}{2}\eta_{\mu\nu}2\overline{\varepsilon}\right)$$

With $\Box$ the usual d'Alambertian.

Thus, you will have to have the coefficient before U/V to get the value of κ usual.

$$\frac{1}{L^2}\gamma^2 = 4\rho\left(\frac{\omega}{\mu}\right)^2 \times \frac{U}{V} = 4\rho\left(\frac{\omega}{\mu}\right)^2 \times T$$

$$\frac{1}{L^2}\gamma^2 => \Box\left(2\varepsilon_{\mu\nu} + \frac{1}{2}\eta_{\mu\nu}2\overline{\varepsilon}\right)$$

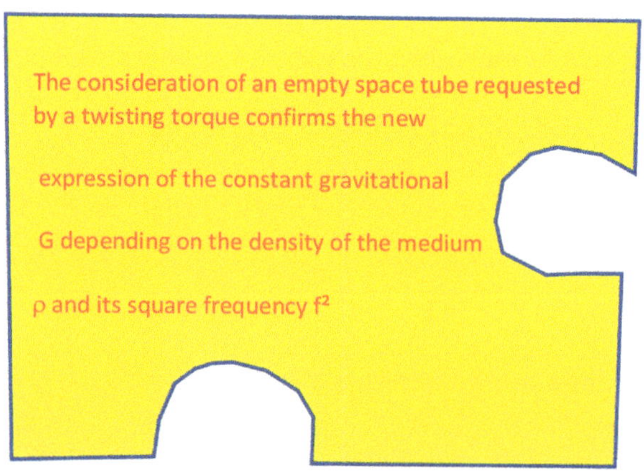

11th piece of the puzzle

Chapter 14

New Mechanical Expression of Einstein's Constant κ

Chapter Summary

Considering a single arm of the interferometer (elongation and shortening deformations)

$$G = \frac{\pi f^2}{\rho}$$

$$\kappa = \frac{8\pi G}{c^4} = 2\rho\left(\frac{\omega}{E}\right)^2 = 2\rho\left(\frac{\omega}{Y}\right)^2$$

Considering the two arms of the interferometer simultaneously (elongation and shortening deformations)

With $\nu = 1$

$$\kappa = \frac{8\pi G}{c^4} = (1+\nu)\rho\left(\frac{\omega}{E}\right)^2 = (1+\nu)\rho\left(\frac{\omega}{Y}\right)^2$$

In pure torsion of a space-time tube

$$\kappa = \frac{8\pi G}{c^4} = 2\rho\left(\frac{\omega}{\mu}\right)^2$$

$$\mu = \frac{E}{2(1+\nu)} = \frac{Y}{2(1+\nu)}$$

DOI: 10.1051/978-2-7598-2573-8.c014

14.1 Steps to Obtain the Mechanical Conversion of κ

In the previous chapters we have deduced from the parallelism between the elasticity theory and general relativity a new mechanical expression of the gravitational constant G.

$$G = \frac{\pi f^2}{\rho}$$

In accordance with the elasticity theory and everything that has been seen in previous chapters, space is therefore proposed to have a quantified microstructure with a characteristic f-associated frequency and a density ρ. G is the macroscopic manifestation of said frequency.

We will therefore use this result to formulate a new expression of Einstein's constant κ based on the strain measurements of Ligo and Virgo and this new expression of G *via* mechanical characteristics associated with the space structure.

14.2 Case where We Consider Only One Interferometer Arm

Taking into account the following:

(a) Since the movements of laser mirrors in interferometers in the case of polarization A_+ are limited to forward or backward displacements, we consider in this chapter only the associated traction/compression waves as representative of these actual data.

(b) The circular frequency ω take the usual expression:

$$\omega = 2\pi f$$

(c) The relationship between Young's modulus Y and speed c is written:

$$E = Y = \rho c^2$$

We get (see chapter 11) the new mechanical expression of Einstein's constant κ function of the vacuum's Eigen circular frequency ω, its equivalent density ρ, and Young's modulus E:

$$\boldsymbol{\kappa = \frac{8\pi G}{c^4} = \frac{8\pi \times \frac{\pi f^2}{\rho}}{c^4} = \frac{8\pi \times \frac{\pi\left(\frac{\omega}{2\pi}\right)^2}{\rho}}{\frac{E^2}{\rho^2}} = 2\rho\left(\frac{\omega}{E}\right)^2 = 2\rho\left(\frac{\omega}{Y}\right)^2}$$

We carefully check the equation to the dimensions of κ:

$$\rho\left(\frac{\omega}{E}\right)^2 = \rho\left(\frac{\omega}{Y}\right)^2 \rightarrow \frac{\mathrm{kg}}{\mathrm{m}^3}\left(\frac{1}{\mathrm{s} \times \frac{\mathrm{kg}}{\mathrm{ms}^2}}\right)^2 = \frac{\mathrm{s}^2}{\mathrm{kgm}} = \frac{1}{N}$$

14.3 Cases where the Two Arms of the Interferometer and Poisson's Ratio are Considered

Taking into account the following:

(a) Since the movements of laser mirrors in interferometers in the case of polarization A_+ are limited to forward or backward displacements, we consider in this chapter only the associated traction/compression waves as representative of these actual data.

(b) Eigen circular frequency takes the usual expression:

$$\omega = 2\pi f$$

We get (see chapter 12):

$$\kappa = \frac{8\pi G}{c^4} = (1+\nu)\rho\left(\frac{\omega}{E}\right)^2 = (1+\nu)\rho\left(\frac{\omega}{Y}\right)^2$$

(c) The relationship between Young's modulus $E(=Y)$ and speed c is written:

$$E = Y = \rho c^2$$

Note: We confirm well the parallelism demonstrated previously with the elasticity theory (see §6.4), in fact is indeed proportional to $(1 + \nu)$. In addition, with $\nu = 1$ we get the same expression of Einstein's constant κ as in the previous paragraph:

$$\left[\left(\varepsilon_{ij} + \frac{\nu}{1-2\nu}\varepsilon_{kk}\delta_{ij}\right)\right]\varepsilon^{ij} = \frac{2(1+\nu)}{Y}U_{ij} = \frac{(1+\nu)}{Y}\sigma_{ij}\varepsilon^{ij} \rightarrow f_{(\varepsilon_{ij}{}^2)} = \boldsymbol{K}_{\left(\frac{1+\nu}{Y}\right)}U$$

14.4 Case of a Pure Torsion of Vacuum Space Tube

Taking into account the following:

(a) We consider the case of polarization $A_\times$.

(b) Eigen circular frequency takes the usual expression.

$$\omega = 2\pi f$$

(c) The relationship between the speed c and the elastic medium shear modulus μ is written:

$$c_{\text{shear}} = \sqrt{\frac{\mu}{\rho}} = \sqrt{\frac{E}{2(1+\nu)\rho}} = \sqrt{\frac{Y}{2(1+\nu)\rho}}$$

We get (see chapter 13):

$$\kappa = \frac{8\pi G}{c^4} = 2\rho\left(\frac{\omega}{\mu}\right)^2$$

With:

$$\mu = \frac{E}{2(1+\nu)} = \frac{Y}{2(1+\nu)}$$

The shear modulus μ plays the role of Young's modulus E.

Note: If constant G depends on a density ρ of this dark elastic material of space, it would be interesting to study how this density varies in the case of rotating galaxy disks. Indeed, any elastic disc reamed in rotation thins near the center and thickens at the periphery at the same time as its radius increases causing variations in the distribution of material and density along this radius. If this is the case, constant G, a function of ρ of space, would vary along the radius of the disc (see figures 1.2a and 1.2b). This could have correlations with the speeds of stars which remain constant at the periphery of the galaxy as held in place by an increase in gravity supposed to come from gravitating dark matter of unknown nature.

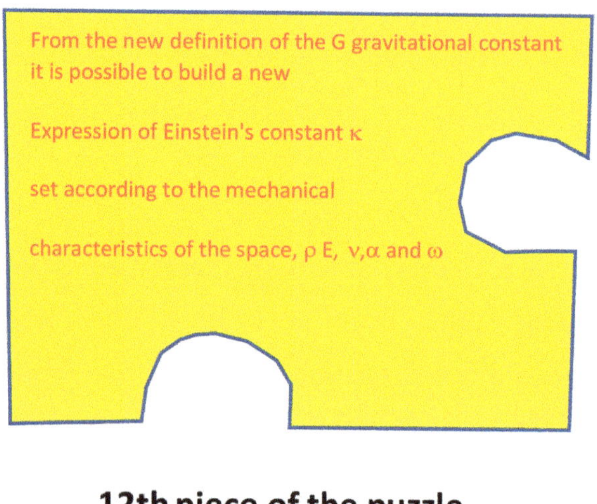

12th piece of the puzzle

Chapter 15

Vacuum Data

Chapter Summary

$$E_{\text{vacuum}} = \frac{1}{2}hf = \frac{1}{2}h\frac{\omega}{2\pi} = \frac{1}{2}\hbar\omega \Rightarrow \frac{2E_{\text{vacuum}}}{\hbar} = \omega$$

$$\rho_{\text{vacuum}} = \frac{m}{V} = \frac{1}{2} \times \frac{\hbar\omega}{c^2V}$$

15.1 Physical Approach or Mathematical Artifact?

Are all the formulas obtained in the previous chapters a physical reality, or are they just a mathematical coincidence? Let us take a look at this about the vacuum characteristics. Casimir's force [62] in relation to quantum field theory, shows that vacuum has a non-zero fundamental state, a non-zero energy and therefore according to special relativity (a non-zero equivalent mass). Space is therefore a physical object and is therefore not empty. The measurement of Casimir's force in a vacuum confirms that $E = mc^2$:

(1) Thanks to the Heisenberg's uncertainty principle, virtual particles and antiparticles are created and annihilated spontaneously, generating a force that brings together two parallel metal plates placed in the vacuum (Casimir's effect). There is an energy in the vacuum.
(2) There is a fundamental state different from 0 of the quantified vacuum (QED).
(3) There is a scalar field (Higgs field) in the empty space.
(4) The vacuum is not nothingness.

DOI: 10.1051/978-2-7598-2573-8.c015

15.2 The Vacuum Energy

In addition, the vacuum energy was measured and calculated:

- According to measurements of cosmological constants: $\rho = 1\times\ 10^{-29}$ g/cm^3 (1×10^{-26} kg/m^3).
- According to quantum field theory: $\rho = 1.11 \times 10^{93}$ g/cm^3.
- According to measurements of cosmological constants: $T_{\text{vacuum}} = 8.987551787 \times 10^{-10}$ kg m^2/s^2/m^3.
- According to quantum field theory: $E_{\text{vacuum}} = 1\times\ 10^{113}$ kg m^2/s^2/m^3.

Even if the values are so different, the vacuum energy is not zero. That is the fundamental point here.

The best way to know what value is right is to measure Young's modulus $E = (Y)$ *via* what we call the new Casimir's test (see chapter 20). Thus, following the quantum field theory, the vacuum energy E_v is quantified to the fundamental state:

Starting from the harmonic oscillator we get for energy:

$$E_n = \hbar\omega\left(n + \frac{1}{2}\right)$$

Which gives the fundamental state with $n = 0$

$$E_v = \frac{1}{2}hf = \frac{1}{2}h\frac{\omega}{2\pi} = \frac{1}{2}\hbar\omega$$

where E_v is the energy of the vacuum, h is Planck's constant and ω is the circular frequency. In addition, the laws of special relativity apply to the vacuum. The energy of the vacuum at rest is:

$$E_v = mc^2$$

We infer from previous expressions a formula of the equivalent mass presented in the vacuum:

$$\frac{1}{2} \times \frac{\hbar\omega}{c^2} = m$$

Dividing by volume V results in the expression of density ρ, we obtain:

$$\frac{1}{2} \times \frac{\hbar\omega}{c^2 V} = \frac{m}{V} = \rho$$

15.3 Consistency of Results with Vacuum Data

It therefore seems consistent that the formulas seen in chapter 14 are all related to density ρ and specific circular frequency ω (*via* f) which may be, in this case, those of the vacuum (*via* the vacuum energy). However, knowing the Einstein's gravitational constant, the circular frequency in link with the density (vacuum energy), volume V is the unknown that will have to be calibrated to satisfy all the physics constants.

For an interferometer arm:

$$\kappa = \frac{8\pi G}{c^4} = 2\rho\left(\frac{\omega}{E}\right)^2$$

For two interferometer arms:

With $\nu = 1$

$$\kappa = \frac{8\pi G}{c^4} = (1+\nu)\rho\left(\frac{\omega}{E}\right)^2$$

For a twisted space cylinder:

$$\kappa = \frac{8\pi G}{c^4} = 2\rho\left(\frac{\omega}{\mu}\right)^2$$

The vacuum elasticity is then characterized by the elasticity modulus E and Poisson's ratio ν.

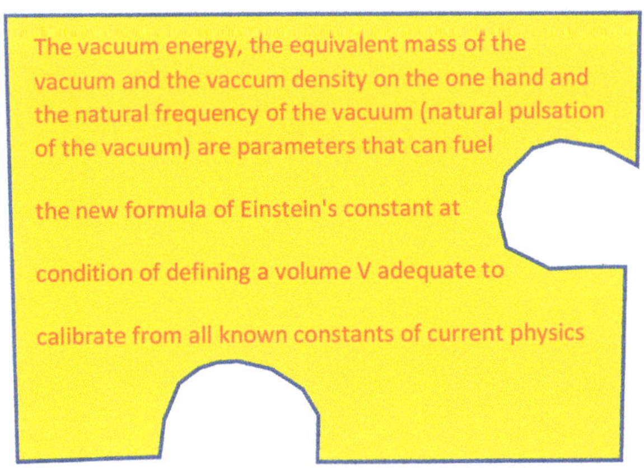

13th piece of the puzzle

Chapter 16

Calibrating the New Mechanical Expression of κ with the Vacuum Data

Chapter Summary

Expression of all formulas based on volume V unknown determined by iterations

$$E = Y = \rho c^2 = \frac{G}{4\pi}\left(\frac{h}{cV}\right)^2$$

$$E = Y = 2E_v^{\frac{3}{2}} \frac{c}{h}\sqrt{\frac{\pi}{GV}}$$

$$f = \frac{Gh}{2\pi V c^2}$$

$$\omega = \frac{Gh}{Vc^2}$$

$$\rho = \frac{Gh^2}{4\pi V^2 c^4} = \frac{E_v}{Vc^2}$$

$$G = \frac{\pi f^2}{\rho}$$

DOI: 10.1051/978-2-7598-2573-8.c016

16.1 Numerical Application to Vacuum Energy – Longitudinal Waves in Interferometric Tubes

16.1.1 Theoretical Development

T is noted as a vacuum energy density measured in J/m^3 with V a volume of spatial elastic matter to be determined.

$$T \times V = E_v = mc^2 = \rho \times V \times c^2$$

So, we can extract the density ρ of the volume:

$$\rho = \frac{E_v}{Vc^2}$$

From the following formula:

$$E_v = \frac{1}{2}hf = \frac{1}{2}h\frac{\omega}{2\pi} = \frac{1}{2}\hbar\omega$$

The expression of the frequency f is extracted:

$$f = \frac{2E_v}{h}$$

By dividing by V the expression below, we obtain $\frac{E_v}{V} = T$ according to the Eigen circular frequency ω and Planck's constant h:

$$E_v = \frac{1}{2}hf = \frac{1}{2}h\frac{\omega}{2\pi} = \frac{1}{2}\hbar\omega$$

So:

$$\frac{E_v}{V} = T = \frac{h\omega}{4\pi V}$$

From the above expression we can extract the Eigen circular frequency ω:

$$\omega = \frac{4\pi E_v}{h}$$

From:

$$\rho = \frac{E_v}{Vc^2}$$

$$\omega = \frac{4\pi E_v}{h}$$

Which is reported in:

$$\kappa = \frac{8\pi G}{c^4} = 2\rho\left(\frac{\omega}{E}\right)^2$$

We get:

$$\frac{8\pi G}{c^4} = 2\frac{E_v}{Vc^2}\left(\frac{\frac{4\pi E_v}{h}}{E}\right)^2$$

$$E^2 = \frac{c^4}{8\pi G} \times 2 \times \frac{E_v}{Vc^2}\left(\frac{4\pi E_v}{h}\right)^2$$

$$E^2 = \frac{4c^2\pi}{G} \times \frac{E_v}{V}\left(\frac{E_v}{h}\right)^2$$

$$E = Y = 2E_v^{3/2}\frac{c}{h}\sqrt{\frac{\pi}{GV}}$$

By introducing the above expression in the following formula:

$$\frac{1}{c^2} = \frac{\rho}{E}$$

and taking into account that:

$$\rho = \frac{E_v}{Vc^2}$$

We get:

$$\frac{1}{c^2} = \frac{\frac{E_v}{Vc^2}}{2E_v^{3/2}\frac{c}{h}\sqrt{\frac{\pi}{GV}}}$$

$$1 = \frac{E_v}{2\mathrm{V}E_v^{3/2}\frac{c}{h}\sqrt{\frac{\pi}{GV}}}$$

$$2\mathrm{V}E_v^{3/2}\frac{c}{h}\sqrt{\frac{\pi}{GV}} = E_v$$

$$4V^2E_v^3\frac{c^2}{h^2} \times \frac{\pi}{GV} = E_v^2$$

$$4VE_v\frac{c^2}{h^2}\times\frac{\pi}{G}=1$$

$$E_v=\rho Vc^2=\frac{Gh^2}{4\pi Vc^2}$$

From the expression above, we deduce the expression of density ρ:

$$\rho=\frac{Gh^2}{4\pi V^2c^4}$$

From the following expression of Young's modulus $E(=Y)$ and taking into account of the new formula of density ρ we obtain:

$$E=\rho c^2=\frac{G}{4\pi}\left(\frac{h}{cV}\right)^2$$

From:

$$f=\frac{2E_v}{h}$$

And from E:

$$E_v=\rho Vc^2=\frac{Gh^2}{4\pi Vc^2}$$

A new equation of the frequency f is obtained:

$$f=\frac{Gh}{2\pi Vc^2}$$

With $\omega=2\pi f$, we get a new formula of the circular frequency ω:

$$\omega=\frac{Gh}{Vc^2}$$

It is therefore necessary to verify the hypothesis of small volumes (quantum approach) allowing to satisfy simultaneously all these equations. According to quantum field theory, the vacuum density ρ is 1.11×10^{93} g/cm^3 and the vacuum energy density is $T=1.00\times10^{113}$ J/m^3. In addition, the fundamental physical constants are:

$G=6.67408\times10^{-11}$ m^3/(kg.s^2)
$h=6.62607004\times10^{-34}$ m^2 kg/s
$c=299\ 792\ 458$ m/s
$\kappa=2.0766\times10^{-43}$ N^{-1}

As all formulas are V-volume functions, we are now looking for the value of this volume V that can satisfy all physical constants. From this volume V, it will then be possible to determine the r dimension of the fibers of the space fabric and Young's modulus E of the space material.

16.1.2 Intensity Obtained for the New G Parameters Based on Vacuum Energy

The question, then, is what volume V takes to simultaneously satisfy the following formula?:

$$\kappa = \frac{8\pi G}{c^4}$$
$$\frac{1}{c^2} = \frac{\rho}{E}$$
$$G = \frac{\pi f^2}{\rho}$$

Once all the iterative calculations have been completed, the results are tabulated in table 16.1.

The proof of these formula is given in appendix I. Indeed, with all these formulas we find by numerical iterations the results of T.G Tenev and M.F Horstemeyer [18] table 1 about the quantum definition of the space time Young's modulus Y.

16.2 Numerical Application to Vacuum Energy – Global Approach by Twist Wave

16.2.1 Theoretical Development

The approach is the same as at section 16.1, but Young's modulus E is replaced throughout the equations by the shear modulus μ. This result is:

$$\mu = \frac{E}{2(1+\nu)} = 2E_v^{3/2}\frac{c}{h}\sqrt{\frac{\pi}{GV}}$$

With the definition of the shear modulus μ we get for the Young's modulus:

$$E = Y = 4(1+\nu)E_v^{\frac{3}{2}}\frac{c}{h}\sqrt{\frac{\pi}{GV}}$$

The energy of the vacuum E_v is depending on volume V considered:

$$E_v = \rho V c^2 = \mu V = \frac{Gh^2}{4\pi V c^2}$$

And density ρ of the vacuum is written.

$$\rho = \frac{\mu}{c^2} = \frac{Gh^2}{4\pi V^2 c^4}$$

TAB. 16.1 – Numerical values (compression wave/traction case), Young's modulus approach.

Parameter	Physical object With $v = 1$ Case of longitudinal waves (pure compression/traction)	Units	Values with $E_{\text{vacuum}} = 1\times 10^{113}$ J/m^3
Volume	V	m^3	1.61×10^{-104}
Radius associated with V (link with string theory/LQG and Planck's length)	r if $V = \frac{4}{3}\pi r^3$	m	1.566×10^{-35}
Vacuum energy	$E_v = \rho V c^2 = \frac{Gh^2}{4\pi V c^2}$	$\frac{\text{kgm}^2}{\text{s}^2} = (\text{J})$	1.61×10^{9}
Vacuum energy density (link with quantum field theory)	$T = \frac{E_v}{V}$	$\frac{\text{kg}}{\text{s}^2\text{m}} = \frac{\text{J}}{\text{m}^3}$	1.00×10^{113}
Density	$\rho = \frac{Gh^2}{4\pi V^2 c^4}$	$\frac{\text{kg}}{\text{m}^3}$	1.11×10^{96}
Young's modulus (link to elasticity theory)	$E = Y = \rho c^2 = \frac{G}{4\pi}\left(\frac{h}{cV}\right)^2$ $E = Y = 2E_v^{\frac{3}{2}} \frac{c}{h}\sqrt{\frac{\pi}{GV}}$	$\frac{\text{kg}}{\text{s}^2\text{m}} = \text{Pa}$	1.00×10^{113}
Speed of light (link with special relativity)	$c = \sqrt{\frac{E}{\rho}}$	$\frac{\text{m}}{\text{s}}$	299 792 458
Frequency	$f = \frac{Gh}{2\pi V c^2}$	1/s	4.861×10^{42}
Period (link to Planck's time)	$T = \frac{1}{f}$	s	2.05684×10^{-43}
Gravitation constant (link to Newton's gravitation)	$G = \pi f^2 \frac{1}{\rho}$	$\frac{\text{m}^3}{\text{kgs}^2}$	6.67408×10^{-11}
Circular frequency	$\omega = \frac{Gh}{Vc^2} = 2\pi f = \frac{4\pi E_v}{h}$	1/s	3.05×10^{43}
Einstein's constant (link with general relativity)	$\kappa = 2\rho\left(\frac{\omega}{E}\right)^2$	$\frac{1}{\text{Newton}}$	2.07658×10^{-43}

And the shear modulus μ is written:

$$\mu = \frac{E}{2(1+\nu)} = \frac{G}{4\pi}\left(\frac{h}{cV}\right)^2$$

and given the relationship between the Young's modulus E and the shear modulus μ:

$$E = \frac{(1+\nu)G}{2\pi}\left(\frac{h}{cV}\right)^2$$

From:

$$f = \frac{2E_v}{h}$$

$$E_v = \rho V c^2 = \mu V = \frac{Gh^2}{4\pi V c^2}$$

The expression of the frequency f is obtained:

$$f = \frac{2\left(\frac{Gh^2}{4\pi V c^2}\right)}{h} = \frac{Gh}{2\pi V c^2}$$

And from the expression connecting the frequency f to the Eigen circular frequency ω, we get:

$$\omega = \frac{Gh}{Vc^2}$$

The space is thus quantified and considering a granular volume of fibers of radius r:

$$V = \frac{4}{3}\pi r^3$$

From:

$$\mu V = \frac{Gh^2}{4\pi V c^2}$$

Taking into account the definition of volume V:

$$V^2 = \frac{1}{4\pi} \times \frac{G}{\mu c^2} h^2$$

$$\left(\frac{4}{3}\pi r^3\right)^2 = \frac{1}{4\pi} \times \frac{G}{\mu c^2} h^2$$

So:

$$r = \left(\frac{9}{64\pi^3} \times \frac{G}{\mu c^2}\right)^{1/6} h^{1/3}$$

Or with the new definition of G:

$$r = \left(\frac{9}{64\pi^2} \times \frac{f^2}{\rho\mu c^2}\right)^{1/6} h^{1/3}$$

And the diameter d is:

$$d = 2r = \left(\frac{9}{\pi^2} \times \frac{f^2}{\rho\mu c^2}\right)^{1/6} h^{1/3} = \left(\frac{9}{\pi^2\rho\mu}\right)^{1/6} \left(\frac{fh}{c}\right)^{1/3}$$

16.2.2 Intensities Obtained for New G Parameters Based on Vacuum Energy

The question is therefore, what volume V does take to simultaneously satisfy the following expressions:

$$G = \frac{\pi f^2}{\rho}$$

$$\kappa = \frac{8\pi G}{c^4}$$

$$c_{\text{shear}} = \sqrt{\frac{\mu}{\rho}} = \sqrt{\frac{E}{2(1+\nu)\rho}}$$

$$\rho c^2 = \mu$$

Once all the iterative calculations have been completed, the results are obtained in table 16.2.

The proof of these formula is given in appendix I. Indeed, with all these formulas we find by numerical iterations the results of T.G Tenev and M.F Horstemeyer [18] table 1 about the quantum definition of the space time Young's modulus Y.

The results are identical to table 16.1 except for the Young 's modulus Y, which is multiplied by 4.

TAB. 16.2 – Numerical values (case of shear waves), shear modulus μ approach.

Parameter	Physical object with $v=1$ Case of shear waves in (twisting)	Units	Values with $E_{\text{vacuum}} = 1\times 10^{113}$ J/m^3
Volume	V	m^3	1.61×10^{-104}
Radius associated with V (link with string theory/LQG and Planck's length)	r if $V = \frac{4}{3}\pi r^3$	m	1.566×10^{-35}
Vacuum energy	$E_v = \rho V c^2 = \mu V = \frac{Gh^2}{4\pi V c^2}$	$\frac{\text{kgm}^2}{\text{s}^2} = (\text{J})$	1.61×10^{9}
Vacuum energy density (link with quantum field theory)	$T = \frac{E_v}{V}$	$\frac{\text{kg}}{\text{s}^2\text{m}} = \frac{\text{J}}{\text{m}^3}$	1.00×10^{113}
Density	$\rho = \frac{Gh^2}{4\pi V^2 c^4}$	$\frac{\text{kg}}{\text{m}^3}$	1.11×10^{96}
Young's modulus (link to elasticity theory)	$E = Y = 2\rho c^2(1+v) = \frac{(1+v)G}{2\pi}\left(\frac{h}{cV}\right)^2$ $\mu = \frac{G}{4\pi}\left(\frac{h}{cV}\right)^2$ $E = Y = 4(1+v)E_v^{3/2}\frac{c}{h}\sqrt{\frac{\pi}{GV}}$	$\frac{\text{kg}}{\text{s}^2\text{m}} = \text{Pa}$	4.00×10^{113}
Speed of light (link with special relativity)	$c = \sqrt{\frac{\mu}{\rho}} = \sqrt{\frac{E}{2(1+v)\rho}}$	$\frac{\text{m}}{\text{s}}$	299 792 458
Frequency	$f = \frac{Gh}{2\pi V c^2}$	1/s	4.861×10^{42}
Period (link to Planck's time)	$T = \frac{1}{f}$	S	2.05684×10^{-43}
Gravitation constant (link to Newton's gravitation)	$G = \pi f^2 \frac{1}{\rho}$	$\frac{\text{m}^3}{\text{kgs}^2}$	6.67408×10^{-11}
Circular frequency	$\omega = \frac{Gh}{Vc^2} = 2\pi f = \frac{4\pi E_v}{h}$	1/s	3.05×10^{43}
Einstein's constant (link with general relativity)	$\kappa = 2\rho\left(\frac{\omega}{\mu}\right)^2$	$\frac{1}{\text{Newton}}$	2.07658×10^{-43}

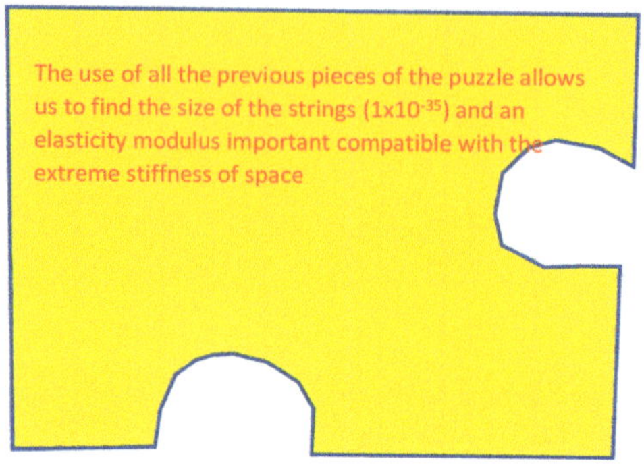

14th piece of the puzzle

Chapter 17

Let's Go Back to the Time Components Based on the New Results

Chapter Summary

The temporal component of the 4-dimensional deformation tensor

$$\frac{1}{R^2}\varepsilon_{00}{}^2 \approx \frac{8\pi G}{c^4} \times \frac{U}{V}$$

The radius of a fiber that makes up space

$$r = \frac{1}{2}\left[\frac{18}{\pi^3} \times \frac{G(1+\nu)}{Ec^2}\right]^{1/6} h^{1/3}$$

$$r = \frac{1}{2}\left[\frac{9}{\pi^2} \times \frac{f^2}{\rho\mu c^2}\right]^{1/6} h^{1/3} \approx \frac{1}{2}\left[\frac{f^2 h^2}{\rho\mu c^2}\right]^{1/6}$$

Quantum time

$$t_q = \left[\frac{9}{\pi^2} \times \frac{f^2}{\rho\mu c^8}\right]^{1/6} h^{1/3} \approx \left[\frac{f^2 h^2}{\rho\mu c^8}\right]^{1/6}$$

$$\boldsymbol{t_q \approx \left[\frac{1}{\rho\mu c^8}\right]^{\frac{1}{6}} E_t^{\frac{1}{3}}}$$

With $E_t = hf$

DOI: 10.1051/978-2-7598-2573-8.c017

17.1 Introduction

In this book, we found an expression of the Einstein's constant κ based on the spatial part only of the gravitational field tensors. To do this, we used the two-dimensional stress tensor and Poisson's ratio ν. We recalled that Einstein established his coupling constant κ using the temporal part of his gravitational field tensor in weak fields (Poisson's Formula and Newton's Gravitational Field Equation). It comes naturally from the idea of extending the stress tensor derived from the four-dimensional elastic theory and deducing information about the potential elastic characteristics of time. We will show in this chapter that it is possible to link a time lapse to the thickness of space fibers with elastic time. Since the fabric is elastic, time itself becomes elastic as demonstrated by special relativity and general relativity.

A similar approach is studied in the publication of Tenev and Horstemeyer {205} and [18] I quote paragraph 2.4.3 "The Lapse Rate postulate relates the flow of proper time to the geometry of the cosmic fabric. All matter-matter interactions are mediated by signals propagating in the fabric as shear waves".

17.2 Impact on the Time of this Search

17.2.1 Time Behaviour as an Elastic Material

In this book, we have so far focused on the spatial part of Einstein's tensor. On the basis of the new definition of G, we can now return to the temporal part of Einstein's tensor (see also T. Tenev and M.F. Horstemeyer, chapter 2.4.3 and 2.6) [18]. Einstein demonstrated that a mass-energy density in space $T_{\mu\nu}$ curves space but also time at the location of this mass *via* a proportionality factor κ. This space-time deformation, mathematically characterized by metric variation $g_{\mu\nu}$ is gravitation. Gravitation is therefore a deformation of the space time and not a force. Near a black hole, the curvature of time is such that time expands to infinity giving the impression that it stops. So, the question is: can time be considered elastic?

For memory some measured facts:

(a) At high speed, time lengthens: the time of a clock placed in a fast-moving aircraft (t') slowed down compared to a clock left on Earth (t) (special relativity principle):

$$\Delta t' = \frac{1}{\sqrt{1-\left(\frac{v}{c}\right)^2}}\Delta t$$

(b) At high speed distances contract:

$$\Delta z' = \sqrt{1-\left(\frac{v}{c}\right)^2}\,\Delta z$$

Therefore, time behaves elastically, lengthens, or shortens, and time behaves opposite to space (dilatation = negative contraction).

The metric is connected at the interval, ds^2. Thus, in a non-inertial frame of reference for example (if the coordinates according to x, y, z do not vary), we have time alone that advances (space-time interval):

$$ds^2 = g_{00}\, c^2\, dt^2$$

or in the general case, the unknowns of the Einstein equation are the 10 components of the $g_{\mu v}$ tensor:

$$\begin{aligned} ds^2 &= g_{00}\, c^2\, dt^2 + 2g_{10}\, dxcdt + 2g_{20}\, dycdt + 2g_{30}\, dzcdt + g_{11}\, dx^2 + 2g_{12}\, dxdy \\ &\quad + 2g_{13}\, dxdz + g_{22}\, dy^2 + 2g_{23}\, dydz + g_{33}\, dz^2 \end{aligned}$$

$$g_{\mu} = \begin{bmatrix} g_{00} & g_{01} & g_{02} & g_{03} \\ g_{10} & g_{11} & g_{12} & g_{13} \\ g_{20} & g_{21} & g_{22} & g_{23} \\ g_{30} & g_{31} & g_{32} & g_{33} \end{bmatrix}$$

To study time, we must therefore focus only the temporal components of the metric. By considering a small variation h_{00} in the temporal component, taking into account the relationship $h_{\mu v} = 2\varepsilon_{\mu v}$ and Poisson's equation applied to the temporal component, we obtain with ϕ the gravitational potential (see also formulas 2.10, 2.11 and 3.2, 3.3 of {205} [18]).

$$g_{00} = \eta_{00} + h_{00} = 1 + 2\varepsilon_{00} \approx 1 + \frac{2\phi}{c^2}$$

Thus, the temporal perturbation $h_{\mu v}$ of the metric for time (index 00) is written:

$$h_{00} = 2\varepsilon_{00} \approx \frac{2\phi}{c^2}$$

The equivalent elastic deformation of time is so:

$$\varepsilon_{00} \approx \frac{\phi}{c^2}$$

The gravitational potential of a sphere of mass M and radius R is written:

$$\phi = \frac{GM}{R} = \frac{\frac{\mathrm{m}^3}{\mathrm{kg\,s}^2} \times \mathrm{kg}}{\mathrm{m}} = \frac{\mathrm{m}^2}{\mathrm{s}^2}$$

By introducing the expression of this gravitational potential into the expression of the elastic deformation of time we obtain:

$$\varepsilon_{00} \approx \frac{GM}{Rc^2}$$

Taking into account the new definition of gravitation constant G that we have demonstrated for the spatial components, we obtain with density ρ of the space substance and f its Eigen vibration frequency:

$$\varepsilon_{00} \approx \frac{\pi f^2 M}{\rho R c^2}$$

By elevating this expression squared we obtain:

$$\varepsilon_{00}{}^2 \approx \frac{\pi^2 f^4 M^2}{\rho^2 R^2 c^4}$$

We have shown in elasticity theory that:

$$\left[\left(\varepsilon_{ij} + \frac{\nu}{1-2\nu}\varepsilon_{kk}\delta_{ij}\right)\right]\varepsilon^{ij} = \frac{2(1+\nu)}{E}U_{ij} = \frac{(1+\nu)}{E}\sigma_{ij}\varepsilon^{ij}$$

So, the time component must also be proportional to $(1+\nu)$:

$$\varepsilon_{00}{}^2 \approx (1+\nu)\frac{\pi^2 f^4 M^2}{\rho^2 R^2 c^4}$$

In this expression, we bring out the new definition of G and we get:

$$\varepsilon_{00}{}^2 \approx (1+\nu)\pi\frac{\pi f^2}{\rho} \times \frac{1}{c^4} \times \frac{f^2 M^2}{R^2 \rho}$$

We multiply each side of the equation by $1/R^2$:

$$\frac{1}{R^2}\varepsilon_{00}{}^2 \approx (1+\nu)\pi\frac{\pi f^2}{\rho} \times \frac{1}{c^4} \times \frac{f^2 M^2}{R^3 \rho R}$$

For memory volume V of a sphere is:

$$V = \frac{4}{3}\pi R^3$$

We introduce volume V *via* R^3 in the above equation of time deformation:

$$\frac{1}{R^2}\varepsilon_{00}{}^2 \approx (1+\nu)\pi\frac{\pi f^2}{\rho}\frac{1}{c^4}\frac{f^2 M^2}{\frac{3V}{4\pi}\rho R}$$

We get after simplification:

$$\frac{1}{R^2}\varepsilon_{00}{}^2 \approx \frac{4}{3}\pi^2(1+\nu)\frac{\pi f^2}{\rho} \times \frac{1}{c^4} \times \frac{f^2 M^2}{V \rho R}$$

Noting that $\rho V = M$ with $d = c/f$ a length and noting that the last term to the dimension of an energy density U/V:

$$\frac{f^2 M}{R} = \frac{f^2 R \times d^2 \times \rho}{R} = \rho \times f^2 \times d^2 = \rho \times c^2 = \frac{mc^2}{V} = \frac{\mathrm{kg}}{\mathrm{m\,s^2}} = \frac{\frac{\mathrm{kg\,m^2}}{\mathrm{s^2}}}{\mathrm{m^3}} = \frac{U}{V}$$

We get:

$$\frac{1}{R^2}{\varepsilon_{00}}^2 \approx \frac{4}{3}\pi \times \pi(1+\nu)\frac{\pi f^2}{\rho} \times \frac{1}{c^4} \times \frac{U}{V}$$

By noting that $3.14 \approx 3$ we get:

$$\frac{1}{R^2}{\varepsilon_{00}}^2 \approx 4\pi(1+\nu)\frac{\pi f^2}{\rho} \times \frac{1}{c^4} \times \frac{U}{V}$$

where $(\nu = 1)$

$$\frac{1}{R^2}{\varepsilon_{00}}^2 \approx 8\pi \times \frac{\pi f^2}{\rho} \times \frac{1}{c^4} \times \frac{U}{V}$$

Either with the new G definition:

$$\frac{\mathbf{1}}{R^2}{\varepsilon_{00}}^2 \approx \frac{\mathbf{8\pi G}}{c^4} \times \frac{U}{V}$$

The temporal component of the metric perturbation gives an expression similar to that obtained by considering a stretch of tube in compression/pure traction. Thus, time behaves like an elastic material. It is so logical to add it with the spacial stress tensor to obtain a 4 dimensional stress tensor.

17.2.2 Relating the Time Intervals with the Thickness Fibers of Spatial Space Sheets

For memory, some measured facts:

(a) Time slows down when a clock is immersed in a gravitational field (see general relativity).

Therefore:

(1) Time is "curved" by gravity.
(2) The more gravitation, the more curvature and the more tension there is in the material and the longer the slowing time expands.

(b) Time acts with a different sign of space,

$$ds^2 = d\tau^2 - dx^2 - dy^2 - dz^2$$

For a gravitational wave, the passage of time is a succession of snapshots following the z direction. Figure 17.1 shows an instantaneous photo of space-time taken during the passage of a gravitational wave.

In the figure 17.1, R_{xy} is the radius of a circle of particles in the plane xy, n is the number of quanta of distance depending on the time $c.t$, $2r$ is the thickness of a sheet of space (r is the radius of the space fiber $= 1 \times 10^{-35}$ m).

The passage of time is the succession of snapshot photos following z. We notice that:

(1) The perturbation $h_{\mu v}$ of the metric depends on the variable, where $ct - z$ are at the same level, but with an opposite sign.
(2) In the matrix $h_{\mu v}$, we see that all time terms and z terms are correlated to zero (in red in the formula below).

$$h_{\mu v} = A_+ \cos\left(\frac{\omega}{c}(ct - z)\right) \begin{bmatrix} 0 & 0 & 0 & 0 \\ 0 & +\mathbf{1} & \mathbf{0} & 0 \\ 0 & \mathbf{0} & -\mathbf{1} & 0 \\ 0 & 0 & 0 & 0 \end{bmatrix}$$

In addition, as the z axis is confused with the axis of time, we therefore have thin sheets of space plane xy, of thickness $2r$ (2×10^{-35} m) that follow each other over time along the z axis. We propose to link the time lapse to the thickness of the fibers that would constitute space (following z in the case of gravitational waves). As the gravitational wave moves from one plane of fibers to another (see figure 17.1), the different time intervals can be counted by these successive passages from one spatial diameter of the fiber to another (quantum of time). Time and space seem to be somewhat quantified:

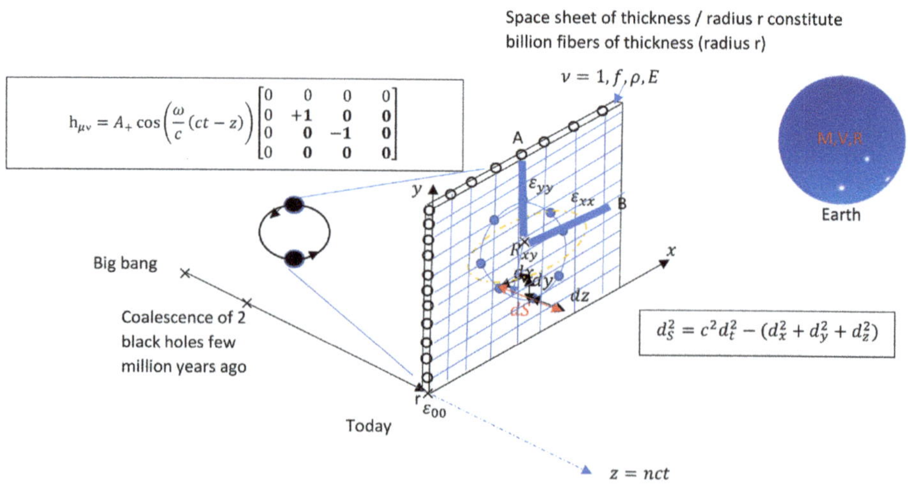

FIG. 17.1 – A snapshot (taken at the speed of light) of the xy plane of space deformed by the passage of a gravitational wave propagating in the z direction.

$$t_q = \frac{2r}{c}$$

Time has a minimum duration corresponding to the time it takes to transmit information at the c speed of one fiber plane to another (multi sandwich sheets).

Based on the following formula, we will get a quantified time frame t_q:

$$\mu = \frac{E}{2(1+\nu)}; c^2 = \frac{\mu}{\rho}$$

$$E_v = \rho V c^2 = \mu V = \frac{Gh^2}{4\pi V c^2}$$

$$V = \frac{4}{3}\pi r^3$$

We get:

$$\mu V = \frac{Gh^2}{4\pi V c^2}$$

Introducing volume V and μ, the above formula becomes:

$$\frac{E}{2(1+\nu)} \times \frac{4}{3}\pi r^3 = \frac{Gh^2}{4\pi \frac{4}{3}\pi r^3 c^2}$$

The radius of the fiber to the power 6 is therefore worth:

$$r^6 = (1+\nu)\frac{18\,Gh^2}{2^6 E\pi^3 c^2}$$

$$r = \left[(1+\nu)\frac{18\,Gh^2}{2^6 E\pi^3 c^2}\right]^{1/6}$$

$$r = \frac{1}{2}\left[\frac{18}{\pi^3} \times \frac{G(1+\nu)}{Ec^2}\right]^{1/6} h^{1/3}$$

So, we can calculate the time frame:

$$t_q = \frac{2r}{c}$$

$$t_q = \frac{2 \times \frac{1}{2}\left[\frac{18}{\pi^3} \times \frac{G(1+\nu)}{Ec^2}\right]^{1/6} h^{1/3}}{c}$$

$$t_q = 2 \times \frac{1}{2}\left[\frac{18}{\pi^3} \times \frac{G(1+\nu)}{Ec^2c^6}\right]^{1/6} h^{1/3}$$

So:

$$t_q = \frac{2r}{c} = \left[\frac{18}{\pi^3} \times \frac{G(1+\nu)}{Ec^8}\right]^{1/6} h^{1/3}$$

With the new definition of G and taking into account the definition of μ we get:

$$\mu = \frac{E}{2(1+\nu)}$$

$$t_q = \frac{2r}{c} = \left[\frac{18}{\pi^3} \times \frac{\pi f^2 E}{\rho E c^8 2\mu}\right]^{1/6} h^{1/3}$$

So:

$$t_q = \left[\frac{9}{\pi^2} \times \frac{f^2}{\rho\mu c^8}\right]^{1/6} h^{1/3} \approx \left[\frac{f^2 h^2}{\rho\mu c^8}\right]^{1/6}$$

$$t_q = \left[\frac{\left(\frac{\text{kgm}^2}{\text{s}^2} \times \text{s}\right)^2}{\text{s}^2 \times \frac{\text{kg}}{\text{m}^3} \times \frac{\text{kgm}}{\text{s}^2\text{m}^2} \times \frac{\text{m}^8}{\text{s}^8}}\right]^{1/6} = \text{s}$$

With the values of table 16.2 results, we get:
$t_q = 1.045 \times 10^{-43}$ s that is more or less the Plank's time.

Proof: With:

$$G = \frac{\pi f^2}{\rho}$$

$$t_q = \left[\frac{9}{\pi^2} \times \frac{f^2 h^2}{\rho\mu c^8}\right]^{1/6} = \left[\frac{9}{\pi^3} \times \frac{4\pi^2 \hbar^2 G}{\mu c^8}\right]^{1/6}$$

With:

$$t_q = \left[\frac{9}{\pi^3} \times \frac{4\pi^2 \hbar^2 G}{\mu c^8}\right]^{1/6} = \left[\frac{36}{\pi} \times \frac{\hbar^2 G}{\mu c^8}\right]^{1/6} = \left[\frac{144}{\pi} \times \frac{\hbar^2 G}{Y c^8}\right]^{1/6}$$

With

$$G = \frac{4\pi c^2 f^2}{Y} = \frac{\pi f^2 c^2}{\mu}$$

We eliminate Y:

$$t_q = \left[\frac{144}{4\pi} \times \frac{\hbar^2 G^2}{\pi c^2 f^2 c^8}\right]^{1/6} = \left[\frac{144}{4\pi^2} \times \frac{\hbar^2 G^2}{f^2 c^{10}}\right]^{1/6}$$

With the definition of the circular frequency:

$$\omega = 2\pi f$$

$$t_q = \left[144 \frac{\hbar^2 G^2}{\omega^2 c^{10}}\right]^{1/6} = 2.28942849 \left[\frac{\hbar^2 G^2}{\omega^2 c^{10}}\right]^{1/6}$$

With the definition of Planck's time:

$$t_p = \sqrt{\frac{\hbar G}{c^5}}$$

$$t_q = \left[144 \frac{\hbar^2 G^2}{\omega^2 c^{10}}\right]^{1/6} = 2.28942849 \left[\frac{t_p^4}{\omega^2}\right]^{1/6}$$

If we consider Planck's circular frequency:

$$\omega_p = \frac{1}{t_p}$$

Indeed we obtain (see table 16.2):
$\omega = 3.05 \times 10^{43}\ \mathrm{s}^{-1}$
and:
$\omega_p = 1.855 \times 10^{43}$ rad/s
So, $\omega = 1.644\ \omega_p$

$$t_q = 2.28942849 \left[\frac{t_p^4}{1.644^2 \left(\frac{1}{t_p}\right)^2}\right]^{1/6} = 1.939 t_p = 1.939 \times 5.39124760 \times 10^{-44}$$
$$= 1.045 \times 10^{-43}\ \mathrm{s}$$

Which is the value of the Planck's time rounded generally (see Wikipedia in French at "temps de Planck").

This length of time is compatible with Planck's time:
$t_p \approx 1.0000 \times 10^{-43}$ s.

If we define the energy of time as the engine of time in terms of energy, we get with $E_t = hf$:

$$t_q \approx \left[\frac{1}{\rho\mu c^8}\right]^{\frac{1}{6}} E_t^{\frac{1}{3}}$$

We confirm that if the frequency f is zero, the space material disappears, and time also disappears. Gravitation can bend time because it bends the fibers of the space fabric that expands. Time passes step by step through each of its fibers, one after the other to transmit information, and therefore it expands like these fibers. This elasticity of time is characteristic of its elastic behavior defined by deformation ε_{00}.

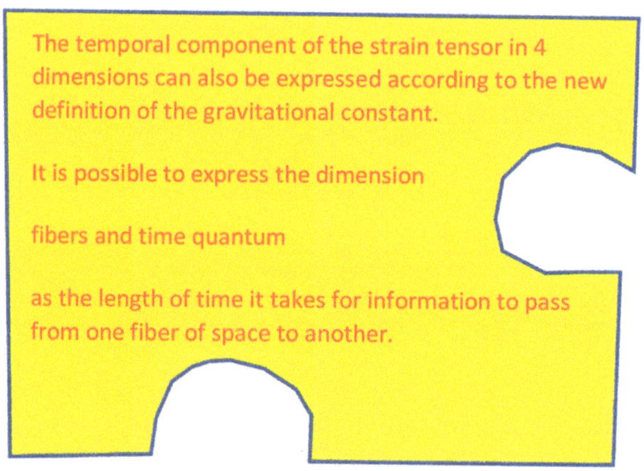

15th piece of the puzzle

Chapter 18

Analogy of Mohr's Circle with Graviton Spin

Chapter Summary

Mohr's circle in elasticity theory is an image of graviton spin in quantum field theory.

18.1 Possible Constitution of Space Material

This vision of the space medium is a new vision of a relativistic Ether without any correlation with the luminescent Ether that does not exist. This medium can be composed of small infinity-sized (quantum) beams in perpetual oscillations forming a three-dimensional deformable frame of strong rigidity density rather than strings without stiffness in bending of 1×10^{-35} m length (see result of chapter 16 and figure F.5). This three-dimensional fabric can be characterized by an equivalent dark elastic material itself, characterized by Young's modulus $E(=Y)$, Poisson's ratio ν and density ρ and eventually the expansion coefficient α. In this case, we characterize space by a kind of elastic substance, in the fundamental state (minimum vacuum energy E_v). Numerically we got the following values:

(1) The fiber length constituting the texture of the space fabric substance is 1.566×10^{-35} m compatible with the string dimension defined in string theory or LQG quantified space and in agreement with the Planck's length $1.61625518 \times 10^{-35}$ postulated by T. Tenev and M.F. Horstemeyer in chapter 3.4 of [18].

(2) Young's modulus $E = Y = 1.00 \times 10^{113}$ Pa (from the longitudinal wave) and 4.00×10^{113} Pa (from the shear wave) is consistent with the results of T. Tenev and M.F. Horstemeyer (4.4×10^{113} Pa, see 3.13) [18] and with the K.T. McDonald's results [13] 5.0×10^{113} Pa. This value is compatible with the extreme rigidity of the space.

DOI: 10.1051/978-2-7598-2573-8.c018

(3) Density $\rho = 1.11 \times 10^{96}$ kg/m^3 is also compatible with the results of T. Tenev and M.F. Horstemeyer (1.30×10^{96} kg m^3, see §3.4, formula 3.14) [18] and the K.T. McDonald's results [13] 5.0×10^{96} kg/m^3.

18.2 Analogy of Mohr's Circle with Graviton Spin

If we take the analogy of a simple beam in pure bending according to Timoshenko's strength of material, a simplification of the elasticity theory, it is possible to understand the fundamental concept used in Einstein's general relativity (curvature space-time $= \kappa \times$ energy density) on the one hand, and assimilate Mohr's stress circle in elasticity (see figure 18.1) to the spin of 2 of graviton in quantum field theory on the other hand. Indeed, when the facet bearing the stresses performs a complete rotation of one turn on the real elastic model, the facet bearing the stresses on Mohr's circle performs two turns. R. Feynman in his "lectures on gravitation" also explains this point see – conference 3, paragraph 3.4, figures 3.3 and 3.4. The Mohr's circle in elasticity stresses therefore has a behavior similar to a spin 2 of a possible graviton.

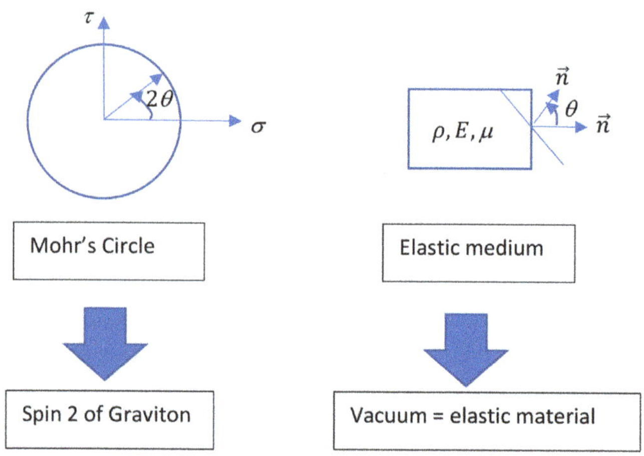

FIG. 18.1 – Analogy between Mohr's circle as an image of the spin of the graviton.

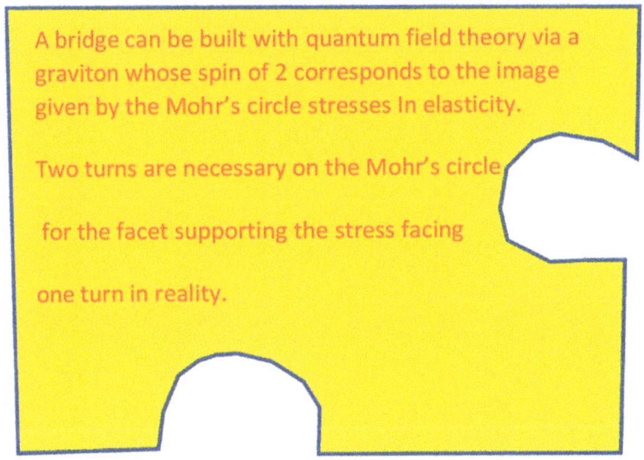

16th piece of the puzzle

Chapter 19

What if We Gave Up the Constant Character of G?

Chapter Summary

$$F = G\frac{Mm}{r^2} \rightarrow G_{\mu\nu} = -\frac{8\pi G}{c^4}T_{\mu\nu} \rightarrow G_{\mu\nu} = -\frac{8\pi^2 f^2}{c^4 \rho}T_{\mu\nu} = -\frac{8\pi^2 \rho f^2}{E^2}T_{\mu\nu}$$
$$= -2\rho\frac{\omega^2}{E^2}T_{\mu\nu} = -2\rho\frac{\omega^2}{Y^2}T_{\mu\nu}$$

(1) The gravitational constant G does not appear to be a stable universal constant as it appears to depend on the characteristics of the material that make up the spatial elastic fabric (ρ, f) calculated from the properties of the vacuum.
(2) G could therefore have varied over time as the density or frequency of the vacuum.
(3) As C. Liegeois Physicist suggested to me: The speed limit of light could find an explanation *via* the limited ratio of light penetration within the elastic space environment (the analogy of water that becomes a high-speed concrete wall).

Newton's concept of gravitation as proposed by Einstein's general relativity, is so not enough. This gravitational force being only an illusion, it seems logical to also abandon the universal and constant character of G created with this force. We show in this book that if Newton and his gravitational force had not existed, Einstein might have found his constant coupling κ without going through the temporal component of his theory (μ, $\nu = 0$), but going through the mechanical and spatial parts of his theory (μ, $\nu = 1, 2, 3$) *via* the elasticity theory, quantum field theory and data of string theory/LQG at the Planck's scale. If this had been the case, we would not have had the idea of forcing, to a universal constant, the parameters $\frac{\pi f^2}{\rho}$ related to the space-making substance that Newton unknowingly called G in his equation. We also show that by proceeding in this way and calculating the G parameter according to the data of the vacuum *via* quantum field theory, we find information

DOI: 10.1051/978-2-7598-2573-8.c019

about the infinitesimal dark substance constituting space: the size of its granulometry 1×10^{-35} m compatible with string theory/LQG; a high Young's modulus required given the very small curvatures generated in space by extremely massive bodies; $Y = E = 1 \times 10^{113}$ Pa; Poisson's ratio equal to 1. The "metal" constituting space is 5×10^{101} times harder than steel at the speed of light (and rather fluid at very low speed)! To further advance the elastic approach to space, we believe it is necessary to carry out measures of Young's modulus (see Casimir test) and shear modulus (measurement of the lateral motion of the laser beam in the large interferometer). This will also help to resolve the paradox of the energies of the vacuum resulting from the cosmological constant Λ (possible dilatation effect of the space material?) on the one hand, and quantum field theory on the other hand by quantifying once and for all the vacuum energy that must be taken into account in the calculation. In summary, if G can vary, probably many of today's physics problems will disappear in turn.

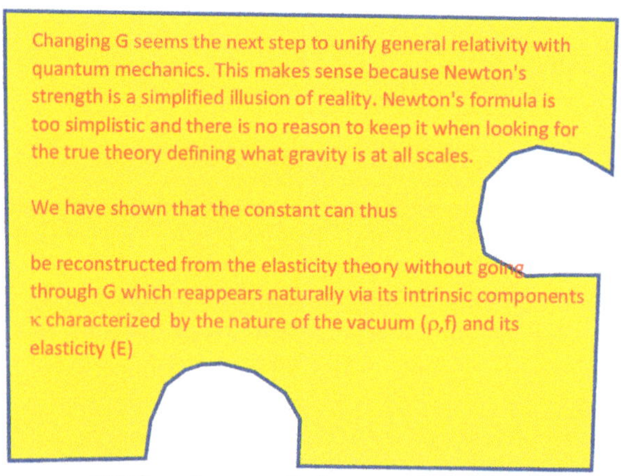

17th piece of the puzzle

Chapter 20

How to Test the New Theory?

20.1 Experimental Test of Young's Modulus of the Elastic Space Medium

An experimental method to establish Young's modulus $E(=Y)$ of the vacuum is to perform Casimir's test. This test consists of taking two metal plates spaced at a very short distance L_0, and positioning them in the vacuum of space by simultaneously measuring F-force and horizontal displacements (see figure 20.1). Generally,

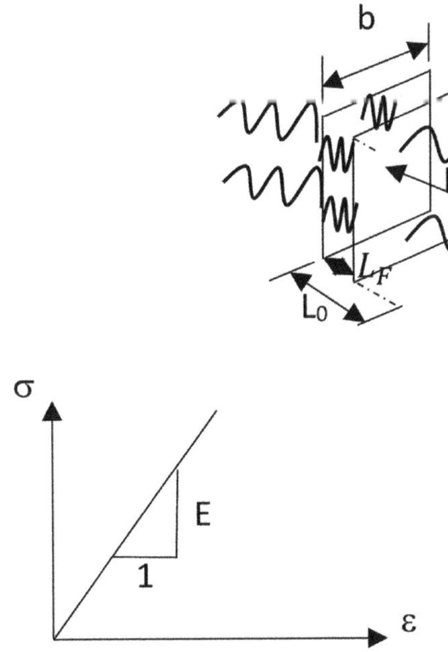

$$S = bh$$

$$\sigma = \frac{F}{S} = \varepsilon E$$

$$\varepsilon = \frac{L_F - L_o}{L_0}$$

FIG. 20.1 – Test to measure Young's modulus $E = Y$ of the spatial medium.

DOI: 10.1051/978-2-7598-2573-8.c020

strength alone is measured, and displacement calculated in the Casimir experience. The result of the test is to draw the stress/strain curve. Of course, the slope of the stress/strain curve corresponds to Young's modulus $E(=Y)$ of the space medium.

20.2 Experimental Test of Pure Space Shear Behavior

In addition, we have shown that we could consider a new type of wave, space torsion waves, which can generate longitudinal and shear waves. It would therefore be extremely interesting to use current interferometers (or future with LISA for example) to measure possible shear deformations (angle), *i.e.* use the potential lateral movements of the laser beam in the interferometers to determine the angular shear strain γ and so the shear modulus μ of space. This will confirm that the approach of general relativity by the elasticity theory is reasonable on the one hand, and that there is indeed an elastic material filling all the space on the other.

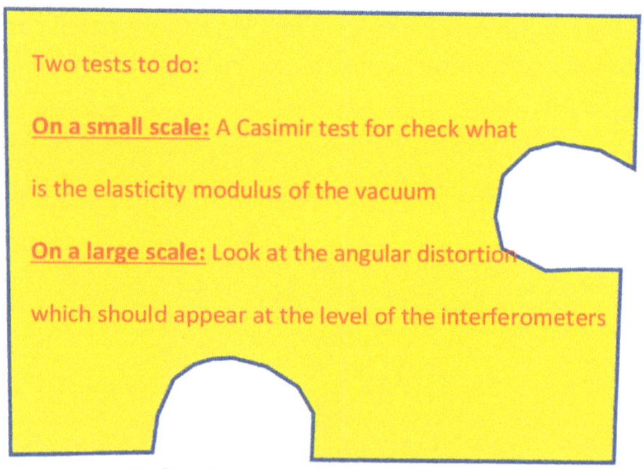

18th piece of the puzzle

Chapter 21

Other Points in Link with the Strength of Material

21.1 An Analysis of the Vibrations of the Space Medium at the Time of the Big Bang

We have shown in tables 16.1 and 16.2 that it is possible to recalculate Einstein's constant κ based on elasticity theory and volume wave theory. Research conducted by Ringermacher and L.R. Mead [26] seems to show that the Universe can also ring. The analysis of these "spatial frequencies" is also an open door to explore and obtain information about the space fabric structure (dynamic behavior of its frame). It would be interesting to compare the successive f ratios of its vibration frequencies with a theory representative of the space medium at the time of the big bang (eigen vibrations and eigen modes of the space structure).

21.2 The Plastic Behavior of the Space Medium in Strong Fields

In this study, we focused on the elastic approach of space in weak fields. Since the presence of a black hole is confirmed, it seems logical that gravitation in a strong field should be closer to the plasticity of the space material, in which case it would be interesting to study what becomes of Einstein's constant in strong fields using the theory of plasticity derived from the strength of materials. A black hole could be seen as a plastic hinge of the space fabric. A plastic flow of fibers then forms. See T. Damour's studies on the Navier–Stokes equation hidden in general relativity in 1979 and appendix F of this book.

DOI: 10.1051/978-2-7598-2573-8.c021

Chapter 22

Conclusions

In light of the previous chapters and appendices of this book, of the author's publication to the Indian Academy of Sciences, *via* the peer-reviewed physics journal Pramana of August 13, 2020 "Mechanical Conversion of Einstein's Gravitational Constant κ" [2], and in Hal Archives ouvertes {59}, the T. Damour's book "If Einstein was told me" [70], and of the papers [13, 18], {205, 213, 215}, it seems logical to propose a new mechanical and physical expression of Einstein's constant κ based on the characteristics of an elastic dark material filling the space by taking into account:

(1) The elastic behaviour of space and time.
(2) The assumption that space is a substance, an elastic dark material characterized by its characteristic $f$frequency, its elastic properties $E = Y$, Poisson's ratio $\nu - 1$, its density (or energy density *via*) $E/V = mc^2/V = \rho c^2$ and dilatation coefficient.
(3) The similarity of the wave equation in an elastic medium and the gravitational wave equation (see appendix G).
(4) The parallelism between the elasticity theory and general relativity in stresses ($\sigma_{\mu\nu} \Rightarrow T_{\mu\nu}$ and in strains $2\varepsilon_{\mu\nu} \Rightarrow h_{\mu\nu}$).
(5) The perfect correlation between the two gravitational wave polarizations ($h_{\mu\nu(A+)}, h_{\mu\nu(A\times)}$) with the two compression/tension and shear tensors of the space medium according to the facet considered in elasticity (case of a medium space cylinder twisted by rotation of two massive objects merging) and the link between spin 2 graviton with the image of Mohr's circle in terms of rotation.
(6) The intrinsic quantum characteristics of this elastic spatial medium with a fundamental state different from 0 (ρ/energy of vacuum and ω circular frequency).
(7) The spatial microstructure quantified with a fiber dimension of Planck's scale and a quantified time interval.
(8) The new definition of G (macroscopic manifestation of the said $f$frequency of the space medium) resulting from the parallelism between the Einstein's gravitation formalism and the Elasticity theory/strength of material of Timoshenko.

Table 22.1 provides a summary of the different formulas obtained. They allow the space medium to be associated with a dark elastic space material.

DOI: 10.1051/978-2-7598-2573-8.c022

TAB. 22.1 – Expressions of modified general relativity.

Parameters	Formulas	Formulas
Strains	$\varepsilon_{ii} = \frac{\delta L}{L}$	$\varepsilon_{ij} = \frac{\gamma}{2}$
Polarized wave $h_{\mu\nu}$	$A_+ \cos\left(\frac{\omega}{c}(ct - z)\right) \begin{bmatrix} 0 & 0 & 0 & 0 \\ 0 & +1 & \mathbf{0} & \mathbf{0} \\ 0 & \mathbf{0} & -1 & \mathbf{0} \\ 0 & \mathbf{0} & \mathbf{0} & \mathbf{0} \end{bmatrix}$	$A_\times \cos\left(\frac{\omega}{c}(ct - z)\right) \begin{bmatrix} 0 & 0 & 0 & 0 \\ 0 & \mathbf{0} & +1 & \mathbf{0} \\ 0 & +1 & \mathbf{0} & \mathbf{0} \\ 0 & \mathbf{0} & \mathbf{0} & \mathbf{0} \end{bmatrix}$
Stress tensor associated	$\sigma_{xy} = \begin{bmatrix} \sigma & 0 & 0 \\ 0 & -\sigma & 0 \\ 0 & 0 & 0 \end{bmatrix}$	$\sigma_{xy} = \begin{bmatrix} 0 & \tau = \sigma & 0 \\ \tau = \sigma & 0 & 0 \\ 0 & 0 & 0 \end{bmatrix}$
Associated elastic wave in the interferometric tube	Longitudinale compression/traction $c = \sqrt{\frac{E}{\rho}}$	Transversal shear $c = \sqrt{\frac{\mu}{\rho}}$
Space curvature = $K \times$ energy density	$\frac{1}{L^2}(\varepsilon)^2 = 4\pi(1+v)\left(\frac{\pi f^2}{\rho}\right)\frac{1}{c^4} \times \frac{U}{V}$	$\frac{1}{L^2}(\gamma)^2 = 16\pi\frac{\pi f^2}{\rho} \times \frac{1}{c^4} \times \frac{U}{V}$
Einstein's gravitational field	$G^{\mu\nu} = -\frac{8\pi G}{c^4} T^{\mu\nu}$	$\partial^\lambda \partial_\lambda \overline{h}_{ij} = \Box \overline{h}_{ij} = -\frac{16\pi G}{c^4} T_{ij}$
New gravitationnal field equation proposed by D. Izabel with mechanical parameters	$G^{\mu\nu} = -(1+v)\rho\left(\frac{\omega}{E}\right)^2 T^{\mu\nu}$ $G^{\mu\nu} = -2\rho\left(\frac{\omega}{E}\right)^2 T^{\mu\nu}$	$G^{\mu\nu} = -2\rho\left(\frac{\omega}{\mu}\right)^2 T^{\mu\nu}$
New expression of the G gravitational constant expressed as a function of mechanical and physical parameters	$G = \frac{\pi f^2 c^2}{E}$ $G = \frac{\pi f^2}{\rho} = \frac{\omega^2}{4\pi\rho}$	$G = \frac{\pi f^2 c^2}{\mu}$ $G = \frac{\pi f^2}{\rho} = \frac{\omega^2}{4\pi\rho}$
Poisson's ratio (elastic space homogeneous)	$(v = 1)$	$(v = 1)$
Young's modulus $E = Y$	$E = Y = \rho c^2$	$E = Y = 2\rho c^2(1+v)$ $E = Y = 4\rho c^2$
Speed c	$c = \frac{1}{f}\sqrt{\frac{GE}{\pi}}$	$c = \frac{1}{f}\sqrt{\frac{GE}{2\pi(1+v)}}$
Frequency f	$f = \frac{Gh}{2\pi V c^2}$	$f = \frac{Gh}{2\pi V c^2}$
Einstein's constant κ	$\kappa = -(1+v)\rho\left(\frac{\omega}{E}\right)^2$ $\kappa = -2\rho\left(\frac{\omega}{E}\right)^2$	$\kappa = -8(1+v)^2\rho\left(\frac{\omega}{E}\right)^2$ $\kappa = -2\rho\left(\frac{\omega}{\mu}\right)^2$

In view of the plane deformations of the particles deposited on a circle (lengthening/shortening of the structure of space) and the associated plane stresses (in the plane perpendicular to the propagation direction of gravitational waves), the space appears to constitute a multisandwich of thin sheets.

Numerically we get the following values:

(1) The length/thickness of fibers constituting the texture of the substance of the space fabric is 1.566×10^{-35} m [2] obtained after iterations compatible with the string dimension defined in string theory, or the dimension of the quantified space in loop quantum gravity and in agreement with the Planck length $1.61625518 \times 10^{-35}$ m postulated by T. Tenev and M.F. Horstemeyer in chapter 3.4 of [18].

(2) The space time quantum Young's modulus $E = Y = 1.00 \times 10^{113}$ Pa (from the longitudinal wave) and 4.00×10^{113} Pa (from the shear wave) [2] compatible with the results. T.G. Tenev and M.F. Horstemeyer (4.4×10^{113} Pa, see 3.4, formula 3.13) [18] and with the result of Kirk T. McDonald [13] (5.0×10^{113} Pa see formula 6). This value is compatible with the extreme rigidity of the space. See also D. Izabel {59}.

(3) Density ρ of 1.11×10^{96} kg/m^3 [2] (vacuum energy density) is also consistent with the results of T.G. Tenev and M.F. Horstemeyer (1.30×10^{96} kg/m^3, see 3.4, formula 3.14) [18] and with the result of Kirk T. McDonald [13] (5.0×10^{96} kg/m^3 see formula 8).

(4) Poisson ratio $\nu = 1$ [2] of the space time material is consistent with the results of T.G. Tenev and M.F. Horstemeyer (see 3.3, formula 3.2) [18].

(5) The new expression of ($G = \frac{\pi c^2 f^2}{Y}$) of D. Izabel demonstrated in this book is in correlation with the K. Mc Donald/R. Weiss approach [13] (see formula (5) of [13] $G = \frac{\pi c^2 f^2}{4Y}$).

(6) The quantum approach of Young's modulus $\boldsymbol{Y} = \frac{6c^7}{2.20422881\pi\hbar G^2}$ of D. Izabel obtained by an iterative approach (see appendix I) is also compatible with the results of T.G. Tenev and M.F. Horstemeyer $\boldsymbol{Y} = \frac{6c^7}{2\pi\hbar G^2}$ [18] table 1.

(7) The lapse of time to pass information from one elastic sheet of space to another is related to Planck's time (1×10^{-43} s).

Thus, to the question of this book "in what is space-time made of?", Our proposed answer is as follows: Space could be made of an equivalent elastic material consisting of at least a three-dimensional sandwich of thin sheets of thickness $t = 1.56 \times 10^{-35}$ m, with a density ρ of 1.1×10^{96} kg/m^3 and Young's modulus $Y = 4 \times 10^{113}$ Pa (or 10^{20} Y_{steel} following Gravitational Wave approach) and Poisson's ratio $\nu = 1$. At slow speed, its texture is rather like a fluid and at the speed of light it becomes a hyper resistant steel, an impassable wall for the speed of light thus blocked at 300 000 km/s inside its structure. Gravitational waves and therefore dynamic deformations of this structure under a loading within it (example the coalescence of two black holes), inherently propagate in this structural fabric at the speed of light. When loaded to the extreme, in the case of intense loading concentrated in an infinitesimal point, as for a black hole, its fibers stretch and plasticize. The synthesis of

TAB. 22.2 – State of the art – synthesis of the mechanical characteristics of the elastic material potentially constituting the elastic space-time {194–219}.

Author	G ($m^3/(kg.s^2)$)		Density ρ Kg/m^3	Y	Young's modulus intensity Y (Pa) Quantum (*)/classical (**) (GW)	Space fiber dimension (m)	Poisson's ratio v
	Formula	Way to obtain		Quantum approach/classical approach			
T.G. Tenev and M.F. Horstemeyer [18] and appendix I	–	Elasticity in 4 dimensions	1.3×10^{96}	$Y = \frac{6}{2\pi}\frac{c^7}{\hbar G^2}$ –	4.4×10^{113} –	1.62×10^{-35} (Planck Length assumed)	1
K.T. McDonald [13] and {215}	$G = \frac{c^2 f^2}{Y}$	Dimensional analysis	5×10^{96}	$Y \approx \frac{c^7}{\hbar G^2}$ $Y = \frac{c^2 f^2}{G}$	5.00×10^{113} 4.5×10^{31}	–	–
R. Weiss/ K. McDonald [13] {215} see appendix J Hamilton lecture 2018 and appendix J	$G = \frac{\pi c^2 f^2}{4Y}$	Strain energy densities of waves in steel (strength of material) and in plane gravitational waves	–	– $Y = \frac{\pi c^2 f^2}{4G}$	– $10^{20} \times Y_{steel}$	–	–

TAB. 22.2 – (continued).

D. Izabel [2] {59} see appendices I and J	$G = \frac{\pi c^2 f^2}{Y}$	Elasticity in plane/interferometer as strain gauge	1.11×10^{96}	$Y = \frac{6}{8.82\pi} \times \frac{c^7}{\hbar G^2}$	2.0×10^{20} Y_{steel}	1.57×10^{-35} (Obtained by iterations)	–
				$Y = \frac{\pi c^2 f^2}{G}$	1.0×10^{113}		
D. Izabel [2] see appendices I and J	$G = \frac{\pi f^2 c^2}{\mu}$	Torsional shear wave of an elastic cylinder clamped at the foot	1.11×10^{96}	$Y = \frac{6}{2.20\pi} \times \frac{c^7}{\hbar G^2}$	8.0×10^{20} Y_{steel}	1.57×10^{-35} (Obtained by iterations)	1
	$G = 2(1+v)\frac{\pi f^2 c^2}{Y}$			$Y = 4\frac{\pi f^2 c^2}{G}$	4.01×10^{113}		

(*) Obtained by considering the quantum energy of the vacuum.

(**) Obtained by considering a Ligo/Virgo gravitational wave frequency of 100 Hz.

$Y_{\text{steel}} = 210\ 000\ 000\ 000\ \text{N/m}^2$ or (Pa) or $210\ 000\ \text{N/mm}^2$ or (MPa).

the results of some of the different authors who have studied this elastic material hypothesis is given in table 22.2. Time also becomes elastic; it is in a way characterized by the transmission of information from a fiber from the structure of space to another. As these fibers are elastic, it therefore takes more or less time to transmit information giving the illusion of time that is lengthening or shortening in duration. Gravity being the deformation of space, the duration of time becomes therefore correlated with this space deformation and gravitation. A lot of gravitation involves a lot of space deformations, a lot of elongation of the space fibers, more time to transmit the deformation from one end to the other (or within the thickness of the fiber) and therefore a time which lengthens in the presence of gravitation. This elastic material thus becomes a new kind of deformable ether of quantum dimension.

We have therefore proposed a characterization of the elastic medium that can constitute space in correlation with the information given to us by gravitational waves whose existence is certain since they were measured in 2014. From the point of view of quantum field theory, we have shown that Mohr's circle behaves like the image of a spin 2 of the graviton. The granulometry of this medium being fine and quantum it is correlated with quantum fluctuations in relation to Casimir's force. Time appears to be just a passage of information from one sheet of the multi-sandwich to another. In all approaches, the gravitational constant G is proportional to the speed of light c squared, the frequency f, and is inversely proportional to Young's modulus Y of elastic space (see table 22.2).

The difference between the two values of Young's modulus (quantum/classical) is related to the difference in the measurement of vacuum energy:

- According to measurements of cosmological constants: $\rho = 1 \times 10^{-29}$ g/cm^3 (1×10^{-26} kg/m^3).
- According to quantum field theory: $\rho = 1.11 \times 10^{93}$ g/cm^3.
- According to measurements of cosmological constants: $T_{\text{vacuum}} = 8.987551787 \times 10^{-10}$ kg.m^2/s^2/m^3.
- According to quantum field theory: $E_{\text{vacuum}} = 1 \times 10^{113}$ kg m^2/s^2/m^3.

Other approaches to Young's modulus of space-time exist.

An Author (M. Beau) consider {213} the bulk modulus in link with expansion of the universe and dark energy and obtain:

$$\boldsymbol{K} = \boldsymbol{B} = \boldsymbol{\lambda} + \frac{\mathbf{2}}{\mathbf{3}}\boldsymbol{\mu} = \frac{\boldsymbol{Y}}{\mathbf{3}(\mathbf{1} - \mathbf{2}\boldsymbol{v})} = \mathbf{1.64 \times 10^{109}\ N/m^2}$$

This approach (K also noted B) is also considered in section 3.2 of the publication by Tenev and Horstemeyer [18] and {205} and in appendix A of the note of K. McDonald of [13].

An approach to Young's modulus of vacuum *via* electromagnetic wave propagation in vacuum is developed in [13] by K. McDonald in appendix B. This leads to a High but lower Young's modulus. He obtains $Y = \varepsilon_0 E_{\text{crit}}^2 = 3 \times 10^{25}$ Pa where E_{crit} critical electric field strength and ε_0 the permittivity of the vacuum.

Finally, A.C. Melissinos {215} develops an approach of Young's modulus based on gravitational waves and the TT gauge. He obtains $Y < \frac{\pi c^2 f^2}{4G} \frac{c\Delta\tau}{\Delta z}$ with $\Delta\tau = 1$ s and $\Delta z = 400$ Mpc.

The most important point is that all these approaches describe Young' modulus Y of the elastic space time. This corroborates the hypothesis of an equivalent elastic medium, of a space filled with a certain dark elastic material and therefore the return of a certain dark ether which unlike its luminiferous ancestor, this one is deformable!

To test this model, we proposed two tests, one on a small scale to measure Young's modulus of the material constituting space and the other on a large scale (distortion angle γ measurement) to demonstrate the successive shear of space sheets and so the corelation of the general relativity with the elastic theory (2 polarizations of the gravitational wave in link with the normal and shear stresses and strains of an elastic medium in torsion). Thus we have two ways to verify whether this theory is accurate.

This new reformulation of G and κ thus allows to reinterpret Einstein's field equation from a new angle uniting general relativity with the mechanics of continuous elastic medium bringing out a new kind of deformable ether that like Einstein's constant appears disappearing and reappearing again but differently like a phoenix from its ashes.

The consequence of this approach to space by an elastic material therefore leads to a rewriting of Einstein's gravitational field equation:

$$G_{\mu\nu} = -(1+\nu)\rho\left(\frac{\omega}{Y}\right)^2 T_{\mu\nu}$$

$$G = \frac{\pi f^2}{\rho} = \frac{\pi f^2 c^2}{Y} \rightarrow \kappa = \frac{8\pi G}{c^4} = (1+\nu)\rho\left(\frac{\omega}{Y}\right)^2 = \frac{8\pi^2 f^2}{Yc^2}$$

$$G_{\mu\nu} = -\frac{8\pi^2 f^2}{Yc^2} T_{\mu\nu}$$

Thus, if we saw correct, G would be related to the Eigen frequency of the fine structure of the space fabric as well as to the mechanical characteristics (ρ, ν) of the material constituting the structure of the space. Only measurements of Young's modulus on the one hand and the lateral movements of the arms of the interferometers or the Trihedra of the LISA project on the other hand, will be used to decide whether or not the space-time is indeed made up of an elastic dark material with an extremely fine texture. That is the rule in physics. Experience and theory must agree.

Appendix A

Chronological Order of Progress of the Author's Reflection and Related Discoveries

(1) From 2013 to 2016, I studied quantum mechanics as a self-taught study (polytechnic X-lectures with J. Dalibard), special relativity, then general relativity, then electromagnetism, then cosmology with the e-lectures of R. Taillet teacher-researcher at Savoie Mont blanc university (introduction to general relativity), then quantum mechanics relativistic and then quantum fields theory with Prasanta Tripathy of the Madras University. I also study philosophy and the history of sciences with E. Klein through his genial lectures; general relativity, the black holes, gravitational waves with the lectures of K. Torne, T. Damour; Quantum entanglement, Bell equations with A. Aspect, Quantum gravity with C. Rovelli.

(2) In September 2016 I discovered that the principle of curvature = K energy density is the same in strength of material (which I know from my studies at INSA of Rennes) and in general relativity (which I studied as a self-taught study) thanks to the conferences of T. Damour at the university of all knowledge the 1st Julys 2000, and that the stress tensor is equivalent to the stress-energy tensor (see https://physique.coursgratuits.net/).

(3) In 27 September 2016 N°. 1 attempt to publish in the European Journal of physic: refused.

(4) In October 26, 2016 Attempt N°. 2 publication at Foundation of physic: refused.

(5) On 9 June 2017, I published in Hall "open archives" the principles of equivalent curvature between the strength of material and general relativity and the equivalence between the stress energy tensor with the stress tensor relying on the article of physics "Free lectures". I also showed that the vibration of a beam and then a plate are the same as Schrödinger's equation of a particle in a potential well to one or two dimensions. The strength of material is the simplest way to understand general relativity and quantum field theory.

DOI: 10.1051/978-2-7598-2573-8.c911

(6) On 23 June 2017, I published in Hall "open archives" that if strains are measured *via* gravitational waves, if there are strains, then there is an associated elastic medium. I calculated the space-time rigidity and found that it is enormous. κ is the opposite of a rigidity is therefore very small. I found out what Andrew Ochadlick posted on a video on the internet on January 20, 2016 about the enormous rigidity of space-time (I discovered 3 years later in May 2020 the book of T. Damour "if Einstein was told to me" that confirms this).

(7) On July 7, 2017, I published in Hall "Open Archives" trying to calibrate κ with the parallelism of plate theory. I found what Americans published in 2016 T.G. Tenev and M.F. Horstemeyer (2016) "A Solid Mechanics Perspective on the Theory of General" that said more or less the same thing.
I tried to bring quantum mechanics into Einstein's constant κ *via* the Planck constant (as American T.G. Tenev and M.F. Horstemeyer (2016) mechanics of space time) and then *via* the known components of the vacuum (problem two values exist on the vacuum energy) – first bridge with the quantum vacuum that has a non-zero energy and therefore an equivalent density $\rho = (E_{\text{vacuum}}/c^2)/V$; *Via* the wave events, I confirmed a link between the material density of the space time and the speed of a wave ($\frac{\rho}{E} = \frac{1}{c^2}$). I also showed that Casimir's force can be a test to measure Young's modulus of the space environment and that the vacuum is not empty.

(8) On September 22, 2017, I met a French expert in quantum field theory. The verdict falls: He said to me "Just because it's the same math doesn't mean it's the same physics. The ether is big enough that we are looking for it, we have not found it. There is no ether, In general relativity it is the geometry of space that is distorted. There is no material. Otherwise you have to prove it". Back to square one. Big doubts.

(9) In December 2017, I discovered what Einstein says about a certain Ether. Discussion with the Hebrew center of Jerusalem (February 6, 2018) to be able to quote what Einstein wrote in 1905 with special relativity he abolished the ether but at a Leiden's conference in 1920, following a discussions with Lorentz on June 17, 1916, he recognized the existence of a certain Ether but no longer the Luminiferous ether that does not exist since it is an electromagnetic wave. So, I'm not wrong, there is space is physical object that deforms, it cannot be done with anything.

(10) In April 2018, I had the intuition that G must be made from its units of a product of the inverse of a density ρ by a frequency f squared. That makes it possible in addition to make fit the data of the vacuum ρ and ω in constant κ. $G = \frac{f^2}{\rho}$ and so $\kappa = \frac{2\rho}{\pi}\left(\frac{\omega}{E}\right)^2$.
I also discovered the prob B mission that measured in May 2011 angular deformations of space dragging around the earth (the earth twists space like a spoon spinning in jam). I was convinced. This is the true measure of new elastic ether around the earth. Sure, the space seems to be made of an elastic material. Now we must prove it.

(11) "Science et avenir" cruise on Nordic see May 2018 and meet with physicists. I then submitted a draft article to one of them. Same observation, "math and parallelism are interesting, but it is not because they are the same math that it is the same physics; dark energy is not a material". Same conclusions as with the specialist in quantum field theory. I doubted it again then. Did I go in the right way?

(12) In July 2018, I took a step back and synthesized all my work and made two major advances: I realized that the metric $g_{\mu\nu}$ in low gravity fields (case of gravitational waves propagating in the vacuum far from the sources that created them) come down to a plate metric $\eta_{\mu\nu}$ called Minkowski supplemented by a perturbation that is worth $h_{\mu\nu}$ twice the strain tensor $\varepsilon_{\mu\nu}$. So, I have three bridges between the theory of elasticity and general relativity:

- A common curvature space principle (deformation function) $= K \times$ an energy-mass of deformation equivalent to the work of external forces.
- A correspondence between the stress energy tensor $T_{\mu\nu}$ and the stress tensor $\sigma_{\mu\nu}$.
- A correspondence between the disturbance of the metric $h_{\mu\nu}$ and the strain tensor, $\varepsilon_{\mu\nu}$.

I list all the measurements and discoveries made for one century on space that show without any ambiguity that it is an elastic medium that deforms and not just a geometry, a mathematical and physical model:

- Galaxies are dragged into expanding space all the faster as they are far from us (Hubble Law).
- The earth drags the space around it, the 4 gyroscopes of the satellite probe B have shown, (it is therefore an elastic physical object deformed in torsion).
- The light of the stars follows the curvature of the sun (visible during an Eclipse) and shown by Eddington. When the sun goes away the stars resume their position (elastic response of space).
- Gravitational waves generate strains of space that are measured in interferometers (so it is an elastic physical object that vibrates).
- Space resonates (Ringermacher).

Finally, whether it is through Higgs's field, Casimir's force, the quantum field theory with a fundamental state of non-zero energy (the fundamental energy state of the harmonic oscillator) or the measurement of the energy of the vacuum, it is clear that the vacuum is not empty and does have an intrinsic energy. Stephen Hawking also used it to show that blackholes radiated (the creation of a pair of antiparticle/particles in the vacuum near a black hole, one is absorbed by the black hole and the other materializes) the vacuum is empty on average but does not have nothingness. So, if there is an energy with $E = mc^2$ there is an equivalent mass and therefore an apparent density of the vacuum and then since there is a quantum energy there is a quantum frequency of vibration with $E = hf$. So, everything to constitute a material of vacuum.

(13) July 24, 2018 Attempt N°. 3 to publish international journal of modern physic D: refused.
(14) September 22, 2018 N°. 4 attempt to publish to Russian Physic journal: no response.
(15) October 14, 2018 attempt N°. 5 publication living review in physic: refused.
(16) January 10, 2019 attempt N°. 6 scientific report: refused.
(17) February 4, 2019 attempt N°. 7 European Physical Journal C: refused.
(18) March 2019: great progress:
I decided to write in tensorial version the principle curvature = $k \times$ energy density (directly from Hooke's laws and the theory of elasticity) and find an equation with the equivalent of a κ in theory of elasticity and use for this the bridges in stresses and strains with the metric and stress energy tensors of general relativity.
I had a flash, a vision: I saw the Ligo and Virgo interferometers no longer as interferometers but as gigantic strain gauges placed on the elastic medium space (like the one used in strength of material when studying strains of beams or truss or when making photo elastic studies).
By writing down the dynamic deformation of the tubes of empty space trapped in the arms of interferometers set in motion by the passage of gravitational waves, I discovered through mathematics the formulation of G which reappears naturally in the equations with the parameters that I had envisaged, but corrected by a factor π. The excitation is at its peak. Sure, the elasticity theory approach told us that if there is an elastic space material that makes up the vacuum space, then $G = \frac{\pi f^2}{\rho}$.
By this approach, I can find exactly a tensor formalism of general relativity in elasticity theory in two dimensions by considering the arms deformations of the interferometers.

$$\begin{bmatrix} \frac{1}{L^2} & 0 \\ 0 & \frac{1}{L^2} \end{bmatrix} \begin{bmatrix} (\varepsilon_{xx})^2 & 0 \\ 0 & (\varepsilon_{yy})^2 \end{bmatrix} = \frac{8\pi G}{c^4} \begin{bmatrix} T_{xx} & 0 \\ 0 & T_{yy} \end{bmatrix}$$

(19) On March 10, 2019 at 10 p.m. I discovered in a conference of T. Damour "association of friends of the laboratory Arago a Banyuls on the sea of October 7, 2017 two fundamental points that intersect my discoveries found in self-taught":
T. Damour writes:

- "Einstein's theory: space-time is an elastic structure that is distorted by the presence within it of mass energy: space = jelly".
- In the question session he explains that the Luminiferous ether does not exist but that another kind of ether exists as Einstein himself said in 1920 at the Leiden conference.

I am therefore on the right track. Of course, I will absolutely have to find a scientific journal willing to publish me so that I will finally be part of the seraglio and my work will finally be recognized.

(20) March 19, 2019 Attempt N°. 8 to publish a foundation of physics: refused.

(21) 25 May 2019 attempt N°. 9 to publish a Quantum studies mathematics and foundation: refused.

(22) 3 May 2019 attempt N°. 10 to Publish in Pramana an Indian physics journal Peer reviewed.

(23) June 2019 attempt N°. 11 Indian journal of physic: refused.

(24) 5 September 2019 finally a response from Pramana's positive article could be published but there is a request for a major revision of the manuscript in particular:
- Bring out curvature = $k \times$ energy related to external and non-inner forces so that the parallelism is total.
- Consider Poisson's ratio.
- Consider the good value of the vacuum energy.

(25) In October 2019 second attempt to publish a Pramana. False track of the auxetic material (negative Poisson's ratio).

(26) In December 2019 new ideas and a third version of the article that will be the right one. All the pieces of the puzzle finally come together at once. The intellectual pleasure is intense. The two theoretical polarizations gravitational waves through the relationship $g_{\mu\nu} = \eta_{\mu\nu} + h_{\mu\nu} = 2\varepsilon_{\mu\nu}$ can be seen, and perfectly correspond to the two expressions of the space-time deformation tensor twisted by the rotation of the two black holes. The space behaves like an elastic medium. The elasticity theory works well. T. Damour is right to write in his book "If Einstein was told to me" and to announce in his lectures "space-time is an elastic structure that is distorted by the presence within its mass energy". Discovery of a K. Thorne conference that explains how black holes twist space-time and generate compression/traction deformations.
The two deformation tensors correspond to two stress tensors that are exactly what it takes when a medium is twisted. Following one of the facets we see compression traction and at 45° of this facet we see pure shears and shear strains (angular distortions).
As these deformation fields are perpendiculars to the propagation of the wave, the space must consist of sheets (a sandwich or each sheet is 10^{-35} m thick according to our numerical applications) that are successively sheared during the passage of the twist wave. We are faced with a problem of flat stresses.
Poisson's ration ν is 1. This is verified by looking at the particle deformation on a circle and looking at the antisymmetric oscillations of the interferometer arms.
Development from each strain tensor of the stress tensor associated with each polarization by Hooke's law (compression/traction of the arms) and on a facet at 45° (arm shear). I understand why there are two polarizations. It all depends on the orientation with which we observe the facet of the space material distorted in twisting by the passage of gravitational waves. The

interferometers should therefore, depending on their orientation on the Earth, either see compression traction or angular movements related to the shear of the sheets.
Angular movements that are not yet measured today.
By studying the dynamic deformation in compression traction of each arm of the interferometer, I reconstruct the spatial part of Einstein's equation of gravitation.
I find for each stress tensors (compression traction) or twist (pure shear) the formula of $G = \frac{\pi f^2}{\rho}$ with f the specific frequency of the material of the space fabric and its apparent ρ density mass (Casimir's vacuum energy or quantum field theory gives a mass to the rest by $E = mc^2$ gives a density mass by dividing E/c^2 by a volume).
Having the right expression of κ, with the data of the vacuum, I find the supposed granulometry of the vacuum 10^{-35} m established in string theory or Loop Quantum gravity.
I find by taking the time component of tensor the right formula for G.
I rewrite what history would have been if Newton had not existed and if Einstein had calibrated his equations not from the temporal component of his tensor and into weak fields *via* the equation of Newton but using the spatial components of his tensors.
In this case, G no longer must be a constant. It is simply calibrated on the mechanical and elastic characteristics of the material that makes up the space: its own frequency and density. κ becomes set according to the elastic characteristics of the space fabric (Young's modulus E) and the characteristics of the vacuum (its own pulsation ω and density ρ). $\kappa = 2\rho\left(\frac{\omega}{E}\right)^2$.
We also understand why the shape of spiral galaxies is reminiscent of the shape of hurricanes or eddies in water during the tsunami in Japan in 2011, space is filled with a fluid (liquid at low speed and solid at high speed) that shapes these geometric shapes. T. Damour also demonstrated that the dynamics of black holes concealed the viscous fluid formulations of Navier–Stokes.
As this material is present in space, it disrupts particle shots in Young's experiments, where particles follow ripples in the quantum vacuum. Quantum objects are particles traveling in a soup of fluctuating 10^{-35} dimension material. This material is solid at the speed of light and liquid otherwise. Hence this speed limit $\left(\frac{\rho}{E} = \frac{1}{c^2}\right)$.

(27) Pramana accepts my publication on February 18, 2020. Victory!

(28) During Christmas 2019, I discovered the work of R. Feynman in his book "lesson on gravitation" in chapter 3.4 that he also speaks of Mohr's circle and the interpretation of polarizations as stress tensors. Indeed, Mohr's circle behaves like a spin of 2 specific to the graviton: it is necessary to make 2 complete turns on Mohr's circle so the facet makes a complete turn in reality. Graviton and quantum field theory begin to connect with general relativity through continuum mechanics (elasticity theory). I discovered during an

internet conference the work of T. Damour who demonstrated that near black holes Einstein's equations hides Navier–Stokes equations of fluids.

(29) On 24 April 2020 the final publication is validated.

(30) On 15 May 2020, I discover and buy T. Damour's book "if Einstein was told to me, chapter 3 elastic space-time" I discover in:

  (a) That T. Damour proposes a formulation of $D_{(g)} = \kappa T$ Einstein's equation according to the theory of elasticity, Hooke's Law: with D a deformation tensor measuring how different a space-time having a geometry described by g of a Minkowski's space-time and T a tension tensor.

  (b) That Einstein never approached his theory from the point of view of elasticity. So, it could not recalibrate G like I did see References [16] from T. Damour's book I quote "note that Einstein never used the phrase, introduced in this book of 'law of elasticity of space-time'. However, we do not think we are betraying the central idea of his theory, but rather clearing it up using that image".

The innovative nature of my work therefore results in the fact that I pushed T. Damour's idea of elastic space-time to the end to characterize exactly κ in elasticity in the case of space deformations measured in the arms of the interferometers Ligo and Virgo and thus to have discovered the hidden parameters that encompass the G gravitational constant.

(31) Publication went live on August 13, 2020.
https://link.springer.com/article/10.1007/s12043-020-01954-5#citeas
https://www.ias.ac.in/describe/article/pram/094/0119

(32) On August 24, 2020, I contacted by email a famous Nobel Prize winner, specialist in this question of the Young modulus of space-time, therefore concerned by my work, and he found this approach to the dynamics of relativity general by the theory of elasticity described in my interesting Pramana Paper and as a nice addition to look at the dynamics of general relativity through the window of elasticity: great victory.

(33) On August 30, I decided to write a book recounting all the ups and downs of this fascinating work, I chose EDP, science for its seriousness.

Appendix B

Measurements of Space-Time Material Deformations (Strains and Angles)

B.1 The Distortion of Space-Time by Lense–Thirring Effect (Frame Dragging) {142}

It has been demonstrated by the Gravity probe B mission (2005 and results in 2011) that the earth by its rotation distorts the space-time around it by two effects: the warping/curvature and the twisting of space-time (see figure B.1 and table B.1). We are talking about:

- Frame dragging effect.
- Geodetic effect.

B.2 The Distortion of Space-Time by the Sun

As space-time is curved by the presence of a mass, we can calculate and measure the change in trajectory of a beam of light that will follow the curvature of space-time. An application is possible in the case of the sun by observing the apparent position variation of the light of a star behind the sun during a total eclipse of it (figure B.2).

Thus in 1919, Eddington observed the apparent position of the stars during an eclipse of the sun then repeated the experiment with the sun far from the night stars and measured by difference the difference in the position of the stars. He deduced an angle of variation of position stars in the presence and absence of the gravitational effect of the sun on vacuum space in accordance with calculations made according to General Relativity (table B.2).

(*) Detail of the numerical application.
$G = 6.67 \times 10^{-11}$ m^3.kg^{-1}.s^{-2}
$M =$ Mass of the Sun $= 2 \times 10^{30}$ kg
$r_0 =$ 700 000 km

DOI: 10.1051/978-2-7598-2573-8.c912

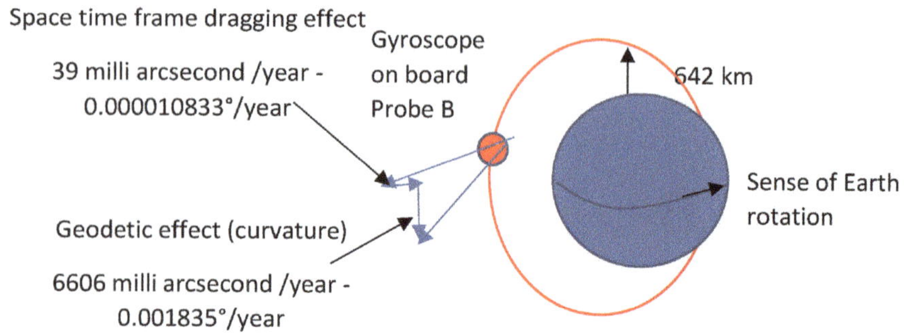

FIG. B.1 – Distortion of space-time around the earth (see {142}).

TAB. B.1 – Comparison of the results of general relativity with those measured by the gravity prob B satellite see table 3.2 {142}.

Effect considered	Prediction of general relativity in milli second arc/year	Measure gravity probe B in milli second arc/year	Error %
Geodetic effect	−6606.1	−6601.8/−18.3	0.28
Frame dragging effect	−39.2	−37.2/−7.2	19

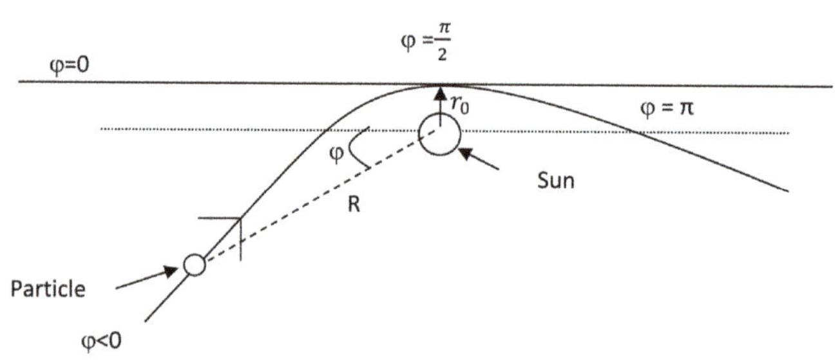

FIG. B.2 – Curvature of a beam of light nearing the sun.

$c = 300\ 000$ km/s

$\Delta\varphi = \frac{2a}{r_0} = \frac{4GM}{r_0 c^2} = \frac{4\times 6.67\times 10^{-11}\times 2\times 10^{30}}{700\,000\,000\times(300\,000\,000)^2} = 8.4698E - 06$ radiant

Which can be expressed in degrees:

Pi radiants => 180°

8.4698E−06 radiants => 0.00048529°

TAB. B.2 – Comparison of the result of general relativity with the Eddington measurements.

Effect considered	Prediction of general relativity (in arc minutes)	Eddington's measurements in 1919 (in arc minutes (source Wikipedia))
Curvature of the beam of a light of a star passing near the sun	$\varphi = \frac{4GM}{r_0 c^2} = 1.747030168$	– Sobral Experience (Brazil): $\phi \sim 1.98 \pm 0.12$ – Principle experiment (Gulf of Guinea): $\phi \sim 1.61 \pm 0.31$

Knowing:
That an arc minute is a 60th of a degree
That an arc second is a 60th of a minute arc
$(1/60)/60°$ => one arc second
$0.00048529°$ => 1.747044 arc seconds

B.3 Space-Time Strains Measured in the Interferometers Ligo and Virgo

Interferometers measure strains of 10^{-21} in 4 km long tubes of vacuum space resulting from the passage of gravitational waves.

Appendix C

History of Physics and Related Formulas

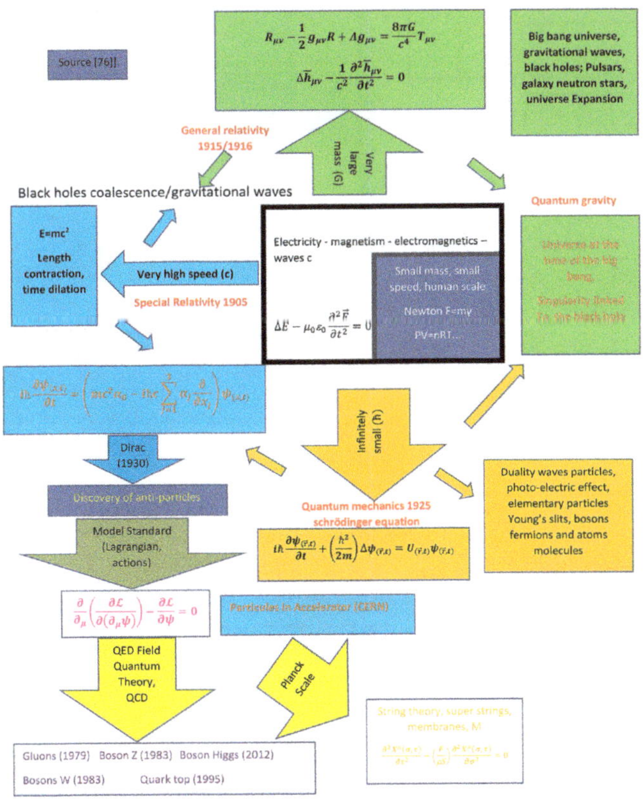

FIG. C.1 – Summarizes the different stages of current physics.

DOI: 10.1051/978-2-7598-2573-8.c913

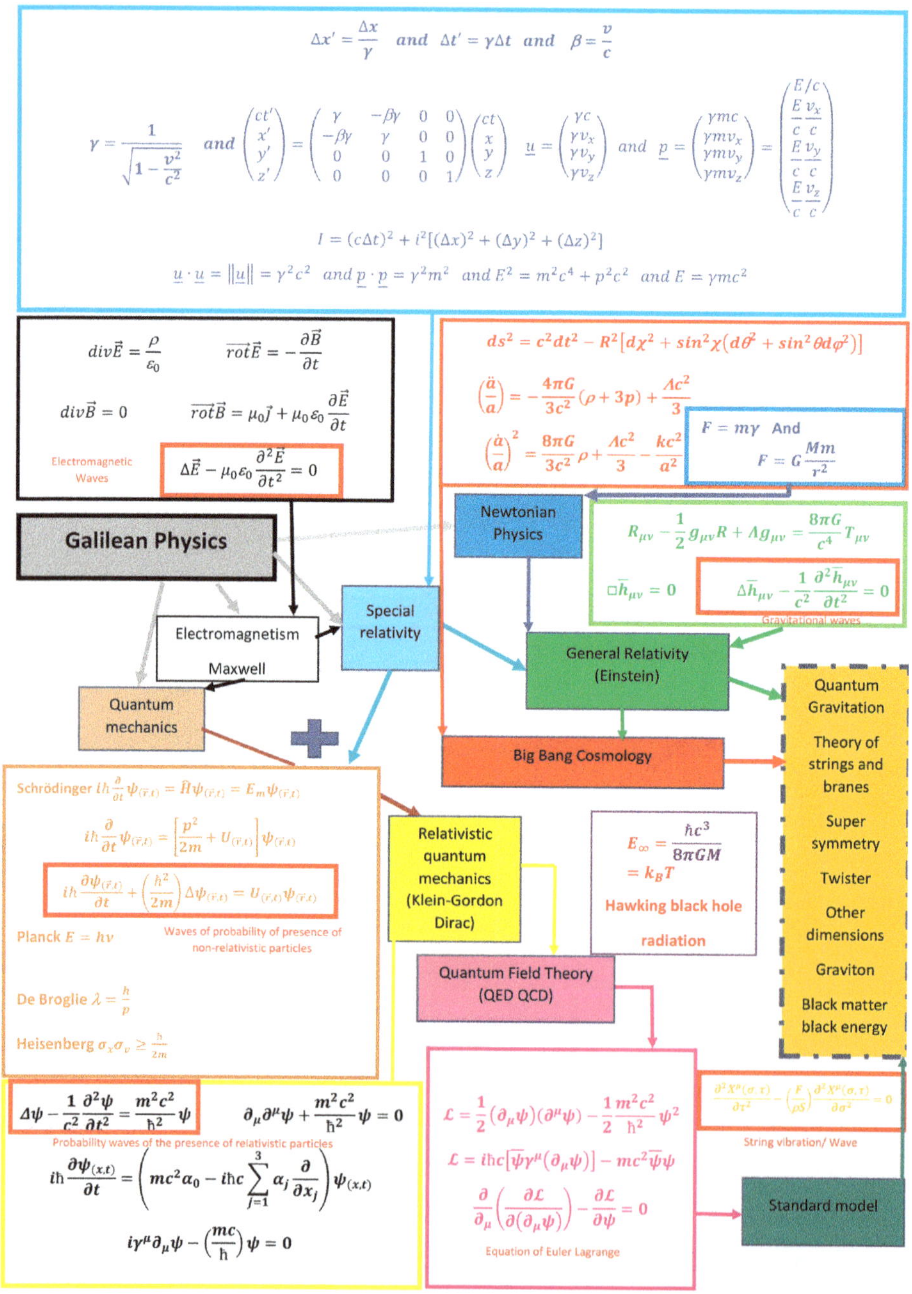
$\Delta x' = \frac{\Delta x}{\gamma}$ and $\Delta t' = \gamma \Delta t$ and $\beta = \frac{v}{c}$
$\gamma = \frac{1}{\sqrt{1-\frac{v^2}{c^2}}}$ and $\begin{pmatrix} ct' \\ x' \\ y' \\ z' \end{pmatrix} = \begin{pmatrix} \gamma & -\beta\gamma & 0 & 0 \\ -\beta\gamma & \gamma & 0 & 0 \\ 0 & 0 & 1 & 0 \\ 0 & 0 & 0 & 1 \end{pmatrix} \begin{pmatrix} ct \\ x \\ y \\ z \end{pmatrix}$ $\underline{u} = \begin{pmatrix} \gamma c \\ \gamma v_x \\ \gamma v_y \\ \gamma v_z \end{pmatrix}$ and $\underline{p} = \begin{pmatrix} \gamma mc \\ \gamma m v_x \\ \gamma m v_y \\ \gamma m v_z \end{pmatrix} = \begin{pmatrix} E/c \\ \frac{E}{c}\frac{v_x}{c} \\ \frac{E}{c}\frac{v_y}{c} \\ \frac{E}{c}\frac{v_z}{c} \end{pmatrix}$
$I = (c\Delta t)^2 + i^2[(\Delta x)^2 + (\Delta y)^2 + (\Delta z)^2]$
$\underline{u} \cdot \underline{u} = \|\underline{u}\| = \gamma^2 c^2$ and $\underline{p} \cdot \underline{p} = \gamma^2 m^2$ and $E^2 = m^2c^4 + p^2c^2$ and $E = \gamma mc^2$
$div\vec{E} = \frac{\rho}{\varepsilon_0}$ $\overrightarrow{rot}\vec{E} = -\frac{\partial \vec{B}}{\partial t}$
$div\vec{B} = 0$ $\overrightarrow{rot}\vec{B} = \mu_0 \vec{J} + \mu_0 \varepsilon_0 \frac{\partial \vec{E}}{\partial t}$
Electromagnetic Waves
$\Delta\vec{E} - \mu_0\varepsilon_0 \frac{\partial^2 \vec{E}}{\partial t^2} = 0$
$ds^2 = c^2dt^2 - R^2[d\chi^2 + sin^2\chi(d\theta^2 + sin^2\theta d\varphi^2)]$
$\left(\frac{\ddot{a}}{a}\right) = -\frac{4\pi G}{3c^2}(\rho + 3p) + \frac{\Lambda c^2}{3}$
$\left(\frac{\dot{a}}{a}\right)^2 = \frac{8\pi G}{3c^2}\rho + \frac{\Lambda c^2}{3} - \frac{kc^2}{a^2}$
$F = m\gamma$ And $F = G\frac{Mm}{r^2}$
Galilean Physics
Newtonian Physics
$R_{\mu\nu} - \frac{1}{2}g_{\mu\nu}R + \Lambda g_{\mu\nu} = \frac{8\pi G}{c^4}T_{\mu\nu}$
$\Box \bar{h}_{\mu\nu} = 0$
$\Delta \bar{h}_{\mu\nu} - \frac{1}{c^2}\frac{\partial^2 \bar{h}_{\mu\nu}}{\partial t^2} = 0$
Gravitational waves
Special relativity
Electromagnetism
Maxwell
General Relativity (Einstein)
Quantum mechanics
Big Bang Cosmology
Quantum Gravitation
Theory of strings and branes
Super symmetry
Twister
Other dimensions
Graviton
Black matter black energy
Schrödinger $i\hbar\frac{\partial}{\partial t}\psi_{(\vec{r},t)} = \hat{H}\psi_{(\vec{r},t)} = E_m\psi_{(\vec{r},t)}$
$i\hbar\frac{\partial}{\partial t}\psi_{(\vec{r},t)} = \left[\frac{p^2}{2m} + U_{(\vec{r},t)}\right]\psi_{(\vec{r},t)}$
$i\hbar\frac{\partial\psi_{(\vec{r},t)}}{\partial t} + \left(\frac{\hbar^2}{2m}\right)\Delta\psi_{(\vec{r},t)} = U_{(\vec{r},t)}\psi_{(\vec{r},t)}$
Planck $E = h\nu$
Waves of probability of presence of non-relativistic particles
De Broglie $\lambda = \frac{h}{p}$
Heisenberg $\sigma_x\sigma_v \geq \frac{\hbar}{2m}$
Relativistic quantum mechanics (Klein-Gordon Dirac)
$E_\infty = \frac{\hbar c^3}{8\pi GM} = k_B T$
Hawking black hole radiation
Quantum Field Theory (QED QCD)
$\Delta\psi - \frac{1}{c^2}\frac{\partial^2\psi}{\partial t^2} = \frac{m^2c^2}{\hbar^2}\psi$
$\partial_\mu\partial^\mu\psi + \frac{m^2c^2}{\hbar^2}\psi = 0$
Probability waves of the presence of relativistic particles
$i\hbar\frac{\partial\psi_{(x,t)}}{\partial t} = \left(mc^2\alpha_0 - i\hbar c\sum_{j=1}^{3}\alpha_j\frac{\partial}{\partial x_j}\right)\psi_{(x,t)}$
$i\gamma^\mu\partial_\mu\psi - \left(\frac{mc}{\hbar}\right)\psi = 0$
$\mathcal{L} = \frac{1}{2}(\partial_\mu\psi)(\partial^\mu\psi) - \frac{1}{2}\frac{m^2c^2}{\hbar^2}\psi^2$
$\mathcal{L} = i\hbar c[\bar{\psi}\gamma^\mu(\partial_\mu\psi)] - mc^2\bar{\psi}\psi$
$\frac{\partial}{\partial_\mu}\left(\frac{\partial\mathcal{L}}{\partial(\partial_\mu\psi)}\right) - \frac{\partial\mathcal{L}}{\partial\psi} = 0$
Equation of Euler Lagrange
$\frac{\partial^2 X^\mu(\sigma,\tau)}{\partial\tau^2} - \left(\frac{F}{\rho S}\right)\frac{\partial^2 X^\mu(\sigma,\tau)}{\partial\sigma^2} = 0$
String vibration/ Wave
Standard model

Appendix D

Calculating the Scalar Curvature *R* of a Sphere

D.1 Introduction

In this appendix, we will calculate the scalar curvature R intervening in Einstein's formulation in the simplified case of a two-dimensional sphere. The case of a space-time 3 sphere will be dealt with in appendix E. We demonstrate in this appendix all the results given in the excellent video of science-click on the mathematic of general relativity 5/8 d'Alessandro Roussel (figure D.1). https://www.youtube.com/watch?v=_iZ9cl0CCA8&list=PLK3v6vxQgv6SYzxABG4AcrYYAOgKTcIv3&index=5.

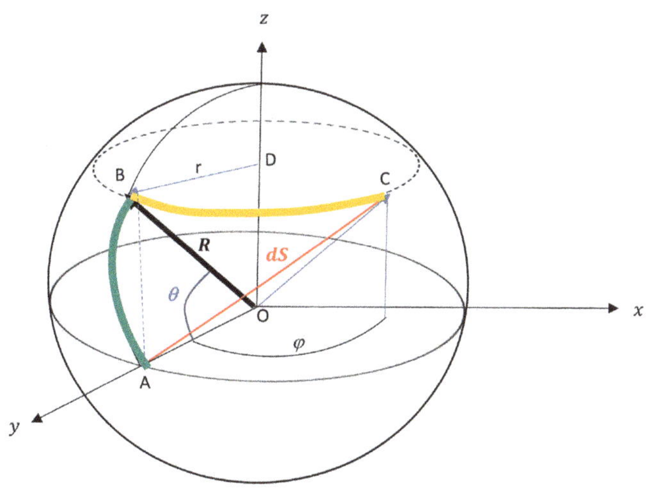

FIG. D.1 – Spherical coordinates.

DOI: 10.1051/978-2-7598-2573-8.c914

D.2 Metric Definition

The arc of the circle is worth $\widehat{AB} = R\theta$

In the triangle $\widehat{OAB}$ we can express the r radius:

$$\cos\theta = \frac{r}{R}$$

So:

$$r = R\cos\theta$$

The length of the bow $\widehat{BC}$ is therefore: $\widehat{BC} = R\varphi\cos\theta$

If we now consider small variations in each of the lengths and angles, the arcs of circles can be likened to small straight distances and one can make a Pythagoras:

We begin by writing the interval in polar coordinates:

$$\widehat{dS}^2 = \widehat{dAB}^2 + d\widehat{BC}^2$$

$$ds^2 = R^2 d\theta^2 + R^2 \cos\,\theta^2 d\varphi^2$$

The metric corresponds to the terms in front of $d\theta^2$ and $d\varphi^2$

$$g_{11} = g_{\theta\theta} = R^2$$

$$g_{22} = g_{\varphi\varphi} = R^2\cos\theta^2$$

$$g_{12} = g_{21} = g_{\theta\varphi} = g_{\varphi\theta} = 0$$

So:

$$g_{\mu\nu} = \begin{bmatrix} R^2 & 0 \\ 0 & R^2\cos\theta^2 \end{bmatrix} => g_{\mu\nu} = \begin{bmatrix} g_{\theta\theta} & g_{\theta\varphi} \\ g_{\varphi\theta} & g_{\varphi\varphi} \end{bmatrix}$$

And by definition we also have:

$$g^{\mu\nu} = \begin{bmatrix} \frac{1}{R^2} & 0 \\ 0 & \frac{1}{R^2\cos\theta^2} \end{bmatrix} => g^{\mu\nu} = \begin{bmatrix} g^{\theta\theta} & g^{\theta\varphi} \\ g^{\varphi\theta} & g^{\varphi\varphi} \end{bmatrix}$$

And the derivatives are made in relation to the following variables:
$\boldsymbol{x^\theta} = \theta$ and $\boldsymbol{x^\varphi} = \varphi$

D.3 Calculation of the First Derivatives of the Metric

All derivatives of the metric are systematically calculated.

$$\frac{dg_{\theta\theta}}{d\theta} = 0$$

$$\frac{dg_{\theta\theta}}{d\varphi} = 0$$

$$\frac{dg_{\theta\varphi}}{d\theta} = 0; \frac{dg_{\varphi\theta}}{d\theta} = 0$$

$$\frac{dg_{\theta\varphi}}{d\varphi} = 0; \frac{dg_{\varphi\theta}}{d\varphi} = 0$$

$$\frac{dg_{\varphi\varphi}}{d\varphi} = 0$$

Using that $(u^n)' = nu^{n-1}u'$ and $(\cos\theta)' = -\sin\theta$

$$\frac{dg_{\varphi\varphi}}{d\theta} = -2R^2\cos\theta\sin\theta$$

D.4 Calculation of the Different Christoffel Symbols

By definition we have:

$$\Gamma^{\lambda}_{\mu\nu} = \frac{1}{2} g^{\lambda\rho}\left\{\frac{\partial g_{\mu\rho}}{\partial x^{\nu}} + \frac{\partial g_{\nu\rho}}{\partial x^{\mu}} - \frac{\partial g_{\mu\nu}}{\partial x^{\rho}}\right\}$$

With $\lambda = 1, 2$ either $\lambda = (\theta, \varphi)$
With θ which will give a 2×2 matrix:

$$\Gamma^{\theta}_{\mu\nu} = \frac{1}{2} g^{\theta\rho}\left\{\frac{\partial g_{\mu\rho}}{\partial x^{\nu}} + \frac{\partial g_{\nu\rho}}{\partial x^{\mu}} - \frac{\partial g_{\mu\nu}}{\partial x^{\rho}}\right\}$$

With φ which will give a 2×2 matrix:

$$\Gamma^{\varphi}_{\mu\nu} = \frac{1}{2} g^{\varphi\rho}\left\{\frac{\partial g_{\mu\rho}}{\partial x^{\nu}} + \frac{\partial g_{\nu\rho}}{\partial x^{\mu}} - \frac{\partial g_{\mu\nu}}{\partial x^{\rho}}\right\}$$

Step 1 $\Gamma^{\theta}_{\mu\nu}$

Two possible values for ρ, either θ, either φ
First value:

$$\Gamma^{\theta}_{\mu\nu} = \frac{1}{2} g^{\theta\theta}\left\{\frac{\partial g_{\mu\theta}}{\partial x^{\nu}} + \frac{\partial g_{\nu\theta}}{\partial x^{\mu}} - \frac{\partial g_{\mu\nu}}{\partial x^{\theta}}\right\}$$

So, we have to calculate this for all the components of μ and v possible.

With $\mu = \theta$, $v = \theta$

$$\Gamma^{\theta}_{\mu v}=\frac{1}{2}g^{\theta\theta}\left\{\frac{\partial g_{\mu\theta}}{\partial x^{v}}+\frac{\partial g_{v\theta}}{\partial x^{\mu}}-\frac{\partial g_{\mu v}}{\partial x^{\theta}}\right\}$$

$$\Gamma^{\theta}_{\theta\theta}=\frac{1}{2}g^{\theta\theta}\left\{\frac{\partial g_{\theta\theta}}{\partial x^{\theta}}+\frac{\partial g_{\theta\theta}}{\partial x^{\theta}}-\frac{\partial g_{\theta\theta}}{\partial x^{\theta}}\right\}$$

$$g_{\mu v}=\begin{bmatrix} R^2 & 0 \\ 0 & R^2\cos\theta^2 \end{bmatrix} => g_{\mu v}=\begin{bmatrix} g_{\theta\theta} & g_{\theta\varphi} \\ g_{\varphi\theta} & g_{\varphi\varphi} \end{bmatrix}$$

$$g^{\mu v}=\begin{bmatrix} \frac{1}{R^2} & 0 \\ 0 & \frac{1}{R^2\cos\theta^2} \end{bmatrix} => g^{\mu v}=\begin{bmatrix} g^{\theta\theta} & g^{\theta\varphi} \\ g^{\varphi\theta} & g^{\varphi\varphi} \end{bmatrix}$$

$g_{\theta\theta}$ is independent θ

$$\Gamma^{\theta}_{\theta\theta}=0$$

With $\mu = \theta, v = \varphi$

$$\Gamma^{\theta}_{\mu v}=\frac{1}{2}g^{\theta\theta}\left\{\frac{\partial g_{\mu\theta}}{\partial x^{v}}+\frac{\partial g_{v\theta}}{\partial x^{\mu}}-\frac{\partial g_{\mu v}}{\partial x^{\theta}}\right\}$$

$$\Gamma^{\theta}_{\theta\varphi}=\frac{1}{2}g^{\theta\theta}\left\{\frac{\partial g_{\theta\theta}}{\partial x^{\varphi}}+\frac{\partial g_{\varphi\theta}}{\partial x^{\theta}}-\frac{\partial g_{\theta\varphi}}{\partial x^{\theta}}\right\}$$

$$g_{\mu v}=\begin{bmatrix} R^2 & 0 \\ 0 & R^2\cos\theta^2 \end{bmatrix} => g_{\mu v}=\begin{bmatrix} g_{\theta\theta} & g_{\theta\varphi} \\ g_{\varphi\theta} & g_{\varphi\varphi} \end{bmatrix}$$

$$g^{\mu v}=\begin{bmatrix} \frac{1}{R^2} & 0 \\ 0 & \frac{1}{R^2\cos\theta^2} \end{bmatrix} => g^{\mu v}=\begin{bmatrix} g^{\theta\theta} & g^{\theta\varphi} \\ g^{\varphi\theta} & g^{\varphi\varphi} \end{bmatrix}$$

$g_{\theta\theta}$ independent φ

$$g_{\theta\varphi}=g_{\varphi\theta}=0$$

$$\Gamma^{\theta}_{\theta\varphi}=0$$

With $\mu = \varphi, v = \theta$

$$\Gamma^{\theta}_{\mu v}=\frac{1}{2}g^{\theta\theta}\left\{\frac{\partial g_{\mu\theta}}{\partial x^{v}}+\frac{\partial g_{v\theta}}{\partial x^{\mu}}-\frac{\partial g_{\mu v}}{\partial x^{\theta}}\right\}$$

$$\boldsymbol{\Gamma^{\theta}_{\varphi\theta}} = \frac{\mathbf{1}}{\mathbf{2}} g^{\theta\theta} \left\{ \frac{\boldsymbol{\partial g_{\varphi\theta}}}{\boldsymbol{\partial x^{\theta}}} + \frac{\boldsymbol{\partial g_{\theta\theta}}}{\boldsymbol{\partial x^{\varphi}}} - \frac{\boldsymbol{\partial g_{\varphi\theta}}}{\boldsymbol{\partial x^{\theta}}} \right\}$$

$$g_{\mu\nu} = \begin{bmatrix} R^2 & 0 \\ 0 & R^2 \cos\theta^2 \end{bmatrix} => g_{\mu\nu} = \begin{bmatrix} g_{\theta\theta} & g_{\theta\varphi} \\ g_{\varphi\theta} & g_{\varphi\varphi} \end{bmatrix}$$

$$g^{\mu\nu} = \begin{bmatrix} \frac{1}{R^2} & 0 \\ 0 & \frac{1}{R^2 \cos\theta^2} \end{bmatrix} => g^{\mu\nu} = \begin{bmatrix} g^{\theta\theta} & g^{\theta\varphi} \\ g^{\varphi\theta} & g^{\varphi\varphi} \end{bmatrix}$$

$g_{\theta\theta}$ independent φ

$$g_{\theta\varphi} = g_{\varphi\theta} = 0$$

$$\boldsymbol{\Gamma^{\theta}_{\varphi\theta} = 0}$$

With $\boldsymbol{\mu = \varphi, \nu = \varphi}$

$$\boldsymbol{\Gamma^{\theta}_{\mu\nu}} = \frac{\mathbf{1}}{\mathbf{2}} g^{\theta\theta} \left\{ \frac{\boldsymbol{\partial g_{\mu\theta}}}{\boldsymbol{\partial x^{\nu}}} + \frac{\boldsymbol{\partial g_{\nu\theta}}}{\boldsymbol{\partial x^{\mu}}} - \frac{\boldsymbol{\partial g_{\mu\nu}}}{\boldsymbol{\partial x^{\theta}}} \right\}$$

$$\boldsymbol{\Gamma^{\theta}_{\varphi\varphi}} = \frac{\mathbf{1}}{\mathbf{2}} g^{\theta\theta} \left\{ \frac{\boldsymbol{\partial g_{\varphi\theta}}}{\boldsymbol{\partial x^{\varphi}}} + \frac{\boldsymbol{\partial g_{\varphi\theta}}}{\boldsymbol{\partial x^{\varphi}}} - \frac{\boldsymbol{\partial g_{\varphi\varphi}}}{\boldsymbol{\partial x^{\theta}}} \right\}$$

$$g_{\mu\nu} = \begin{bmatrix} R^2 & 0 \\ 0 & R^2 \cos\theta^2 \end{bmatrix} => g_{\mu\nu} = \begin{bmatrix} g_{\theta\theta} & g_{\theta\varphi} \\ g_{\varphi\theta} & g_{\varphi\varphi} \end{bmatrix}$$

$$g^{\mu\nu} = \begin{bmatrix} \frac{1}{R^2} & 0 \\ 0 & \frac{1}{R^2 \cos\theta^2} \end{bmatrix} => g^{\mu\nu} = \begin{bmatrix} g^{\theta\theta} & g^{\theta} \\ g^{\varphi\theta} & g^{\varphi\varphi} \end{bmatrix}$$

$$g_{\theta\varphi} = g_{\varphi\theta} = 0$$

$$\frac{dg_{\varphi\varphi}}{d\theta} = -2R^2 \cos\theta \sin\theta$$

So:

$$\boldsymbol{\Gamma^{\theta}_{\varphi\varphi}} = \frac{\mathbf{1}}{\mathbf{2}} g^{\theta\theta} \left\{ \frac{\boldsymbol{\partial g_{\varphi\theta}}}{\boldsymbol{\partial x^{\varphi}}} + \frac{\boldsymbol{\partial g_{\varphi\theta}}}{\boldsymbol{\partial x^{\varphi}}} - \frac{\boldsymbol{\partial g_{\varphi\varphi}}}{\boldsymbol{\partial x^{\theta}}} \right\} = \frac{\mathbf{1}}{\mathbf{2}} \frac{1}{R^2} \left\{ 0 + 0 - \left(-2R^2 \cos\theta \sin\theta \right) \right\}$$

$$\boldsymbol{\Gamma^{\theta}_{\varphi\varphi} = (\cos\theta\sin\theta)}$$

Second value:

$$\boldsymbol{\Gamma^{\theta}_{\mu\nu}} = \frac{\mathbf{1}}{\mathbf{2}} g^{\theta\varphi} \left\{ \frac{\boldsymbol{\partial g_{\mu\varphi}}}{\boldsymbol{\partial x^{\nu}}} + \frac{\boldsymbol{\partial g_{\nu\varphi}}}{\boldsymbol{\partial x^{\mu}}} - \frac{\boldsymbol{\partial g_{\mu\nu}}}{\boldsymbol{\partial x^{\varphi}}} \right\}$$

So, we have to calculate this for all the components of μ and ν possible.
With $\boldsymbol{\mu = \theta, \nu = \theta}$

$$\boldsymbol{\Gamma^{\theta}_{\mu\nu}} = \frac{\mathbf{1}}{\mathbf{2}} g^{\theta\varphi} \left\{ \frac{\boldsymbol{\partial g_{\mu\varphi}}}{\boldsymbol{\partial x^{\nu}}} + \frac{\boldsymbol{\partial g_{\nu\varphi}}}{\boldsymbol{\partial x^{\mu}}} - \frac{\boldsymbol{\partial g_{\mu\nu}}}{\boldsymbol{\partial x^{\varphi}}} \right\}$$

$$\boldsymbol{\Gamma^{\theta}_{\theta\theta}} = \frac{\mathbf{1}}{\mathbf{2}} g^{\theta\varphi} \left\{ \frac{\boldsymbol{\partial g_{\theta\varphi}}}{\boldsymbol{\partial x^{\theta}}} + \frac{\boldsymbol{\partial g_{\theta\varphi}}}{\boldsymbol{\partial x^{\theta}}} - \frac{\boldsymbol{\partial g_{\theta\theta}}}{\boldsymbol{\partial x^{\varphi}}} \right\}$$

$$g_{\mu\nu} = \begin{bmatrix} R^2 & 0 \\ 0 & R^2\cos\theta^2 \end{bmatrix} => g_{\mu\nu} = \begin{bmatrix} g_{\theta\theta} & g_{\theta\varphi} \\ g_{\varphi\theta} & g_{\varphi\varphi} \end{bmatrix}$$

$$g^{\mu\nu} = \begin{bmatrix} \frac{1}{R^2} & 0 \\ 0 & \frac{1}{R^2\cos\theta^2} \end{bmatrix} => g^{\mu\nu} = \begin{bmatrix} g^{\theta\theta} & g^{\theta\varphi} \\ g^{\varphi\theta} & g^{\varphi\varphi} \end{bmatrix}$$

$$g_{\theta\varphi} = g_{\varphi\theta} = 0$$

$g_{\theta\theta}$ independent φ

$$\boldsymbol{\Gamma^{\theta}_{\theta\theta} = 0}$$

With $\boldsymbol{\mu = \theta, \nu = \varphi}$

$$\boldsymbol{\Gamma^{\theta}_{\mu\nu}} = \frac{\mathbf{1}}{\mathbf{2}} g^{\theta\varphi} \left\{ \frac{\boldsymbol{\partial g_{\mu\varphi}}}{\boldsymbol{\partial x^{\nu}}} + \frac{\boldsymbol{\partial g_{\nu\varphi}}}{\boldsymbol{\partial x^{\mu}}} - \frac{\boldsymbol{\partial g_{\mu\nu}}}{\boldsymbol{\partial x^{\varphi}}} \right\}$$

$$\boldsymbol{\Gamma^{\theta}_{\theta\varphi}} = \frac{\mathbf{1}}{\mathbf{2}} g^{\theta\varphi} \left\{ \frac{\boldsymbol{\partial g_{\theta\varphi}}}{\boldsymbol{\partial x^{\varphi}}} + \frac{\boldsymbol{\partial g_{\varphi\varphi}}}{\boldsymbol{\partial x^{\theta}}} - \frac{\boldsymbol{\partial g_{\theta\varphi}}}{\boldsymbol{\partial x^{\varphi}}} \right\}$$

$$g_{\mu\nu} = \begin{bmatrix} R^2 & 0 \\ 0 & R^2\cos\theta^2 \end{bmatrix} => g_{\mu\nu} = \begin{bmatrix} g_{\theta\theta} & g_{\theta\varphi} \\ g_{\varphi\theta} & g_{\varphi\varphi} \end{bmatrix}$$

$$g^{\mu\nu} = \begin{bmatrix} \frac{1}{R^2} & 0 \\ 0 & \frac{1}{R^2\cos\theta^2} \end{bmatrix} => g^{\mu\nu} = \begin{bmatrix} g^{\theta\theta} & g^{\theta\varphi} \\ g^{\varphi\theta} & g^{\varphi\varphi} \end{bmatrix}$$

$$g_{\theta\varphi} = g_{\varphi\theta} = 0$$

$$g^{\theta\varphi} = 0$$

$$\frac{dg_{\varphi\varphi}}{d\theta} = -2R^2\cos\theta\sin\theta$$

$$\boldsymbol{\Gamma^{\theta}_{\theta\varphi}} = \frac{\mathbf{1}}{\mathbf{2}}\boldsymbol{g^{\theta\varphi}}\left\{\frac{\boldsymbol{\partial g_{\theta\varphi}}}{\boldsymbol{\partial x^{\varphi}}} + \frac{\boldsymbol{\partial g_{\varphi\varphi}}}{\boldsymbol{\partial x^{\theta}}} - \frac{\boldsymbol{\partial g_{\theta\varphi}}}{\boldsymbol{\partial x^{\varphi}}}\right\} = \frac{\mathbf{1}}{\mathbf{2}} \times 0\{0 - (2R^2\cos\theta\sin\theta) - 0\}$$

$$\boldsymbol{\Gamma^{\theta}_{\theta\varphi} = 0}$$

With $\boldsymbol{\mu = \varphi, \nu = \theta}$

$$\boldsymbol{\Gamma^{\theta}_{\mu\nu}} = \frac{\mathbf{1}}{\mathbf{2}}\boldsymbol{g^{\theta\varphi}}\left\{\frac{\boldsymbol{\partial g_{\mu\varphi}}}{\boldsymbol{\partial x^{\nu}}} + \frac{\boldsymbol{\partial g_{\nu\varphi}}}{\boldsymbol{\partial x^{\mu}}} - \frac{\boldsymbol{\partial g_{\mu\nu}}}{\boldsymbol{\partial x^{\varphi}}}\right\}$$

$$g_{\mu\nu} = \begin{bmatrix} R^2 & 0 \\ 0 & R^2\cos\theta^2 \end{bmatrix} => g_{\mu\nu} = \begin{bmatrix} g_{\theta\theta} & g_{\theta\varphi} \\ g_{\varphi\theta} & g_{\varphi\varphi} \end{bmatrix}$$

$$g^{\mu\nu} = \begin{bmatrix} \frac{1}{R^2} & 0 \\ 0 & \frac{1}{R^2\cos\theta^2} \end{bmatrix} => g^{\mu\nu} = \begin{bmatrix} g^{\theta\theta} & g^{\theta\varphi} \\ g^{\varphi\theta} & g^{\varphi\varphi} \end{bmatrix}$$

$$g^{\theta\varphi} = 0$$

$$\boldsymbol{\Gamma^{\theta}_{\varphi\theta} = 0}$$

With $\boldsymbol{\mu = \varphi, \nu = \varphi}$

$$\boldsymbol{\Gamma^{\theta}_{\mu\nu}} = \frac{\mathbf{1}}{\mathbf{2}}\boldsymbol{g^{\theta\varphi}}\left\{\frac{\boldsymbol{\partial g_{\mu\varphi}}}{\boldsymbol{\partial x^{\nu}}} + \frac{\boldsymbol{\partial g_{\nu\varphi}}}{\boldsymbol{\partial x^{\mu}}} - \frac{\boldsymbol{\partial g_{\mu\nu}}}{\boldsymbol{\partial x^{\varphi}}}\right\}$$

$$g_{\mu\nu} = \begin{bmatrix} R^2 & 0 \\ 0 & R^2\cos\theta^2 \end{bmatrix} => g_{\mu\nu} = \begin{bmatrix} g_{\theta\theta} & g_{\theta\varphi} \\ g_{\varphi\theta} & g_{\varphi\varphi} \end{bmatrix}$$

$$g^{\mu\nu} = \begin{bmatrix} \frac{1}{R^2} & 0 \\ 0 & \frac{1}{R^2\cos\theta^2} \end{bmatrix} => g^{\mu\nu} = \begin{bmatrix} g^{\theta\theta} & g^{\theta\varphi} \\ g^{\varphi\theta} & g^{\varphi\varphi} \end{bmatrix}$$

$$g^{\theta\varphi} = 0$$

$$\boldsymbol{\Gamma^{\theta}_{\varphi\varphi} = 0}$$

Step 2 $\boldsymbol{\Gamma^{\varphi}_{\mu\nu}}$

Two possible values for $\boldsymbol{\rho}$ either $\boldsymbol{\theta}$ either $\boldsymbol{\varphi}$
First value:

$$\boldsymbol{\Gamma^{\varphi}_{\mu\nu} = \frac{1}{2}g^{\varphi\theta}\left\{\frac{\partial g_{\mu\theta}}{\partial x^{\nu}} + \frac{\partial g_{\nu\theta}}{\partial x^{\mu}} - \frac{\partial g_{\mu\nu}}{\partial x^{\theta}}\right\}}$$

So, we have to calculate this for all the components of μ and ν possible.

With $\boldsymbol{\mu = \theta, \nu = \theta}$

$$\boldsymbol{\Gamma^{\varphi}_{\mu\nu} = \frac{1}{2}g^{\varphi\theta}\left\{\frac{\partial g_{\mu\theta}}{\partial x^{\nu}} + \frac{\partial g_{\nu\theta}}{\partial x^{\mu}} - \frac{\partial g_{\mu\nu}}{\partial x^{\theta}}\right\}}$$

$$\boldsymbol{\Gamma^{\varphi}_{\theta\theta} = \frac{1}{2}g^{\varphi\theta}\left\{\frac{\partial g_{\theta\theta}}{\partial x^{\theta}} + \frac{\partial g_{\theta\theta}}{\partial x^{\theta}} - \frac{\partial g_{\theta\theta}}{\partial x^{\theta}}\right\}}$$

$$g_{\mu\nu} = \begin{bmatrix} R^2 & 0 \\ 0 & R^2\cos\theta^2 \end{bmatrix} => g_{\mu\nu} = \begin{bmatrix} g_{\theta\theta} & g_{\theta\varphi} \\ g_{\varphi\theta} & g_{\varphi\varphi} \end{bmatrix}$$

$$g^{\mu\nu} = \begin{bmatrix} \frac{1}{R^2} & 0 \\ 0 & \frac{1}{R^2\cos\theta^2} \end{bmatrix} => g^{\mu\nu} = \begin{bmatrix} g^{\theta\theta} & g^{\theta\varphi} \\ g^{\varphi\theta} & g^{\varphi\varphi} \end{bmatrix}$$

$$g^{\varphi\theta} = 0$$

$$\boldsymbol{\Gamma^{\varphi}_{\theta\theta} = 0}$$

With $\boldsymbol{\mu = \theta, \nu = \varphi}$

$$\boldsymbol{\Gamma^{\varphi}_{\mu\nu} = \frac{1}{2}g^{\varphi\theta}\left\{\frac{\partial g_{\mu\theta}}{\partial x^{\nu}} + \frac{\partial g_{\nu\theta}}{\partial x^{\mu}} - \frac{\partial g_{\mu\nu}}{\partial x^{\theta}}\right\}}$$

$$g_{\mu\nu} = \begin{bmatrix} R^2 & 0 \\ 0 & R^2\cos\theta^2 \end{bmatrix} => g_{\mu\nu} = \begin{bmatrix} g_{\theta\theta} & g_{\theta\varphi} \\ g_{\varphi\theta} & g_{\varphi\varphi} \end{bmatrix}$$

$$g^{\mu\nu} = \begin{bmatrix} \frac{1}{R^2} & 0 \\ 0 & \frac{1}{R^2\cos\theta^2} \end{bmatrix} => g^{\mu\nu} = \begin{bmatrix} g^{\theta\theta} & g^{\theta\varphi} \\ g^{\varphi\theta} & g^{\varphi\varphi} \end{bmatrix}$$

$$g^{\varphi\theta} = 0$$

$$\boldsymbol{\Gamma^{\varphi}_{\theta\varphi} = 0}$$

With $\mu = \varphi, v = \theta$

$$\Gamma^{\varphi}_{\mu v} = \frac{1}{2} g^{\varphi\theta} \left\{ \frac{\partial g_{\mu\theta}}{\partial x^{v}} + \frac{\partial g_{v\theta}}{\partial x^{\mu}} - \frac{\partial g_{\mu v}}{\partial x^{\theta}} \right\}$$

$$g_{\mu v} = \begin{bmatrix} R^2 & 0 \\ 0 & R^2 \cos\theta^2 \end{bmatrix} => g_{\mu v} = \begin{bmatrix} g_{\theta\theta} & g_{\theta\varphi} \\ g_{\varphi\theta} & g_{\varphi\varphi} \end{bmatrix}$$

$$g^{\mu v} = \begin{bmatrix} \frac{1}{R^2} & 0 \\ 0 & \frac{1}{R^2 \cos\theta^2} \end{bmatrix} => g^{\mu v} = \begin{bmatrix} g^{\theta\theta} & g^{\theta\varphi} \\ g^{\varphi\theta} & g^{\varphi\varphi} \end{bmatrix}$$

$$g^{\varphi\theta} = 0$$

$$\Gamma^{\varphi}_{\varphi\theta} = 0$$

With $\mu = \varphi, v = \varphi$

$$\Gamma^{\varphi}_{\mu v} = \frac{1}{2} g^{\varphi\theta} \left\{ \frac{\partial g_{\mu\theta}}{\partial x^{v}} + \frac{\partial g_{v\theta}}{\partial x^{\mu}} - \frac{\partial g_{\mu v}}{\partial x^{\theta}} \right\}$$

$$g_{\mu v} = \begin{bmatrix} R^2 & 0 \\ 0 & R^2 \cos\theta^2 \end{bmatrix} => g_{\mu v} = \begin{bmatrix} g_{\theta\theta} & g_{\theta\varphi} \\ g_{\varphi\theta} & g_{\varphi\varphi} \end{bmatrix}$$

$$g^{\mu v} = \begin{bmatrix} \frac{1}{R^2} & 0 \\ 0 & \frac{1}{R^2 \cos\theta^2} \end{bmatrix} => g^{\mu v} = \begin{bmatrix} g^{\theta\theta} & g^{\theta\varphi} \\ g^{\varphi\theta} & g^{\varphi\varphi} \end{bmatrix}$$

$$g^{\varphi\theta} = 0$$

$$\Gamma^{\varphi}_{\varphi\varphi} = 0$$

Second value:

$$\Gamma^{\varphi}_{\mu v} = \frac{1}{2} g^{\varphi\varphi} \left\{ \frac{\partial g_{\mu\varphi}}{\partial x^{v}} + \frac{\partial g_{v\varphi}}{\partial x^{\mu}} - \frac{\partial g_{\mu v}}{\partial x^{\varphi}} \right\}$$

So, we have to calculate this for all the components of μ and v possible.

With $\mu = \theta, v = \theta$

$$\Gamma^{\varphi}_{\mu v} = \frac{1}{2} g^{\varphi\varphi} \left\{ \frac{\partial g_{\mu\varphi}}{\partial x^{v}} + \frac{\partial g_{v\varphi}}{\partial x^{\mu}} - \frac{\partial g_{\mu v}}{\partial x^{\varphi}} \right\}$$

$$\boldsymbol{\Gamma^{\varphi}_{\theta\theta}} = \frac{1}{2} g^{\varphi\varphi} \left\{ \frac{\partial g_{\theta\varphi}}{\partial x^{\theta}} + \frac{\partial g_{\theta\varphi}}{\partial x^{\theta}} - \frac{\partial g_{\theta\theta}}{\partial x^{\varphi}} \right\}$$

$$g_{\mu\nu} = \begin{bmatrix} R^2 & 0 \\ 0 & R^2\cos\theta^2 \end{bmatrix} => g_{\mu\nu} = \begin{bmatrix} g_{\theta\theta} & g_{\theta\varphi} \\ g_{\varphi\theta} & g_{\varphi\varphi} \end{bmatrix}$$

$$g^{\mu\nu} = \begin{bmatrix} \frac{1}{R^2} & 0 \\ 0 & \frac{1}{R^2\cos\theta^2} \end{bmatrix} => g^{\mu\nu} = \begin{bmatrix} g^{\theta\theta} & g^{\theta\varphi} \\ g^{\varphi\theta} & g^{\varphi\varphi} \end{bmatrix}$$

$$g_{\theta\varphi} = 0$$

$g_{\theta\theta}$ independent of φ

$$\boldsymbol{\Gamma^{\varphi}_{\theta\theta}} = 0$$

With $\mu = \theta, \nu = \varphi$

$$\boldsymbol{\Gamma^{\varphi}_{\mu\nu}} = \frac{1}{2} g^{\varphi\varphi} \left\{ \frac{\partial g_{\mu\varphi}}{\partial x^{\nu}} + \frac{\partial g_{\nu\varphi}}{\partial x^{\mu}} - \frac{\partial g_{\mu\nu}}{\partial x^{\varphi}} \right\}$$

$$\boldsymbol{\Gamma^{\varphi}_{\theta\varphi}} = \frac{1}{2} g^{\varphi\varphi} \left\{ \frac{\partial g_{\theta\varphi}}{\partial x^{\varphi}} + \frac{\partial g_{\varphi\varphi}}{\partial x^{\theta}} - \frac{\partial g_{\theta\varphi}}{\partial x^{\varphi}} \right\}$$

$$g_{\mu\nu} = \begin{bmatrix} R^2 & 0 \\ 0 & R^2\cos\theta^2 \end{bmatrix} => g_{\mu\nu} = \begin{bmatrix} g_{\theta\theta} & g_{\theta\varphi} \\ g_{\varphi\theta} & g_{\varphi\varphi} \end{bmatrix}$$

$$g^{\mu\nu} = \begin{bmatrix} \frac{1}{R^2} & 0 \\ 0 & \frac{1}{R^2\cos\theta^2} \end{bmatrix} => g^{\mu\nu} = \begin{bmatrix} g^{\theta\theta} & g^{\theta\varphi} \\ g^{\varphi\theta} & g^{\varphi\varphi} \end{bmatrix}$$

$$g_{\theta\varphi} = 0$$

$$\frac{dg_{\varphi\varphi}}{d\theta} = -2R^2\cos\theta\sin\theta$$

$$\boldsymbol{\Gamma^{\varphi}_{\theta\varphi}} = \frac{1}{2} g^{\varphi\varphi} \left\{ \frac{\partial g_{\theta\varphi}}{\partial x^{\varphi}} + \frac{\partial g_{\varphi\varphi}}{\partial x^{\theta}} - \frac{\partial g_{\theta\varphi}}{\partial x^{\varphi}} \right\} = \frac{1}{2} \frac{1}{R^2\cos\theta^2} \{0 + -2R^2\cos\theta\sin\theta - 0\}$$

$$\boldsymbol{\Gamma^{\varphi}_{\theta\varphi} = -\tan\theta}$$

With $\mu = \varphi, v = \theta$

$$\Gamma^{\varphi}_{\mu v} = \frac{1}{2} g^{\varphi\varphi} \left\{ \frac{\partial g_{\mu\varphi}}{\partial x^{v}} + \frac{\partial g_{v\varphi}}{\partial x^{\mu}} - \frac{\partial g_{\mu v}}{\partial x^{\varphi}} \right\}$$

$$\Gamma^{\varphi}_{\varphi\theta} = \frac{1}{2} g^{\varphi\varphi} \left\{ \frac{\partial g_{\varphi\varphi}}{\partial x^{\theta}} + \frac{\partial g_{\theta\varphi}}{\partial x^{\varphi}} - \frac{\partial g_{\varphi\theta}}{\partial x^{\varphi}} \right\}$$

$$g_{\mu v} = \begin{bmatrix} R^2 & 0 \\ 0 & R^2\cos\theta^2 \end{bmatrix} => g_{\mu v} = \begin{bmatrix} g_{\theta\theta} & g_{\theta\varphi} \\ g_{\varphi\theta} & g_{\varphi\varphi} \end{bmatrix}$$

$$g^{\mu v} = \begin{bmatrix} \frac{1}{R^2} & 0 \\ 0 & \frac{1}{R^2\cos\theta^2} \end{bmatrix} => g^{\mu v} = \begin{bmatrix} g^{\theta\theta} & g^{\theta\varphi} \\ g^{\varphi\theta} & g^{\varphi\varphi} \end{bmatrix}$$

$$g_{\theta\varphi} = 0$$

$$\frac{dg_{\varphi\varphi}}{d\theta} = -2R^2\cos\theta\sin\theta$$

$$\Gamma^{\varphi}_{\varphi\theta} = \frac{1}{2} g^{\varphi\varphi} \left\{ \frac{\partial g_{\varphi\varphi}}{\partial x^{\theta}} + \frac{\partial g_{\theta\varphi}}{\partial x^{\varphi}} - \frac{\partial g_{\varphi\theta}}{\partial x^{\varphi}} \right\} = \frac{1}{2} \frac{1}{R^2\cos\theta^2} \left\{ -2R^2\cos\theta\sin\theta + 0 - 0 \right\}$$

$$\Gamma^{\varphi}_{\varphi\theta} = -\tan\theta$$

With $\mu = \varphi, v = \varphi$

$$\Gamma^{\varphi}_{\mu v} = \frac{1}{2} g^{\varphi\varphi} \left\{ \frac{\partial g_{\mu\varphi}}{\partial x^{v}} + \frac{\partial g_{v\varphi}}{\partial x^{\mu}} - \frac{\partial g_{\mu v}}{\partial x^{\varphi}} \right\}$$

$$\Gamma^{\varphi}_{\varphi\varphi} = \frac{1}{2} g^{\varphi\varphi} \left\{ \frac{\partial g_{\varphi\varphi}}{\partial x^{\varphi}} + \frac{\partial g_{\varphi\varphi}}{\partial x^{\varphi}} - \frac{\partial g_{\varphi\varphi}}{\partial x^{\varphi}} \right\}$$

$$g_{\mu v} = \begin{bmatrix} R^2 & 0 \\ 0 & R^2\cos\theta^2 \end{bmatrix} => g_{\mu v} = \begin{bmatrix} g_{\theta\theta} & g_{\theta\varphi} \\ g_{\varphi\theta} & g_{\varphi\varphi} \end{bmatrix}$$

$$g^{\mu v} = \begin{bmatrix} \frac{1}{R^2} & 0 \\ 0 & \frac{1}{R^2\cos\theta^2} \end{bmatrix} => g^{\mu v} = \begin{bmatrix} g^{\theta\theta} & g^{\theta\varphi} \\ g^{\varphi\theta} & g^{\varphi\varphi} \end{bmatrix}$$

None of the terms depend on φ:

$$\boldsymbol{\Gamma}^{\varphi}_{\varphi\varphi} = \mathbf{0}$$

Summary of results (we keep only cases that does not give the four values null):

$$\boldsymbol{\Gamma}^{\theta}_{\mu\nu} = \begin{bmatrix} \boldsymbol{\Gamma}^{\theta}_{\theta\theta} & \boldsymbol{\Gamma}^{\theta}_{\theta\varphi} \\ \boldsymbol{\Gamma}^{\theta}_{\varphi\theta} & \boldsymbol{\Gamma}^{\theta}_{\varphi\varphi} \end{bmatrix} = \begin{bmatrix} \mathbf{0} & \mathbf{0} \\ \mathbf{0} & \mathbf{cos\theta sin\theta} \end{bmatrix}$$

$$\boldsymbol{\Gamma}^{\varphi}_{\mu\nu} = \begin{bmatrix} \boldsymbol{\Gamma}^{\varphi}_{\theta\theta} & \boldsymbol{\Gamma}^{\varphi}_{\theta\varphi} \\ \boldsymbol{\Gamma}^{\varphi}_{\varphi\theta} & \boldsymbol{\Gamma}^{\varphi}_{\varphi\varphi} \end{bmatrix} = \begin{bmatrix} \mathbf{0} & -\mathbf{tan\theta} \\ -\mathbf{tan\theta} & \mathbf{0} \end{bmatrix}$$

D.5 Calculation of the Components of Riemann's Curvature Tensor

The curvature tensor is written:

$$R^{\lambda}_{\mu\nu\alpha} = \boldsymbol{\Gamma}^{\lambda}_{\mu\alpha,\nu} - \boldsymbol{\Gamma}^{\lambda}_{\mu\nu,\alpha} + \boldsymbol{\Gamma}^{\lambda}_{\nu\eta}\boldsymbol{\Gamma}^{\eta}_{\mu\alpha} - \boldsymbol{\Gamma}^{\lambda}_{\alpha\eta}\boldsymbol{\Gamma}^{\eta}_{\mu\nu}$$

For memory, Ricci's tensor is written:

$$R_{\mu\nu} = R^{\lambda}_{\mu\nu\lambda}$$

Or:

$$R_{\mu\nu} = R^{\lambda}_{\mu\lambda\nu}$$

With a contraction on λ:
What we need in the end is contraction ($\alpha \to \lambda$):

$$R^{\lambda}_{\mu\nu\alpha} = \boldsymbol{\Gamma}^{\lambda}_{\mu\alpha,\nu} - \boldsymbol{\Gamma}^{\lambda}_{\mu\nu,\alpha} + \boldsymbol{\Gamma}^{\lambda}_{\nu\eta}\boldsymbol{\Gamma}^{\eta}_{\mu\alpha} - \boldsymbol{\Gamma}^{\lambda}_{\alpha\eta}\boldsymbol{\Gamma}^{\eta}_{\mu\nu}$$

$$R^{\lambda}_{\mu\nu\alpha} = \frac{\partial\boldsymbol{\Gamma}^{\lambda}_{\mu\alpha}}{\partial x^{\nu}} - \frac{\partial\boldsymbol{\Gamma}^{\lambda}_{\mu\nu}}{\partial x^{\alpha}} + \boldsymbol{\Gamma}^{\lambda}_{\nu\eta}\boldsymbol{\Gamma}^{\eta}_{\mu\alpha} - \boldsymbol{\Gamma}^{\lambda}_{\alpha\eta}\boldsymbol{\Gamma}^{\eta}_{\mu\nu}$$

We are going to have to calculate:

$$R^{\lambda}_{\theta\theta\lambda} = R_{\theta\theta} = \boldsymbol{\Gamma}^{\lambda}_{\theta\lambda,\theta} - \boldsymbol{\Gamma}^{\lambda}_{\theta\theta,\lambda} + \boldsymbol{\Gamma}^{\lambda}_{\theta\eta}\boldsymbol{\Gamma}^{\eta}_{\theta\lambda} - \boldsymbol{\Gamma}^{\lambda}_{\lambda\eta}\boldsymbol{\Gamma}^{\eta}_{\theta\theta}$$

$$R^{\lambda}_{\varphi\varphi\lambda} = R_{\varphi\varphi} = \boldsymbol{\Gamma}^{\lambda}_{\varphi\lambda,\varphi} - \boldsymbol{\Gamma}^{\lambda}_{\varphi\varphi,\lambda} + \boldsymbol{\Gamma}^{\lambda}_{\varphi\eta}\boldsymbol{\Gamma}^{\eta}_{\varphi\lambda} - \boldsymbol{\Gamma}^{\lambda}_{\lambda\eta}\boldsymbol{\Gamma}^{\eta}_{\varphi\varphi}$$

$$R^{\lambda}_{\theta\varphi\lambda} = R_{\theta\varphi} = R_{\varphi\theta} = \boldsymbol{\Gamma}^{\lambda}_{\theta\lambda,\varphi} - \boldsymbol{\Gamma}^{\lambda}_{\theta\varphi,\lambda} + \boldsymbol{\Gamma}^{\lambda}_{\varphi\eta}\boldsymbol{\Gamma}^{\eta}_{\theta\lambda} - \boldsymbol{\Gamma}^{\lambda}_{\lambda\eta}\boldsymbol{\Gamma}^{\eta}_{\theta\varphi}$$

$$R^{\lambda}_{\theta\lambda\theta} = \Gamma^{\lambda}_{\theta\theta,\lambda} - \Gamma^{\lambda}_{\theta\lambda,\theta} + \Gamma^{\lambda}_{\lambda\eta}\Gamma^{\eta}_{\theta\theta} - \Gamma^{\lambda}_{\theta\eta}\Gamma^{\eta}_{\theta\lambda}$$

$$R^{\lambda}_{\varphi\lambda\varphi} = \Gamma^{\lambda}_{\varphi\varphi,\lambda} - \Gamma^{\lambda}_{\varphi\lambda,\varphi} + \Gamma^{\lambda}_{\lambda\eta}\Gamma^{\eta}_{\varphi\varphi} - \Gamma^{\lambda}_{\varphi\eta}\Gamma^{\eta}_{\varphi\lambda}$$

(a) Study of the first term

$$R^{\lambda}_{\theta\theta\lambda} = R_{\theta\theta} = \underset{\textbf{T1}}{\Gamma^{\lambda}_{\theta\lambda,\theta}} - \underset{\textbf{T2}}{\Gamma^{\lambda}_{\theta\theta,\lambda}} + \underset{\textbf{T3}}{\Gamma^{\lambda}_{\theta\eta}\Gamma^{\eta}_{\theta\lambda}} - \underset{\textbf{T4}}{\Gamma^{\lambda}_{\lambda\eta}\Gamma^{\eta}_{\theta\theta}}$$

Term T1:

$$\Gamma^{\lambda}_{\theta\lambda,\theta} = \Gamma^{\theta}_{\theta\theta,\theta} + \Gamma^{\varphi}_{\theta\varphi,\theta}$$

$$\Gamma^{\theta}_{\mu\nu} = \begin{bmatrix} \Gamma^{\theta}_{\theta\theta} & \Gamma^{\theta}_{\theta\varphi} \\ \Gamma^{\theta}_{\varphi\theta} & \Gamma^{\theta}_{\varphi\varphi} \end{bmatrix} = \begin{bmatrix} 0 & 0 \\ 0 & \cos\theta\sin\theta \end{bmatrix}$$

$$\Gamma^{\varphi}_{\mu\nu} = \begin{bmatrix} \Gamma^{\varphi}_{\theta\theta} & \Gamma^{\varphi}_{\theta\varphi} \\ \Gamma^{\varphi}_{\varphi\theta} & \Gamma^{\varphi}_{\varphi\varphi} \end{bmatrix} = \begin{bmatrix} 0 & -\tan\theta \\ -\tan\theta & 0 \end{bmatrix}$$

$$\Gamma^{\lambda}_{\theta\lambda,\theta} = \Gamma^{\theta}_{\theta\theta,\theta} + \Gamma^{\varphi}_{\theta\varphi,\theta} = \frac{\partial\Gamma^{\theta}_{\theta\theta}}{\partial\theta} + \frac{\partial\Gamma^{\varphi}_{\theta\varphi}}{\partial\theta} = 0 - \frac{\partial\tan\theta}{\partial\theta} = -\frac{1}{\cos^2\theta}$$

$$T_1 = -\frac{1}{\cos^2\theta}$$

Term T2:

$$\Gamma^{\lambda}_{\theta\theta,\lambda} = \Gamma^{\theta}_{\theta\theta,\theta} + \Gamma^{\varphi}_{\theta\theta,\varphi}$$

$$\Gamma^{\theta}_{\mu\nu} = \begin{bmatrix} \Gamma^{\theta}_{\theta\theta} & \Gamma^{\theta}_{\theta\varphi} \\ \Gamma^{\theta}_{\varphi\theta} & \Gamma^{\theta}_{\varphi\varphi} \end{bmatrix} = \begin{bmatrix} 0 & 0 \\ 0 & \cos\theta\sin\theta \end{bmatrix}$$

$$\Gamma^{\varphi}_{\mu\nu} = \begin{bmatrix} \Gamma^{\varphi}_{\theta\theta} & \Gamma^{\varphi}_{\theta\varphi} \\ \Gamma^{\varphi}_{\varphi\theta} & \Gamma^{\varphi}_{\varphi\varphi} \end{bmatrix} = \begin{bmatrix} 0 & -\tan\theta \\ -\tan\theta & 0 \end{bmatrix}$$

$$\Gamma^{\lambda}_{\theta\theta,\lambda} = \Gamma^{\theta}_{\theta\theta,\theta} + \Gamma^{\varphi}_{\theta\theta,\varphi} = \frac{\partial\Gamma^{\theta}_{\theta\theta}}{\partial\theta} + \frac{\partial\Gamma^{\varphi}_{\theta\theta}}{\partial\varphi} = 0 + 0 = 0$$

$$\boldsymbol{T_2 = 0}$$

Term T3:

$$\Gamma^{\lambda}_{\theta\eta}\Gamma^{\eta}_{\theta\lambda}$$

Attention we have two sums entangled, one on λ and one on η.

$$\Gamma^{\lambda}_{\theta\eta}\Gamma^{\eta}_{\theta\lambda} = \Gamma^{\lambda}_{\theta\theta}\Gamma^{\theta}_{\theta\lambda} + \Gamma^{\lambda}_{\theta\varphi}\Gamma^{\varphi}_{\theta\lambda}$$

$$\Gamma^{\lambda}_{\theta\eta}\Gamma^{\eta}_{\theta\lambda} = \Gamma^{\theta}_{\theta\theta}\Gamma^{\theta}_{\theta\theta} + \Gamma^{\varphi}_{\theta\theta}\Gamma^{\theta}_{\theta\varphi} + \Gamma^{\theta}_{\theta\varphi}\Gamma^{\varphi}_{\theta\theta} + \Gamma^{\varphi}_{\theta\varphi}\Gamma^{\varphi}_{\theta\varphi}$$

$$\Gamma^{\lambda}_{\theta\eta}\Gamma^{\eta}_{\theta\lambda} = 0 \times 0 + 0 \times 0 + 0 \times 0 - \tan\theta \times -\tan\theta$$

$$\boldsymbol{\Gamma^{\theta}_{\mu\nu} = \begin{bmatrix} \Gamma^{\theta}_{\theta\theta} & \Gamma^{\theta}_{\theta\varphi} \\ \Gamma^{\theta}_{\varphi\theta} & \Gamma^{\theta}_{\varphi\varphi} \end{bmatrix} = \begin{bmatrix} 0 & 0 \\ 0 & \cos\theta\sin\theta \end{bmatrix}}$$

$$\boldsymbol{\Gamma^{\varphi}_{\mu\nu} = \begin{bmatrix} \Gamma^{\varphi}_{\theta\theta} & \Gamma^{\varphi}_{\theta\varphi} \\ \Gamma^{\varphi}_{\varphi\theta} & \Gamma^{\varphi}_{\varphi\varphi} \end{bmatrix} = \begin{bmatrix} 0 & -\tan\theta \\ -\tan\theta & 0 \end{bmatrix}}$$

$$\boldsymbol{T_3 = \tan^2\theta}$$

Term T4:

$$\Gamma^{\lambda}_{\lambda\eta}\Gamma^{\eta}_{\theta\theta}$$

Attention we have two sums entangled, one on λ and one on η:

$$\Gamma^{\lambda}_{\lambda\eta}\Gamma^{\eta}_{\theta\theta} = \Gamma^{\theta}_{\theta\eta}\Gamma^{\eta}_{\theta\theta} + \Gamma^{\varphi}_{\varphi\eta}\Gamma^{\eta}_{\theta\theta}$$

$$\Gamma^{\lambda}_{\lambda\eta}\Gamma^{\eta}_{\theta\theta} = \Gamma^{\theta}_{\theta\theta}\Gamma^{\theta}_{\theta\theta} + \Gamma^{\theta}_{\theta\varphi}\Gamma^{\varphi}_{\theta\theta} + \Gamma^{\varphi}_{\varphi\theta}\Gamma^{\theta}_{\theta\theta} + \Gamma^{\varphi}_{\varphi\varphi}\Gamma^{\varphi}_{\theta\theta}$$

$$\Gamma^{\lambda}_{\lambda\eta}\Gamma^{\eta}_{\theta\theta} = 0 \times 0 + 0 \times 0 - \tan\theta \times 0 + 0 \times 0 = 0$$

$$\Gamma^{\theta}_{\mu\nu} = \begin{bmatrix} \Gamma^{\theta}_{\theta\theta} & \Gamma^{\theta}_{\theta\varphi} \\ \Gamma^{\theta}_{\varphi\theta} & \Gamma^{\theta}_{\varphi\varphi} \end{bmatrix} = \begin{bmatrix} 0 & 0 \\ 0 & \cos\theta\sin\theta \end{bmatrix}$$

$$\Gamma^{\varphi}_{\mu\nu} = \begin{bmatrix} \Gamma^{\varphi}_{\theta\theta} & \Gamma^{\varphi}_{\theta\varphi} \\ \Gamma^{\varphi}_{\varphi\theta} & \Gamma^{\varphi}_{\varphi\varphi} \end{bmatrix} = \begin{bmatrix} 0 & -\tan\theta \\ -\tan\theta & 0 \end{bmatrix}$$

$$T_4 = 0$$

Summary

$$T_1 = -\frac{1}{\cos^2\theta}, T_2 = 0, T_3 = \tan^2\theta, T_4 = 0$$

$$R^{\lambda}_{\theta\theta\lambda} = R_{\theta\theta} = \underset{\mathbf{T1}}{\Gamma^{\lambda}_{\theta\lambda,\theta}} - \underset{\mathbf{T2}}{\Gamma^{\lambda}_{\theta\theta,\lambda}} + \underset{\mathbf{T3}}{\Gamma^{\lambda}_{\theta\eta}\Gamma^{\eta}_{\theta\lambda}} - \underset{\mathbf{T4}}{\Gamma^{\lambda}_{\lambda\eta}\Gamma^{\eta}_{\theta\theta}}$$

$$R^{\lambda}_{\theta\theta\lambda} = -\frac{1}{\cos^2\theta} - 0 + \tan^2\theta - 0$$

$$R^{\lambda}_{\theta\theta\lambda} = -\frac{1}{\cos^2\theta} + \frac{\sin^2\theta}{\cos^2\theta} = \frac{-\cos^2\theta - \sin^2\theta}{\cos^2\theta} + \frac{\sin^2\theta}{\cos^2\theta} = -1$$

$$R^{\lambda}_{\theta\theta\lambda} = -1$$

(b) Study of the second term

$$R^{\lambda}_{\varphi\varphi\lambda} = R_{\varphi\varphi} = \underset{\mathbf{T1}}{\Gamma^{\lambda}_{\varphi\lambda,\varphi}} - \underset{\mathbf{T2}}{\Gamma^{\lambda}_{\varphi\varphi,\lambda}} + \underset{\mathbf{T3}}{\Gamma^{\lambda}_{\varphi\eta}\Gamma^{\eta}_{\varphi\lambda}} - \underset{\mathbf{T4}}{\Gamma^{\lambda}_{\lambda\eta}\Gamma^{\eta}_{\varphi\varphi}}$$

Term T1:

$$\Gamma^{\lambda}_{\varphi\lambda,\theta} = \Gamma^{\theta}_{\varphi\theta,\varphi} + \Gamma^{\varphi}_{\varphi\varphi,\varphi}$$

$$\Gamma^{\theta}_{\mu\nu} = \begin{bmatrix} \Gamma^{\theta}_{\theta\theta} & \Gamma^{\theta}_{\theta\varphi} \\ \Gamma^{\theta}_{\varphi\theta} & \Gamma^{\theta}_{\varphi\varphi} \end{bmatrix} = \begin{bmatrix} 0 & 0 \\ 0 & \cos\theta\sin\theta \end{bmatrix}$$

$$\Gamma^{\varphi}_{\mu\nu} = \begin{bmatrix} \Gamma^{\varphi}_{\theta\theta} & \Gamma^{\varphi}_{\theta\varphi} \\ \Gamma^{\varphi}_{\varphi\theta} & \Gamma^{\varphi}_{\varphi\varphi} \end{bmatrix} = \begin{bmatrix} 0 & -\tan\theta \\ -\tan\theta & 0 \end{bmatrix}$$

$$\Gamma^{\lambda}_{\varphi\lambda,\theta} = \Gamma^{\theta}_{\varphi\theta,\varphi} + \Gamma^{\varphi}_{\varphi\varphi,\varphi} = \frac{\partial \Gamma^{\theta}_{\varphi\theta}}{\partial \varphi} + \frac{\partial \Gamma^{\varphi}_{\varphi\varphi}}{\partial \varphi} = 0 - 0 = 0$$

$$\boldsymbol{T_1 = 0}$$

Term T2:

$$\Gamma^{\lambda}_{\varphi\varphi,\lambda}$$

$$\Gamma^{\lambda}_{\varphi\varphi,\lambda} = \Gamma^{\theta}_{\varphi\varphi,\theta} + \Gamma^{\varphi}_{\varphi\varphi,\varphi}$$

$$\boldsymbol{\Gamma^{\theta}_{\mu\nu} = \begin{bmatrix} \Gamma^{\theta}_{\theta\theta} & \Gamma^{\theta}_{\theta\varphi} \\ \Gamma^{\theta}_{\varphi\theta} & \Gamma^{\theta}_{\varphi\varphi} \end{bmatrix} = \begin{bmatrix} 0 & 0 \\ 0 & \cos\theta\sin\theta \end{bmatrix}}$$

$$\boldsymbol{\Gamma^{\varphi}_{\mu\nu} = \begin{bmatrix} \Gamma^{\varphi}_{\theta\theta} & \Gamma^{\varphi}_{\theta\varphi} \\ \Gamma^{\varphi}_{\varphi\theta} & \Gamma^{\varphi}_{\varphi\varphi} \end{bmatrix} = \begin{bmatrix} 0 & -\tan\theta \\ -\tan\theta & 0 \end{bmatrix}}$$

$$\Gamma^{\lambda}_{\varphi\varphi,\lambda} = \Gamma^{\theta}_{\varphi\varphi,\theta} + \Gamma^{\varphi}_{\varphi\varphi,\varphi} = \frac{\partial \Gamma^{\theta}_{\varphi\varphi}}{\partial \theta} + \frac{\partial \Gamma^{\varphi}_{\varphi\varphi}}{\partial \varphi} = \frac{\partial(\cos\theta\sin\theta)}{\partial \theta} + 0 = -\sin^2\theta + \cos^2\theta$$

$$\boldsymbol{T_2 = \cos^2\theta - \sin^2\theta}$$

Term T3:

$$\Gamma^{\lambda}_{\varphi\eta}\Gamma^{\eta}_{\varphi\lambda}$$

Attention we have two sums entangled, one on λ and one on η.

$$\Gamma^{\lambda}_{\varphi\eta}\Gamma^{\eta}_{\varphi\lambda} = \Gamma^{\lambda}_{\varphi\theta}\Gamma^{\theta}_{\varphi\lambda} + \Gamma^{\lambda}_{\varphi\varphi}\Gamma^{\varphi}_{\varphi\lambda}$$

$$\Gamma^{\lambda}_{\varphi\eta}\Gamma^{\eta}_{\varphi\lambda} = \Gamma^{\theta}_{\varphi\theta}\Gamma^{\theta}_{\varphi\theta} + \Gamma^{\varphi}_{\varphi\theta}\Gamma^{\theta}_{\varphi\varphi} + \Gamma^{\theta}_{\varphi\varphi}\Gamma^{\varphi}_{\varphi\theta} + \Gamma^{\varphi}_{\varphi\varphi}\Gamma^{\varphi}_{\varphi\varphi}$$

$$\Gamma^{\lambda}_{\varphi\eta}\Gamma^{\eta}_{\varphi\lambda} = 0 \times 0 - \tan\theta \times \cos\theta\sin\theta - \cos\theta\sin\theta \times \tan\theta + 0 \times 0$$

$$\Gamma^{\theta}_{\mu\nu} = \begin{bmatrix} \Gamma^{\theta}_{\theta\theta} & \Gamma^{\theta}_{\theta\varphi} \\ \Gamma^{\theta}_{\varphi\theta} & \Gamma^{\theta}_{\varphi\varphi} \end{bmatrix} = \begin{bmatrix} 0 & 0 \\ 0 & \cos\theta\sin\theta \end{bmatrix}$$

$$\Gamma^{\varphi}_{\mu\nu} = \begin{bmatrix} \Gamma^{\varphi}_{\theta\theta} & \Gamma^{\varphi}_{\theta\varphi} \\ \Gamma^{\varphi}_{\varphi\theta} & \Gamma^{\varphi}_{\varphi\varphi} \end{bmatrix} = \begin{bmatrix} 0 & -\tan\theta \\ -\tan\theta & 0 \end{bmatrix}$$

$$T_3 = -2\sin^2\theta$$

Term T4:

$$\Gamma^{\lambda}_{\lambda\eta}\Gamma^{\eta}_{\varphi\varphi}$$

Attention we have two sums entangled, one on λ and one on η:

$$\Gamma^{\lambda}_{\lambda\eta}\Gamma^{\eta}_{\varphi\varphi} = \Gamma^{\theta}_{\theta\eta}\Gamma^{\eta}_{\varphi\varphi} + \Gamma^{\varphi}_{\varphi\eta}\Gamma^{\eta}_{\varphi\varphi}$$

$$\Gamma^{\lambda}_{\lambda\eta}\Gamma^{\eta}_{\varphi\varphi} = \Gamma^{\theta}_{\theta\theta}\Gamma^{\theta}_{\varphi\varphi} + \Gamma^{\theta}_{\theta\varphi}\Gamma^{\varphi}_{\varphi\varphi} + \Gamma^{\varphi}_{\varphi\theta}\Gamma^{\theta}_{\varphi\varphi} + \Gamma^{\varphi}_{\varphi\varphi}\Gamma^{\varphi}_{\varphi\varphi}$$

$$\Gamma^{\lambda}_{\lambda\eta}\Gamma^{\eta}_{\varphi\varphi} = 0 \times \cos\theta\sin\theta + 0 \times 0 - \tan\theta \times \cos\theta\sin\theta + 0 \times 0 = -\sin^2\theta$$

$$\Gamma^{\theta}_{\mu\nu} = \begin{bmatrix} \Gamma^{\theta}_{\theta\theta} & \Gamma^{\theta}_{\theta\varphi} \\ \Gamma^{\theta}_{\varphi\theta} & \Gamma^{\theta}_{\varphi\varphi} \end{bmatrix} = \begin{bmatrix} 0 & 0 \\ 0 & \cos\theta\sin\theta \end{bmatrix}$$

$$\Gamma^{\varphi}_{\mu\nu} = \begin{bmatrix} \Gamma^{\varphi}_{\theta\theta} & \Gamma^{\varphi}_{\theta\varphi} \\ \Gamma^{\varphi}_{\varphi\theta} & \Gamma^{\varphi}_{\varphi\varphi} \end{bmatrix} = \begin{bmatrix} 0 & -\tan\theta \\ -\tan\theta & 0 \end{bmatrix}$$

$$T_4 = -\sin^2\theta$$

Summary:

$$T_1 = 0, T_2 = \cos^2\theta - \sin^2\theta, T_3 = -2\sin^2\theta, T_4 = -\sin^2\theta$$

$$R^{\lambda}_{\varphi\varphi\lambda} = R_{\varphi\varphi} = \underset{\textbf{T1}}{\Gamma^{\lambda}_{\varphi\lambda,\varphi}} - \underset{\textbf{T2}}{\Gamma^{\lambda}_{\varphi\varphi,\lambda}} + \underset{\textbf{T3}}{\Gamma^{\lambda}_{\varphi\eta}\Gamma^{\eta}_{\varphi\lambda}} - \underset{\textbf{T4}}{\Gamma^{\lambda}_{\lambda\eta}\Gamma^{\eta}_{\varphi\varphi}}$$

$$R^{\lambda}_{\varphi\varphi\lambda} = 0 - \cos^2\theta + \sin^2\theta - 2\sin^2\theta + \sin^2\theta = -\cos^2\theta$$

$$\boldsymbol{R^{\lambda}_{\varphi\varphi\lambda} = -\cos^2\theta}$$

(c) Study of the third term

$$R^{\lambda}_{\theta\varphi\lambda} = R_{\theta\varphi} = R_{\varphi\theta} = \underset{\textbf{T1}}{\Gamma^{\lambda}_{\theta\lambda,\varphi}} - \underset{\textbf{T2}}{\Gamma^{\lambda}_{\theta\varphi,\lambda}} + \underset{\textbf{T3}}{\Gamma^{\lambda}_{\varphi\eta}\Gamma^{\eta}_{\theta\lambda}} - \underset{\textbf{T4}}{\Gamma^{\lambda}_{\lambda\eta}\Gamma^{\eta}_{\theta\varphi}}$$

Term T1:

$$\Gamma^{\lambda}_{\theta\lambda,\theta} = \Gamma^{\theta}_{\theta\theta,\varphi} + \Gamma^{\varphi}_{\theta\varphi,\varphi}$$

$$\Gamma^{\theta}_{\mu\nu} = \begin{bmatrix} \Gamma^{\theta}_{\theta\theta} & \Gamma^{\theta}_{\theta\varphi} \\ \Gamma^{\theta}_{\varphi\theta} & \Gamma^{\theta}_{\varphi\varphi} \end{bmatrix} = \begin{bmatrix} 0 & 0 \\ 0 & \cos\theta\sin\theta \end{bmatrix}$$

$$\Gamma^{\varphi}_{\mu\nu} = \begin{bmatrix} \Gamma^{\varphi}_{\theta\theta} & \Gamma^{\varphi}_{\theta\varphi} \\ \Gamma^{\varphi}_{\varphi\theta} & \Gamma^{\varphi}_{\varphi\varphi} \end{bmatrix} = \begin{bmatrix} 0 & -\tan\theta \\ -\tan\theta & 0 \end{bmatrix}$$

$$\Gamma^{\lambda}_{\theta\lambda,\theta} = \Gamma^{\theta}_{\theta\theta,\varphi} + \Gamma^{\varphi}_{\theta\varphi,\varphi} = \frac{\partial\Gamma^{\theta}_{\theta\theta}}{\partial\varphi} + \frac{\partial\Gamma^{\varphi}_{\theta\varphi}}{\partial\varphi} = 0 - 0 = 0$$

$$T_1 = 0$$

Term T2:

$$\Gamma^{\lambda}_{\theta\varphi,\lambda}$$

$$\Gamma^{\lambda}_{\theta\varphi,\lambda} = \Gamma^{\theta}_{\theta\varphi,\theta} + \Gamma^{\varphi}_{\theta\varphi,\varphi}$$

$$\Gamma^{\theta}_{\mu\nu} = \begin{bmatrix} \Gamma^{\theta}_{\theta\theta} & \Gamma^{\theta}_{\theta\varphi} \\ \Gamma^{\theta}_{\varphi\theta} & \Gamma^{\theta}_{\varphi\varphi} \end{bmatrix} = \begin{bmatrix} 0 & 0 \\ 0 & \cos\theta\sin\theta \end{bmatrix}$$

$$\boldsymbol{\Gamma}^{\varphi}_{\mu\nu} = \begin{bmatrix} \boldsymbol{\Gamma}^{\varphi}_{\theta\theta} & \boldsymbol{\Gamma}^{\varphi}_{\theta\varphi} \\ \boldsymbol{\Gamma}^{\varphi}_{\varphi\theta} & \boldsymbol{\Gamma}^{\varphi}_{\varphi\varphi} \end{bmatrix} = \begin{bmatrix} \mathbf{0} & -\mathbf{tan}\boldsymbol{\theta} \\ -\mathbf{tan}\boldsymbol{\theta} & \mathbf{0} \end{bmatrix}$$

$$\Gamma^{\lambda}_{\theta\varphi,\lambda} = \Gamma^{\theta}_{\theta\varphi,\theta} + \Gamma^{\varphi}_{\theta\varphi,\varphi} = \frac{\partial\Gamma^{\theta}_{\theta\varphi}}{\partial\theta} + \frac{\partial\Gamma^{\varphi}_{\theta\varphi}}{\partial\varphi} = 0 - \frac{\partial(\tan\theta)}{\partial\varphi} = 0$$

$$\boldsymbol{T_2} = \mathbf{0}$$

Term T3:

$$\Gamma^{\lambda}_{\varphi\eta}\Gamma^{\eta}_{\theta\lambda}$$

Attention we have two sums entangled, one on λ and one on η.

$$\Gamma^{\lambda}_{\varphi\eta}\Gamma^{\eta}_{\theta\lambda} = \Gamma^{\lambda}_{\varphi\theta}\Gamma^{\theta}_{\theta\lambda} + \Gamma^{\lambda}_{\varphi\varphi}\Gamma^{\varphi}_{\theta\lambda}$$

$$\Gamma^{\lambda}_{\varphi\eta}\Gamma^{\eta}_{\theta\lambda} = \Gamma^{\theta}_{\varphi\theta}\Gamma^{\theta}_{\theta\theta} + \Gamma^{\varphi}_{\varphi\theta}\Gamma^{\theta}_{\theta\varphi} + \Gamma^{\theta}_{\varphi\varphi}\Gamma^{\varphi}_{\theta\theta} + \Gamma^{\varphi}_{\varphi\varphi}\Gamma^{\varphi}_{\theta\varphi}$$

$$\Gamma^{\lambda}_{\varphi\eta}\Gamma^{\eta}_{\theta\lambda} = 0 \times 0 - \tan\theta \times 0 + \cos\theta\sin\theta \times 0 + 0 \times -\tan\theta = 0$$

$$\boldsymbol{\Gamma}^{\theta}_{\mu\nu} = \begin{bmatrix} \boldsymbol{\Gamma}^{\theta}_{\theta\theta} & \boldsymbol{\Gamma}^{\theta}_{\theta\varphi} \\ \boldsymbol{\Gamma}^{\theta}_{\varphi\theta} & \boldsymbol{\Gamma}^{\theta}_{\varphi\varphi} \end{bmatrix} = \begin{bmatrix} \mathbf{0} & \mathbf{0} \\ \mathbf{0} & \mathbf{cos}\boldsymbol{\theta}\mathbf{sin}\boldsymbol{\theta} \end{bmatrix}$$

$$\boldsymbol{\Gamma}^{\varphi}_{\mu\nu} = \begin{bmatrix} \boldsymbol{\Gamma}^{\varphi}_{\theta\theta} & \boldsymbol{\Gamma}^{\varphi}_{\theta\varphi} \\ \boldsymbol{\Gamma}^{\varphi}_{\varphi\theta} & \boldsymbol{\Gamma}^{\varphi}_{\varphi\varphi} \end{bmatrix} = \begin{bmatrix} \mathbf{0} & -\mathbf{tan}\boldsymbol{\theta} \\ -\mathbf{tan}\boldsymbol{\theta} & \mathbf{0} \end{bmatrix}$$

$$\boldsymbol{T_3} = \mathbf{0}$$

Term T4:

$$\Gamma^{\lambda}_{\lambda\eta}\Gamma^{\eta}_{\theta\varphi}$$

Attention we have two sums entangled, one on λ and one on η:

$$\Gamma^{\lambda}_{\lambda\eta}\Gamma^{\eta}_{\theta\varphi} = \Gamma^{\theta}_{\theta\eta}\Gamma^{\eta}_{\theta\varphi} + \Gamma^{\varphi}_{\varphi\eta}\Gamma^{\eta}_{\theta\varphi}$$

$$\Gamma^{\lambda}_{\lambda\eta}\Gamma^{\eta}_{\theta\varphi} = \Gamma^{\theta}_{\theta\theta}\Gamma^{\theta}_{\theta\varphi} + \Gamma^{\theta}_{\theta\varphi}\Gamma^{\varphi}_{\theta\varphi} + \Gamma^{\varphi}_{\varphi\theta}\Gamma^{\theta}_{\theta\varphi} + \Gamma^{\varphi}_{\varphi\varphi}\Gamma^{\varphi}_{\theta\varphi}$$

$$\Gamma^{\lambda}_{\lambda\eta}\Gamma^{\eta}_{\varphi\varphi} = 0 \times 0 + 0 \times -\tan\theta + \tan\theta \times 0 + 0 \times \tan\theta = 0$$

$$\boldsymbol{\Gamma}^{\theta}_{\mu\nu} = \begin{bmatrix} \Gamma^{\theta}_{\theta\theta} & \Gamma^{\theta}_{\theta\varphi} \\ \Gamma^{\theta}_{\varphi\theta} & \Gamma^{\theta}_{\varphi\varphi} \end{bmatrix} = \begin{bmatrix} 0 & 0 \\ 0 & \cos\theta\sin\theta \end{bmatrix}$$

$$\boldsymbol{\Gamma}^{\varphi}_{\mu\nu} = \begin{bmatrix} \Gamma^{\varphi}_{\theta\theta} & \Gamma^{\varphi}_{\theta\varphi} \\ \Gamma^{\varphi}_{\varphi\theta} & \Gamma^{\varphi}_{\varphi\varphi} \end{bmatrix} = \begin{bmatrix} 0 & -\tan\theta \\ -\tan\theta & 0 \end{bmatrix}$$

$$T_4 = 0$$

Summary:

$$T_1 = 0, T_2 = 0, T_3 = 0, T_4 = 0$$

$$R^{\lambda}_{\theta\varphi\lambda} = R_{\theta\varphi} = R_{\varphi\theta} = \underset{\textbf{T1}}{\Gamma^{\lambda}_{\theta\lambda,\varphi}} - \underset{\textbf{T2}}{\Gamma^{\lambda}_{\theta\varphi,\lambda}} + \underset{\textbf{T3}}{\Gamma^{\lambda}_{\theta\eta}\Gamma^{\eta}_{\varphi\lambda}} - \underset{\textbf{T4}}{\Gamma^{\lambda}_{\lambda\eta}\Gamma^{\eta}_{\theta\varphi}}$$

$$R^{\lambda}_{\theta\varphi\lambda} = R_{\theta\varphi} = 0 + 0 - 0 - 0 = 0$$

$$R^{\lambda}_{\theta\varphi\lambda} = R_{\theta\varphi} = 0$$

(d) Fourth-term study

$$R^{\lambda}_{\mu\nu\alpha} = \Gamma^{\lambda}_{\mu\alpha,\nu} - \Gamma^{\lambda}_{\mu\nu,\alpha} + \Gamma^{\lambda}_{\nu\eta}\Gamma^{\eta}_{\mu\alpha} - \Gamma^{\lambda}_{\alpha\eta}\Gamma^{\eta}_{\mu\nu}$$
$$R^{\lambda}_{\theta\nu\theta} = \Gamma^{\lambda}_{\theta\theta,\nu} - \Gamma^{\lambda}_{\theta\nu,\theta} + \Gamma^{\lambda}_{\nu\eta}\Gamma^{\eta}_{\theta\theta} - \Gamma^{\lambda}_{\theta\eta}\Gamma^{\eta}_{\theta\nu}$$

And with the contraction:

$$R^{\lambda}_{\theta\lambda\theta} = \underset{\textbf{T1}}{\Gamma^{\lambda}_{\theta\theta,\lambda}} - \underset{\textbf{T2}}{\Gamma^{\lambda}_{\theta\lambda,\theta}} + \underset{\textbf{T3}}{\Gamma^{\lambda}_{\lambda\eta}\Gamma^{\eta}_{\theta\theta}} - \underset{\textbf{T4}}{\Gamma^{\lambda}_{\theta\eta}\Gamma^{\eta}_{\theta\lambda}}$$

Term T1:

$$\Gamma^{\lambda}_{\theta\theta,\lambda} = \Gamma^{\theta}_{\theta\theta,\theta} + \Gamma^{\varphi}_{\theta\theta,\varphi}$$

$$\boldsymbol{\Gamma^{\theta}_{\mu\nu}} = \begin{bmatrix} \boldsymbol{\Gamma^{\theta}_{\theta\theta}} & \boldsymbol{\Gamma^{\theta}_{\theta\varphi}} \\ \boldsymbol{\Gamma^{\theta}_{\varphi\theta}} & \boldsymbol{\Gamma^{\theta}_{\varphi\varphi}} \end{bmatrix} = \begin{bmatrix} \mathbf{0} & \mathbf{0} \\ \mathbf{0} & \mathbf{cos}\boldsymbol{\theta}\mathbf{sin}\boldsymbol{\theta} \end{bmatrix}$$

$$\boldsymbol{\Gamma^{\varphi}_{\mu\nu}} = \begin{bmatrix} \boldsymbol{\Gamma^{\varphi}_{\theta\theta}} & \boldsymbol{\Gamma^{\varphi}_{\theta\varphi}} \\ \boldsymbol{\Gamma^{\varphi}_{\varphi\theta}} & \boldsymbol{\Gamma^{\varphi}_{\varphi\varphi}} \end{bmatrix} = \begin{bmatrix} \mathbf{0} & -\mathbf{tan}\boldsymbol{\theta} \\ -\mathbf{tan}\boldsymbol{\theta} & \mathbf{0} \end{bmatrix}$$

$$\Gamma^{\lambda}_{\theta\theta,\lambda} = \Gamma^{\theta}_{\theta\theta,\theta} + \Gamma^{\varphi}_{\theta\theta,\varphi} = \frac{\partial \Gamma^{\theta}_{\theta\theta}}{\partial\theta} + \frac{\partial \Gamma^{\varphi}_{\theta\theta}}{\partial\varphi} = 0 + 0 = 0$$

$$\boldsymbol{T_1} = \mathbf{0}$$

Term T2:

$$\Gamma^{\lambda}_{\theta\lambda,\theta} = \Gamma^{\theta}_{\theta\theta,\theta} + \Gamma^{\varphi}_{\theta\varphi,\theta}$$

$$\boldsymbol{\Gamma^{\theta}_{\mu\nu}} = \begin{bmatrix} \boldsymbol{\Gamma^{\theta}_{\theta\theta}} & \boldsymbol{\Gamma^{\theta}_{\theta\varphi}} \\ \boldsymbol{\Gamma^{\theta}_{\varphi\theta}} & \boldsymbol{\Gamma^{\theta}_{\varphi\varphi}} \end{bmatrix} = \begin{bmatrix} \mathbf{0} & \mathbf{0} \\ \mathbf{0} & \mathbf{cos}\boldsymbol{\theta}\mathbf{sin}\boldsymbol{\theta} \end{bmatrix}$$

$$\boldsymbol{\Gamma^{\varphi}_{\mu\nu}} = \begin{bmatrix} \boldsymbol{\Gamma^{\varphi}_{\theta\theta}} & \boldsymbol{\Gamma^{\varphi}_{\theta\varphi}} \\ \boldsymbol{\Gamma^{\varphi}_{\varphi\theta}} & \boldsymbol{\Gamma^{\varphi}_{\varphi\varphi}} \end{bmatrix} = \begin{bmatrix} \mathbf{0} & -\mathbf{tan}\boldsymbol{\theta} \\ -\mathbf{tan}\boldsymbol{\theta} & \mathbf{0} \end{bmatrix}$$

$$\Gamma^{\lambda}_{\theta\lambda,\theta} = \Gamma^{\theta}_{\theta\theta,\theta} + \Gamma^{\varphi}_{\theta\varphi,\theta} = \frac{\partial \Gamma^{\theta}_{\theta\theta}}{\partial\theta} + \frac{\partial \Gamma^{\varphi}_{\theta\varphi}}{\partial\theta} = 0 - \frac{1}{\cos^2\theta} = 0$$

$$\boldsymbol{T_2} = -\frac{\mathbf{1}}{\mathbf{cos}^2\boldsymbol{\theta}}$$

Term T3:

$$\Gamma^{\lambda}_{\lambda\eta}\Gamma^{\eta}_{\theta\theta}$$

Attention we have two sums entangled, one on λ and one on η.

$$\Gamma^{\lambda}_{\lambda\eta}\Gamma^{\eta}_{\theta\theta} = \Gamma^{\theta}_{\theta\eta}\Gamma^{\eta}_{\theta\theta} + \Gamma^{\varphi}_{\varphi\eta}\Gamma^{\eta}_{\theta\theta}$$

$$\Gamma^{\lambda}_{\lambda\eta}\Gamma^{\eta}_{\theta\theta} = \Gamma^{\theta}_{\theta\theta}\Gamma^{\theta}_{\theta\theta} + \Gamma^{\theta}_{\theta\varphi}\Gamma^{\varphi}_{\theta\theta} + \Gamma^{\varphi}_{\varphi\theta}\Gamma^{\theta}_{\theta\theta} + \Gamma^{\varphi}_{\varphi\varphi}\Gamma^{\varphi}_{\theta\theta}$$

$$\Gamma^{\lambda}_{\lambda\eta}\Gamma^{\eta}_{\theta\theta} = 0 \times 0 + 0 \times 0 - \tan\theta \times 0 + 0 \times 0$$

$$\boldsymbol{\Gamma^{\theta}_{\mu\nu}} = \begin{bmatrix} \boldsymbol{\Gamma^{\theta}_{\theta\theta}} & \boldsymbol{\Gamma^{\theta}_{\theta\varphi}} \\ \boldsymbol{\Gamma^{\theta}_{\varphi\theta}} & \boldsymbol{\Gamma^{\theta}_{\varphi\varphi}} \end{bmatrix} = \begin{bmatrix} \mathbf{0} & \mathbf{0} \\ \mathbf{0} & \mathbf{cos\theta sin\theta} \end{bmatrix}$$

$$\boldsymbol{\Gamma^{\varphi}_{\mu\nu}} = \begin{bmatrix} \boldsymbol{\Gamma^{\varphi}_{\theta\theta}} & \boldsymbol{\Gamma^{\varphi}_{\theta\varphi}} \\ \boldsymbol{\Gamma^{\varphi}_{\varphi\theta}} & \boldsymbol{\Gamma^{\varphi}_{\varphi\varphi}} \end{bmatrix} = \begin{bmatrix} \mathbf{0} & \mathbf{-tan\theta} \\ \mathbf{-tan\theta} & \mathbf{0} \end{bmatrix}$$

$$\boldsymbol{T_3 = 0}$$

Term T4:

$$\Gamma^{\lambda}_{\theta\eta}\Gamma^{\eta}_{\theta\lambda}$$

Attention we have two sums entangled, one on λ and one on η:

$$\Gamma^{\lambda}_{\theta\eta}\Gamma^{\eta}_{\theta\lambda} = \Gamma^{\theta}_{\theta\eta}\Gamma^{\eta}_{\theta\theta} + \Gamma^{\varphi}_{\theta\eta}\Gamma^{\eta}_{\theta\varphi}$$

$$\Gamma^{\lambda}_{\theta\eta}\Gamma^{\eta}_{\theta\lambda} = \Gamma^{\theta}_{\theta\theta}\Gamma^{\theta}_{\theta\theta} + \Gamma^{\theta}_{\theta\varphi}\Gamma^{\varphi}_{\theta\theta} + \Gamma^{\varphi}_{\theta\theta}\Gamma^{\theta}_{\theta\varphi} + \Gamma^{\varphi}_{\theta\varphi}\Gamma^{\varphi}_{\theta\varphi}$$

$$\Gamma^{\lambda}_{\lambda\eta}\Gamma^{\eta}_{\theta\theta} = 0 \times 0 + 0 \times 0 + 0 \times 0 - \tan\theta \times -\tan\theta$$

$$\boldsymbol{\Gamma^{\theta}_{\mu\nu}} = \begin{bmatrix} \boldsymbol{\Gamma^{\theta}_{\theta\theta}} & \boldsymbol{\Gamma^{\theta}_{\theta\varphi}} \\ \boldsymbol{\Gamma^{\theta}_{\varphi\theta}} & \boldsymbol{\Gamma^{\theta}_{\varphi\varphi}} \end{bmatrix} = \begin{bmatrix} \mathbf{0} & \mathbf{0} \\ \mathbf{0} & \mathbf{cos\theta sin\theta} \end{bmatrix}$$

$$\Gamma^{\varphi}_{\mu\nu} = \begin{bmatrix} \Gamma^{\varphi}_{\theta\theta} & \Gamma^{\varphi}_{\theta\varphi} \\ \Gamma^{\varphi}_{\varphi\theta} & \Gamma^{\varphi}_{\varphi\varphi} \end{bmatrix} = \begin{bmatrix} 0 & -\tan\theta \\ -\tan\theta & 0 \end{bmatrix}$$

$$T_4 = \tan^2\theta$$

Summary:

$$T_1 = 0, T_2 = -\frac{1}{\cos^2\theta}, T_3 = 0, T_4 = \tan^2\theta$$

$$R^{\lambda}_{\theta\lambda\theta} = \Gamma^{\lambda}_{\theta\theta,\lambda} - \Gamma^{\lambda}_{\theta\lambda,\theta} + \Gamma^{\lambda}_{\lambda\eta}\Gamma^{\eta}_{\theta\theta} - \Gamma^{\lambda}_{\theta\eta}\Gamma^{\eta}_{\theta\lambda}$$

T1 T2 T3 T4

$$R^{\lambda}_{\theta\lambda\theta} = R_{\theta\theta} = 0 + \frac{1}{\cos^2\theta} + 0 - \tan^2\theta$$

$$R^{\lambda}_{\theta\theta\lambda} = R_{\theta\theta} = \frac{1}{\cos^2\theta} - \frac{\sin^2\theta}{\cos^2\theta} = \frac{\cos^2\theta + \sin^2\theta}{\cos^2\theta} - \frac{\sin^2\theta}{\cos^2\theta} = 1$$

$$R^{\lambda}_{\theta\lambda\theta} = R_{\theta\theta} = 1$$

(e) Study of the fifth term

$$R^{\lambda}_{\mu\nu\alpha} = \Gamma^{\lambda}_{\mu\alpha,\nu} - \Gamma^{\lambda}_{\mu\nu,\alpha} + \Gamma^{\lambda}_{\nu\eta}\Gamma^{\eta}_{\mu\alpha} - \Gamma^{\lambda}_{\alpha\eta}\Gamma^{\eta}_{\mu\nu}$$
$$R^{\lambda}_{\varphi\nu\varphi} = \Gamma^{\lambda}_{\varphi\varphi,\nu} - \Gamma^{\lambda}_{\varphi\nu,\varphi} + \Gamma^{\lambda}_{\nu\eta}\Gamma^{\eta}_{\varphi\varphi} - \Gamma^{\lambda}_{\varphi\eta}\Gamma^{\eta}_{\varphi\nu}$$

And with the contraction:

$$R^{\lambda}_{\varphi\lambda\varphi} = \Gamma^{\lambda}_{\varphi\varphi,\lambda} - \Gamma^{\lambda}_{\varphi\lambda,\varphi} + \Gamma^{\lambda}_{\lambda\eta}\Gamma^{\eta}_{\varphi\varphi} - \Gamma^{\lambda}_{\varphi\eta}\Gamma^{\eta}_{\varphi\lambda}$$

T1 T2 T3 T4

Term T1:

$$\Gamma^{\lambda}_{\varphi\varphi,\lambda} = \Gamma^{\theta}_{\varphi\varphi,\theta} + \Gamma^{\varphi}_{\varphi\varphi,\varphi}$$

$$\Gamma^{\theta}_{\mu\nu} = \begin{bmatrix} \Gamma^{\theta}_{\theta\theta} & \Gamma^{\theta}_{\theta\varphi} \\ \Gamma^{\theta}_{\varphi\theta} & \Gamma^{\theta}_{\varphi\varphi} \end{bmatrix} = \begin{bmatrix} 0 & 0 \\ 0 & \cos\theta\sin\theta \end{bmatrix}$$

$$\boldsymbol{\Gamma^{\varphi}_{\mu\nu} = \begin{bmatrix} \Gamma^{\varphi}_{\theta\theta} & \Gamma^{\varphi}_{\theta\varphi} \\ \Gamma^{\varphi}_{\varphi\theta} & \Gamma^{\varphi}_{\varphi\varphi} \end{bmatrix} = \begin{bmatrix} 0 & -\tan\theta \\ -\tan\theta & 0 \end{bmatrix}}$$

$$\Gamma^{\lambda}_{\varphi\varphi,\lambda} = \Gamma^{\theta}_{\varphi\varphi,\theta} + \Gamma^{\varphi}_{\varphi\varphi,\varphi} = \frac{\partial\Gamma^{\theta}_{\varphi\varphi}}{\partial\theta} + \frac{\partial\Gamma^{\varphi}_{\varphi\varphi}}{\partial\varphi} = -\sin^2\theta + \cos^2\theta + 0$$

$$\boldsymbol{T_1 = -\sin^2\theta + \cos^2\theta}$$

Term T2:

$$\Gamma^{\lambda}_{\varphi\lambda,\varphi} = \Gamma^{\theta}_{\varphi\theta,\varphi} + \Gamma^{\varphi}_{\varphi\varphi,\varphi}$$

$$\boldsymbol{\Gamma^{\theta}_{\mu\nu} = \begin{bmatrix} \Gamma^{\theta}_{\theta\theta} & \Gamma^{\theta}_{\theta\varphi} \\ \Gamma^{\theta}_{\varphi\theta} & \Gamma^{\theta}_{\varphi\varphi} \end{bmatrix} = \begin{bmatrix} 0 & 0 \\ 0 & \cos\theta\sin\theta \end{bmatrix}}$$

$$\boldsymbol{\Gamma^{\varphi}_{\mu\nu} = \begin{bmatrix} \Gamma^{\varphi}_{\theta\theta} & \Gamma^{\varphi}_{\theta\varphi} \\ \Gamma^{\varphi}_{\varphi\theta} & \Gamma^{\varphi}_{\varphi\varphi} \end{bmatrix} = \begin{bmatrix} 0 & -\tan\theta \\ -\tan\theta & 0 \end{bmatrix}}$$

$$\Gamma^{\lambda}_{\varphi\lambda,\varphi} = \Gamma^{\theta}_{\varphi\theta,\varphi} + \Gamma^{\varphi}_{\varphi\varphi,\varphi} = \frac{\partial\Gamma^{\theta}_{\varphi\theta}}{\partial\varphi} + \frac{\partial\Gamma^{\varphi}_{\varphi\varphi}}{\partial\varphi} = 0 + 0$$

$$\boldsymbol{T_2 = 0}$$

Term T3:

$$\Gamma^{\lambda}_{\lambda\eta}\Gamma^{\eta}_{\varphi\varphi}$$

Attention we have two sums entangled, one on λ and one on η.

$$\Gamma^{\lambda}_{\lambda\eta}\Gamma^{\eta}_{\varphi\varphi} = \Gamma^{\theta}_{\theta\eta}\Gamma^{\eta}_{\varphi\varphi} + \Gamma^{\varphi}_{\varphi\eta}\Gamma^{\eta}_{\varphi\varphi}$$

$$\Gamma^{\lambda}_{\lambda\eta}\Gamma^{\eta}_{\varphi\varphi} = \Gamma^{\theta}_{\theta\theta}\Gamma^{\theta}_{\varphi\varphi} + \Gamma^{\theta}_{\theta\varphi}\Gamma^{\varphi}_{\varphi\varphi} + \Gamma^{\varphi}_{\varphi\theta}\Gamma^{\theta}_{\varphi\varphi} + \Gamma^{\varphi}_{\varphi\varphi}\Gamma^{\varphi}_{\varphi\varphi}$$

$$\Gamma^{\lambda}_{\lambda\eta}\Gamma^{\eta}_{\theta\theta} = 0 \times \cos\theta\sin\theta + 0 \times 0 - \tan\theta \times \cos\theta\sin\theta + 0 \times 0$$

$$\boldsymbol{\Gamma^{\theta}_{\mu\nu} = \begin{bmatrix} \Gamma^{\theta}_{\theta\theta} & \Gamma^{\theta}_{\theta\varphi} \\ \Gamma^{\theta}_{\varphi\theta} & \Gamma^{\theta}_{\varphi\varphi} \end{bmatrix} = \begin{bmatrix} 0 & 0 \\ 0 & \cos\theta\sin\theta \end{bmatrix}}$$

$$\Gamma^{\varphi}_{\mu\nu} = \begin{bmatrix} \Gamma^{\varphi}_{\theta\theta} & \Gamma^{\varphi}_{\theta\varphi} \\ \Gamma^{\varphi}_{\varphi\theta} & \Gamma^{\varphi}_{\varphi\varphi} \end{bmatrix} = \begin{bmatrix} 0 & -\tan\theta \\ -\tan\theta & 0 \end{bmatrix}$$

$$T_3 = -\sin^2\theta$$

Term T4:

$$\Gamma^{\lambda}_{\varphi\eta}\Gamma^{\eta}_{\varphi\lambda}$$

Attention we have two sums entangled, one on λ and one on η:

$$\Gamma^{\lambda}_{\varphi\eta}\Gamma^{\eta}_{\varphi\lambda} = \Gamma^{\theta}_{\varphi\eta}\Gamma^{\eta}_{\varphi\theta} + \Gamma^{\varphi}_{\varphi\eta}\Gamma^{\eta}_{\varphi\varphi}$$

$$\Gamma^{\lambda}_{\varphi\eta}\Gamma^{\eta}_{\varphi\lambda} = \Gamma^{\theta}_{\varphi\theta}\Gamma^{\theta}_{\varphi\theta} + \Gamma^{\theta}_{\varphi\varphi}\Gamma^{\varphi}_{\varphi\theta} + \Gamma^{\varphi}_{\varphi\theta}\Gamma^{\theta}_{\varphi\varphi} + \Gamma^{\varphi}_{\varphi\varphi}\Gamma^{\varphi}_{\varphi\varphi}$$

$$\Gamma^{\lambda}_{\lambda\eta}\Gamma^{\eta}_{\theta\theta} = 0 \times 0 + \cos\theta\sin\theta \times -\tan\theta - \tan\theta \times \cos\theta\sin\theta + 0 \times 0$$

$$\Gamma^{\theta}_{\mu\nu} = \begin{bmatrix} \Gamma^{\theta}_{\theta\theta} & \Gamma^{\theta}_{\theta\varphi} \\ \Gamma^{\theta}_{\varphi\theta} & \Gamma^{\theta}_{\varphi\varphi} \end{bmatrix} = \begin{bmatrix} 0 & 0 \\ 0 & \cos\theta\sin\theta \end{bmatrix}$$

$$\Gamma^{\varphi}_{\mu\nu} = \begin{bmatrix} \Gamma^{\varphi}_{\theta\theta} & \Gamma^{\varphi}_{\theta\varphi} \\ \Gamma^{\varphi}_{\varphi\theta} & \Gamma^{\varphi}_{\varphi\varphi} \end{bmatrix} = \begin{bmatrix} 0 & -\tan\theta \\ -\tan\theta & 0 \end{bmatrix}$$

$$T_4 = -2\sin^2\theta$$

Summary:

$$T_1 = -\sin^2\theta + \cos^2\theta, T_2 = 0, T_3 = -\sin^2\theta, T_4 = -2\sin^2\theta$$

$$R^{\lambda}_{\varphi\lambda\varphi} = \Gamma^{\lambda}_{\varphi\varphi,\lambda} - \Gamma^{\lambda}_{\varphi\lambda,\varphi} + \Gamma^{\lambda}_{\lambda\eta}\Gamma^{\eta}_{\varphi\varphi} - \Gamma^{\lambda}_{\varphi\eta}\Gamma^{\eta}_{\varphi\lambda}$$

T1 T2 T3 T4

$$R^{\lambda}_{\varphi\lambda\varphi} = R_{\varphi\varphi} = -\sin^2\theta + \cos^2\theta + 0 - \sin^2\theta + 2\sin^2\theta$$

$$R^{\lambda}_{\varphi\lambda\varphi} = R_{\varphi\varphi} = \cos^2\theta$$

D.6 Scalar Curvature Calculation

$$R = g^{\mu\nu} R_{\mu\nu} = g^{\theta\theta} R_{\theta\theta} + g^{\varphi\varphi} R_{\varphi\varphi}$$

With:
Summary:

$$\boldsymbol{R_{\mu\nu} = \begin{bmatrix} 1 & 0 \\ 0 & \cos^2\theta \end{bmatrix} = \begin{bmatrix} R_{\theta\theta} & R_{\theta\varphi} \\ R_{\varphi\theta} & R_{\varphi\varphi} \end{bmatrix}}$$

$$g_{\mu\nu} = \begin{bmatrix} R^2 & 0 \\ 0 & R^2\cos\theta^2 \end{bmatrix} => g_{\mu\nu} = \begin{bmatrix} g_{\theta\theta} & g_{\theta\varphi} \\ g_{\varphi\theta} & g_{\varphi\varphi} \end{bmatrix}$$

$$g^{\mu\nu} = \begin{bmatrix} \frac{1}{R^2} & 0 \\ 0 & \frac{1}{R^2\cos\theta^2} \end{bmatrix} => g^{\mu\nu} = \begin{bmatrix} g^{\theta\theta} & g^{\theta\varphi} \\ g^{\varphi\theta} & g^{\varphi\varphi} \end{bmatrix}$$

$$R = g^{\mu\nu} R_{\mu\nu} = g^{\theta\theta} R_{\theta\theta} + g^{\varphi\varphi} R_{\varphi\varphi} = \frac{1}{R^2} \times 1 + \frac{1}{R^2\cos\theta^2} \times \cos\theta^2 = \frac{2}{R^2}$$

Scalar curvature R of a sphere of radius R is:

$$\boldsymbol{R = \frac{2}{R^2}}$$

D.7 Einstein's Tensor Expression

The formula is the following:

$$G_{\mu\nu} = R_{\mu\nu} - \frac{1}{2} g_{\mu\nu} R$$

$$G_{\mu\nu} = \begin{bmatrix} 1 & 0 \\ 0 & \cos^2\theta \end{bmatrix} - \frac{1}{2} \begin{bmatrix} R^2 & 0 \\ 0 & R^2\cos\theta^2 \end{bmatrix} \frac{2}{R^2} = \begin{bmatrix} 0 & 0 \\ 0 & 0 \end{bmatrix}$$

Appendix E

Application of Einstein's Equation in Cosmology – Demonstration of Friedmann–Lemaitre Equations

– Sources see reference [57] transcription e-lecture about relativity general and cosmology of Richard Taillet teacher-researcher at Savoie Mont blanc university and precise development of formula given in {67} à {69}.

E.1 Expression of Metric and Stress Energy Tensor

According to [57]

The interval and the metric are given by:

$$ds^2 = c^2\,dt^2 - a^2(t)\left\{\frac{dr^2}{1-kr^2} + r^2\left(d\Omega^2\right)\right\} \tag{E.1}$$

$$ds^2 = c^2\,dt^2 - a^2(t)\left\{\frac{dr^2}{1-kr^2} + r^2\left(d\theta^2 + \sin^2\theta d\varphi^2\right)\right\} \tag{E.2}$$

In this expression:

c is the speed of light,

$a_{(t)} = \frac{r_{(t)}}{r_0}$ is the scale factor,

k is the curvature factor in $(1/\mathrm{m}^2)$ it takes the following values: (1 spherical shape, 0 flat universe, -1 hyperbolic shape),

DOI: 10.1051/978-2-7598-2573-8.c915

t, r, θ, φ the 4 space-time coordinates considered (spherical coordinates).

$$ds^2 = g_{\mu\nu} dx^\mu dx^\nu$$

$$ds^2 = \sum_{\mu=0}^{3} \sum_{\nu=0}^{3} g_{\mu\nu} dx^\mu dx^\nu$$

μ and ν vary from 0 to 4 with 0 for time and 1, 2, 3 for space.

The development of $g_{\mu\nu}$ is given below:

$$ds^2 = g_{tt} dt^2 + 2g_{rt} dr dt + 2g_{\theta t} d\theta dt + 2g_{\varphi t} d\varphi dt$$

$$+ g_{rr} dr^2 + 2g_{\theta r} d\theta dr + 2g_{\varphi r} d\varphi dr$$

$$+ g_{\theta\theta} d\theta^2 + 2g_{\varphi\theta} d\varphi d\theta$$

$$+ g_{\varphi\varphi} d\varphi^2$$

Note that there are ten different terms, hence the 10 independent equations that we get later.

In the case of the universe, Friedmann–Lemaitre's metric is a solution in force today. We only have the following terms: all the others are zero:

$$g_{tt} = c^2$$

$$g_{rr} = -\frac{a^2(t)}{1 - kr^2}$$

$$g_{\theta\theta} = -a^2(t) r^2$$

$$g_{\varphi\varphi} = -a^2(t) \sin^2\theta$$

So:

$$g_{\mu\nu} = \begin{bmatrix} c^2 & 0 & 0 & 0 \\ 0 & -\frac{a^2(t)}{1-kr^2} & 0 & 0 \\ 0 & 0 & -a^2(t) r^2 & 0 \\ 0 & 0 & 0 & -a^2(t) r^2 \sin^2\theta \end{bmatrix}$$

This metric has two different aims:
On the one hand, it allows to calculate the intervals ds^2 from the coordinates (mathematical aspect), on the other hand it is the gravitational field.
All gravitational effects altered the relationship between coordinates and distances.

Einstein's equations tell us how to calculate the $a(t)$ that is in this $g_{\mu v}$. So the universe changes over time t and therefore has a history.

$$R_{\mu v} - \frac{1}{2} g_{\mu v} R + \Lambda g_{\mu v} = -\frac{8\pi G}{c^4} T_{\mu v} \quad \text{(E.3)}$$

For memory we have:
$g_{\mu v}$ is the unknown metric,
$R_{\mu v}$ Ricci's tensor built from the curvature tensor (contraction) (made up of first and second derivatives of $g_{\mu v}$),
For memory, Ricci's tensor is written:
With a contraction on λ: $R_{\mu v} = R^{\lambda}_{\mu v \lambda}$.
What we need in the end is therefore the contraction of the Riemann tensor ($\alpha => \lambda$) in the expression below for the curvature tensor.

$$R^{\lambda}_{\mu v \alpha} = \Gamma^{\lambda}_{\mu\alpha,v} - \Gamma^{\lambda}_{\mu v,\alpha} + \Gamma^{\lambda}_{\mu\eta}\Gamma^{\eta}_{v\alpha} - \Gamma^{\lambda}_{\alpha\eta}\Gamma^{\eta}_{\mu v} \quad \text{(E.4)}$$

And with the affine connection that is worth:
$\Gamma^{\lambda}_{\mu v}$ Knowing that $\Gamma^{\lambda}_{\mu v} = \frac{1}{2} g^{\lambda\rho}(g_{\mu\rho,v} + g_{v\rho,\mu} - g_{\mu v,\rho}) = \frac{1}{2} g^{\lambda\rho}\left(\frac{\partial g_{\mu\rho}}{\partial x^{v}} + \frac{\partial g_{v\rho}}{\partial x^{\mu}} - \frac{\partial g_{\mu v}}{\partial x^{\rho}}\right)$
And that for example:

$$\Gamma^{\lambda}_{r\lambda,r} = \Gamma^{t}_{rt,r} + \Gamma^{r}_{rr,r} + \Gamma^{\theta}_{r\theta,r} + \Gamma^{\varphi}_{r\varphi,r} = \Gamma^{\lambda}_{r\lambda,r} = \frac{\partial \Gamma^{t}_{rt}}{\partial r} + \frac{\partial \Gamma^{r}_{rr}}{\partial r} + \frac{\partial \Gamma^{\theta}_{r\theta}}{\partial r} + \frac{\partial^{\varphi}_{r\varphi}}{\partial r}$$

R is the scalar curvature obtained from $R_{\mu v}$,
$\Lambda g_{\mu v}$ is the cosmological constant associated with the dark energy. The theory is consistent (whether to put it or not).
So, the left term is a set of $g_{\mu v}$ and its derivatives,
G Newton's constant,
V the speed of light,

$T_{\mu\nu}$ is the stress energy tensor that characterizes what space-time contains at every point in space-time.

All are fields because they have values at every point in space-time.

As we do cosmology, we rely on the cosmological principle that the universe is homogeneous and isotropic.

So, the above quantities, are the same all over space.

When you have homogeneous and isotropic content,

$$T^{\mu\nu} = -pg^{\mu\nu} + (\rho + p)u^{\mu}u^{\nu}$$

$$T_{\mu\nu} = \begin{bmatrix} \rho c^2 & 0 & 0 & 0 \\ 0 & -p & 0 & 0 \\ 0 & 0 & -p & 0 \\ 0 & 0 & 0 & -p \end{bmatrix} g^{\mu\nu}$$

In cosmology, we can describe what is in the universe, the masses, the planets, the galaxies, the neutrinos by a fluid, that is, there are enough masses in the universe to somehow consider that it is the equivalent of atoms in a gas, p.

In other words, we can take the mass of galaxies, consider it diffuses to satisfy the cosmological principle and suddenly describe this mass distribution as a fluid that would fill the entire universe.

When we want to describe a fluid, ρ is fluid energy density, a volume energy (quantity of energy per unit of volume)

$$\rho = \frac{M}{a^3} \text{ if we have a cube of } a \times a \times a$$

Be careful, we do not make the difference between the covariant tensors and contravavriant.

When the indices are down, the p values are with a negative sign.

Pressure also dimensionally has the value of one energy per unit of volume.

When you have a particle fluid (statistical physics) in the case of perfect gases, you have the following expression that connects the pressure to the speed of the fluid's constituents:

$$p = \frac{1}{3}\rho_m v^2$$

For non-relativistic fluids, it is essentially a density of masses.

$$\rho = \rho_m c^2$$

So, we see that in the 4 terms of the tensor $T_{\mu\nu}$, the first is in c^2 and the others in v^2.

So when you have a non-relativistic fluid, that is to say that the velocities of the constituents are small in front of the speed of light, v^2 becomes small in front of c^2 and therefore the terms of pressure are going to be small before the term of ρ.

The terms of pressure are therefore small in front of fluids, non-relativistic gases.

E.2 Steps 1 Determination of the Affine Connections

According to [52, 53] from the interval and the associated metric the space-time interval is:

$$ds^2 = c^2\,dt^2 - a(t)\left\{\frac{dr^2}{1-kr^2} + r^2\left(d\theta^2 + \sin^2\theta d\varphi^2\right)\right\} \tag{E.5}$$

This gives us the shape of the metric tensor $g_{\mu\nu}$ since $ds^2 = g_{\mu\nu}\,dx^\mu\,dx^\nu$

$$g_{\mu\nu} = \begin{bmatrix} c^2 & 0 & 0 & 0 \\ 0 & -\frac{a^2(t)}{1-kr^2} & 0 & 0 \\ 0 & 0 & -a^2(t)r^2 & 0 \\ 0 & 0 & 0 & -a^2(t)r^2\sin^2\theta \end{bmatrix}$$

Diagonal matrix, spherical symmetry imposes something like this on us.
The inverse matrix is then written with $a = a(t)$:

$$g^{\mu\nu} = \begin{pmatrix} \frac{1}{c^2} & 0 & 0 & 0 \\ 0 & -\frac{(1-kr^2)}{a^2} & 0 & 0 \\ 0 & 0 & -\frac{1}{a^2r^2} & 0 \\ 0 & 0 & 0 & -\frac{1}{a^2r^2\sin^2\theta} \end{pmatrix}$$

We will now calculate them $\Gamma^\lambda_{\mu\nu}$ knowing that $\Gamma^\lambda_{\mu\nu} = \frac{1}{2}g^{\lambda\rho}(g_{\mu\rho,\nu} + g_{\nu\rho,\mu} - g_{\mu\nu,\rho}) = \frac{1}{2}g^{\lambda\rho}(\frac{\partial g_{\mu\rho}}{\partial x^\nu} + \frac{\partial g_{\nu\rho}}{\partial x^\mu} - \frac{\partial g_{\mu\nu}}{\partial x^\rho})$ Then we will have to calculate the $R^\lambda_{\mu\nu\alpha}$.

Step one

$\Gamma^\lambda_{\mu\nu} = \frac{1}{2}g^{\lambda\rho}(g_{\mu\rho,\nu} + g_{\nu\rho,\mu} - g_{\mu\nu,\rho})$ with $\lambda = (0,\,1,\,2,\,3)$ so $\lambda = (t,\,r,\,\theta,\,\varphi)$
$\Gamma^t_{\mu\nu} = \frac{1}{2}g^{t\rho}(g_{\mu\rho,\nu} + g_{\nu\rho,\mu} - g_{\mu\nu,\rho})$ with t, which will give a 4×4 matrix
$\Gamma^r_{\mu\nu} = \frac{1}{2}g^{r\rho}(g_{\mu\rho,\nu} + g_{\nu\rho,\mu} - g_{\mu\nu,\rho})$ with r, which will give a 4×4 matrix
$\Gamma^\theta_{\mu\nu} = \frac{1}{2}g^{\theta\rho}(g_{\mu\rho,\nu} + g_{\nu\rho,\mu} - g_{\mu\nu,\rho})$ with θ, which will give a 4×4 matrix
$\Gamma^\varphi_{\mu\nu} = \frac{1}{2}g^{\varphi\rho}(g_{\mu\rho,\nu} + g_{\nu\rho,\mu} - g_{\mu\nu,\rho})$ with φ, which will give a 4×4 matrix

Study of the first term in *t*

$\Gamma^t_{\mu\nu} = \frac{1}{2}g^{t\rho}(g_{\mu\rho,\nu} + g_{\nu\rho,\mu} - g_{\mu\nu,\rho})$ Which will give a 4×4 matrix.
As the matrix $g^{t\rho}$ is diagonal, only the terms where $t = \rho$ are not zero.
So we can replace it everywhere t by ρ:

$$\Gamma^t_{\mu\nu} = \frac{1}{2}g^{tt}(g_{\mu t,\nu} + g_{\nu t,\mu} - g_{\mu\nu,t})$$

We can therefore replace g^{tt} with its value (for memory it is the inverse matrix of g that must be read):

$$g^{\mu\nu} = \begin{pmatrix} \frac{1}{c^2} & 0 & 0 & 0 \\ 0 & -\frac{(1-kr^2)}{a^2} & 0 & 0 \\ 0 & 0 & -\frac{1}{a^2r^2} & 0 \\ 0 & 0 & 0 & -\frac{1}{a^2r^2\sin^2\theta} \end{pmatrix}$$

$$=> g^{tt} = \frac{1}{c^2}$$

We get:

$$\Gamma^t_{\mu\nu} = \frac{1}{2c^2}(g_{\mu t,\nu} + g_{\nu t,\mu} - g_{\mu\nu,t})$$

$$\begin{matrix} & t & r & \theta & \varphi & \end{matrix}$$

$$g_{\mu\nu} = \begin{pmatrix} c^2 & 0 & 0 & 0 \\ 0 & -\dfrac{a^2}{1-kr^2} & 0 & 0 \\ 0 & 0 & -a^2r^2 & 0 \\ 0 & 0 & 0 & -a^2r^2\sin^2\theta \end{pmatrix} \begin{matrix} t \\ r \\ \theta \\ \varphi \end{matrix} \quad => \quad g_{\mu\nu} = \begin{pmatrix} g_{tt} & g_{tr} & g_{t\theta} & g_{t\varphi} \\ g_{rt} & g_{rr} & g_{r\theta} & g_{r\varphi} \\ g_{\theta t} & g_{\theta r} & g_{\theta\theta} & g_{\theta\varphi} \\ g_{\varphi t} & g_{\varphi r} & g_{\varphi\theta} & g_{\varphi\varphi} \end{pmatrix}$$

We get:

$$\Gamma^t_{\mu\nu} = \frac{1}{2c^2}(g_{\mu t,\nu} + g_{\nu t,\mu} - g_{\mu\nu,t})$$

So, we have to calculate this for all the components of μ and ν possible.

First case μ and $\nu = t$

$\Gamma^t_{tt} = \frac{1}{2c^2}(g_{tt,t} + g_{tt,t} - g_{tt,t}) = 0$ because $g_{tt} = c^2$ independent of t so the derivative is null.

$$\Gamma^t_{\mu\nu} = \frac{1}{2c^2}\begin{pmatrix} \times & & & \\ & & & \\ & & & \\ & & & \end{pmatrix} => \Gamma^t_{\mu\nu} = \frac{1}{2c^2}\begin{pmatrix} 0 & & & \\ & & & \\ & & & \\ & & & \end{pmatrix}$$

Second case $\mu = t$ and $\nu = r$

$\Gamma^t_{\mu\nu} = \frac{1}{2c^2}(g_{\mu t,\nu} + g_{\nu t,\mu} - g_{\mu\nu,t}) => \Gamma^t_{tr} = \frac{1}{2c^2}(g_{tt,r} + g_{rt,t} - g_{tr,t})$

$g_{tt} = c^2$ it doesn't vary in r => So the derivative is null and the term of g outside the diagonal $= 0$

(g_{rt} et g_{tr}) so the derivative is null.

The derivative term in relation to t is zero, remains $\Gamma^t_{tr} = 0$

$$\Gamma^t_{\mu\nu} = \frac{1}{2c^2}\begin{pmatrix} 0 & \times & & \\ & & & \\ & & & \\ & & & \end{pmatrix} => \Gamma^t_{\mu\nu} = \frac{1}{2c^2}\begin{pmatrix} 0 & 0 & & \\ & & & \\ & & & \\ & & & \end{pmatrix}$$

By symmetry (because the $\Gamma^t_{\mu\nu}$) are symmetrical on $\mu\nu$:

$$\Gamma^t_{\mu\nu} = \frac{1}{2c^2}\begin{pmatrix} 0 & 0 & & \\ 0 & & & \\ & & & \\ & & & \end{pmatrix}$$

Third case $\mu = r$ and $\nu = r$

$\Gamma^t_{\mu\nu} = \frac{1}{2c^2}\left(g_{\mu t,\nu} + g_{\nu t,\mu} - g_{\mu\nu,t}\right) => \Gamma^t_{rr} = \frac{1}{2c^2}\left(g_{rt,r} + g_{rt,r} - g_{rr,t}\right)$.

And $(g_{rt,r}) = 0 =>$ derivative null

$$g_{\mu\nu} = \begin{pmatrix} c^2 & 0 & 0 & 0 \\ 0 & -\frac{a^2}{1-kr^2} & 0 & 0 \\ 0 & 0 & -a^2r^2 & 0 \\ 0 & 0 & 0 & -a^2r^2\sin^2\theta \end{pmatrix} \begin{matrix} t \\ r \\ \theta \\ \varphi \end{matrix} => g_{\mu\nu} = \begin{pmatrix} g_{tt} & g_{tr} & g_{t\theta} & g_{t\varphi} \\ g_{rt} & g_{rr} & g_{r\theta} & g_{r\varphi} \\ g_{\theta t} & g_{\theta r} & g_{\theta\theta} & g_{\theta\varphi} \\ g_{\varphi t} & g_{\varphi r} & g_{\varphi\theta} & g_{\varphi\varphi} \end{pmatrix}$$

$g_{rr,t} =>$ varies according to t since $a(t) => g_{rr,t} = -\frac{2a\dot{a}}{1-kr^2}$ because the derivative $(u^n = nu^{n-1}\ u')$.

$$\Gamma^t_{\mu\nu} = \frac{1}{2c^2}\begin{pmatrix} 0 & 0 & & \\ 0 & \times & & \\ & & & \\ & & & \end{pmatrix} => \Gamma^t_{\mu\nu} = \frac{1}{2c^2}\begin{pmatrix} 0 & 0 & & \\ 0 & \frac{2a\dot{a}}{1-kr^2} & & \\ & & & \\ & & & \end{pmatrix}$$

Fourth $\mu = \theta$ and $\nu = \theta$

$$\Gamma^t_{\mu\nu} = \frac{1}{2c^2}\left(g_{\mu t,\nu} + g_{\nu t,\mu} - g_{\mu\nu,t}\right) => \Gamma^t_{\theta\theta} = \frac{1}{2c^2}\left(g_{\theta t,\theta} + g_{\theta t,\theta} - g_{\theta\theta,t}\right)$$

$g_{\theta t,\theta} = 0$ and derivatives $=> 0$

$$g_{\mu\nu} = \begin{pmatrix} c^2 & 0 & 0 & 0 \\ 0 & -\frac{a}{1-kr^2} & 0 & 0 \\ 0 & 0 & -a^2r^2 & 0 \\ 0 & 0 & 0 & -a^2r^2\sin^2\theta \end{pmatrix} \begin{matrix} t \\ r \\ \theta \\ \varphi \end{matrix} => g_{\mu\nu} = \begin{pmatrix} g_{tt} & g_{tr} & g_{t\theta} & g_{t\varphi} \\ g_{rt} & g_{rr} & g_{r\theta} & g_{r\varphi} \\ g_{\theta t} & g_{\theta r} & g_{\theta\theta} & g_{\theta\varphi} \\ g_{\varphi t} & g_{\varphi r} & g_{\varphi\theta} & g_{\varphi\varphi} \end{pmatrix}$$

$g_{\theta\theta,t} =>$ varies according to t since one has been $=> g_{\theta\theta,t} = -2a\dot{a}r^2$.

$$\Gamma^t_{\mu\nu} = \frac{1}{2c^2}\begin{pmatrix} 0 & 0 & & \\ 0 & \times & & \\ & & & \\ & & & \end{pmatrix} => \Gamma^t_{\mu\nu} = \frac{1}{2c^2}\begin{pmatrix} 0 & 0 & & \\ 0 & \frac{2a\dot{a}}{1-kr^2} & & \\ & & 2a\dot{a}r^2 & \\ & & & \end{pmatrix}$$

Ditto with $\boldsymbol{\mu = \varphi}$ **and** $\boldsymbol{v = \varphi}$

$$\Gamma^t_{\mu v} = \frac{1}{2c^2}(g_{\mu t,v} + g_{vt,\mu} - g_{\mu v,t}) => \Gamma^t_{\varphi\varphi} = \frac{1}{2c^2}(g_{\varphi t,\varphi} + g_{\varphi t,\varphi} - g_{\varphi\varphi,t})$$

$g_{\varphi t,\varphi} = 0$ and derivatives => 0.

$$g_{\mu v} = \begin{pmatrix} c^2 & 0 & 0 & 0 \\ 0 & -\frac{a^2}{1-kr^2} & 0 & 0 \\ 0 & 0 & -a^2r^2 & 0 \\ 0 & 0 & 0 & -a^2r^2\sin^2\theta \end{pmatrix} \begin{matrix} t \\ r \\ \theta \\ \varphi \end{matrix} => g_{\mu v} = \begin{pmatrix} g_{tt} & g_{tr} & g_{t\theta} & g_{t\varphi} \\ g_{rt} & g_{rr} & g_{r\theta} & g_{r\varphi} \\ g_{\theta t} & g_{\theta r} & g_{\theta\theta} & g_{\theta\varphi} \\ g_{\varphi t} & g_{\varphi r} & g_{\varphi\theta} & g_{\varphi\varphi} \end{pmatrix}$$

$g_{\varphi\varphi,t}$ => varies according to t since we have been $a(t)$ => $a(t)$=> $g_{\varphi\varphi,t} = -2a\dot{a}r^2\sin^2\theta$.

$$\Gamma^t_{\mu v} = \frac{1}{2c^2}\begin{pmatrix} 0 & 0 & & \\ 0 & \times & & \\ & & & \\ & & & \end{pmatrix} = > \Gamma^t_{\mu v} = \frac{1}{2c^2}\begin{pmatrix} 0 & 0 & & \\ 0 & \frac{2a\dot{a}}{1-kr^2} & & \\ & & 2a\dot{a}r^2 & \\ & & & 2a\dot{a}r^2\sin^2\theta \end{pmatrix}$$

Then, for all cross terms, g is zero so derivatives are zero.

$$\Gamma^t_{\mu v} = \frac{1}{2c^2}\begin{pmatrix} 0 & 0 & 0 & 0 \\ 0 & \frac{2a\dot{a}}{1-kr^2} & 0 & 0 \\ 0 & 0 & 2a\dot{a}r^2 & 0 \\ 0 & 0 & 0 & 2a\dot{a}r^2\sin^2\theta \end{pmatrix} \tag{E.6}$$

$$\Gamma^t_{tt} = 0$$

$$\Gamma^t_{rr} = \frac{a\dot{a}}{c^2(1-kr^2)}$$

$$\Gamma^t_{\theta\theta} = \frac{a\dot{a}r^2}{c^2}$$

$$\Gamma^t_{\varphi\varphi} = \frac{a\dot{a}r^2\sin^2\theta}{c^2}$$

Study of the second term in r

$\Gamma^r_{\mu v} = \frac{1}{2} g^{r\rho}(g_{\mu\rho,v} + g_{v\rho,\mu} - g_{\mu v,\rho})$ Which will give a 4×4 matrix.

As the matrix $g^{r\rho}$ is diagonal, only the terms where $r = \rho$ are no null.

So, we can replace everywhere r par ρ.

$$\Gamma^r_{\mu v} = \frac{1}{2} g^{rr}(g_{\mu r,v} + g_{\mu r,v} - g_{\mu v,r})$$

We can therefore replace g^{rr} with its value (for memory it is the inverse matrix of g that must be read).

$$g_{\mu v} = \begin{pmatrix} c^2 & 0 & 0 & 0 \\ 0 & -\frac{a^2}{1-kr^2} & 0 & 0 \\ 0 & 0 & -a^2r^2 & 0 \\ 0 & 0 & 0 & -a^2r^2\sin^2\theta \end{pmatrix} => g^{rr} = -\frac{(1-kr^2)}{a^2}$$

We get:

$$\Gamma^r_{\mu v} = -\frac{(1-kr^2)}{2a^2}(g_{\mu r,v} + g_{\mu r,v} - g_{\mu v,r})$$

$$\begin{matrix} & t & r & \theta & \varphi & \end{matrix}$$

$$g_{\mu v} = \begin{pmatrix} c^2 & 0 & 0 & 0 \\ 0 & \frac{a^2}{1-kr^2} & 0 & 0 \\ 0 & 0 & -a^2r^2 & 0 \\ 0 & 0 & 0 & -a^2r^2\sin^2\theta \end{pmatrix} \begin{matrix} t \\ r \\ \theta \\ \varphi \end{matrix} \quad => \quad g_{\mu v} = \begin{pmatrix} g_{tt} & g_{tr} & g_{t\theta} & g_{t\varphi} \\ g_{rt} & g_{rr} & g_{r\theta} & g_{r\varphi} \\ g_{\theta t} & g_{\theta r} & g_{\theta\theta} & g_{\theta\varphi} \\ g_{\varphi t} & g_{\varphi r} & g_{\varphi\theta} & g_{\varphi\varphi} \end{pmatrix}$$

We get:

$$\Gamma^r_{\mu v} = -\frac{(1-kr^2)}{2a^2}(g_{\mu r,v} + g_{\mu r,v} - g_{\mu v,r})$$

So, we have to calculate this for all the components of μ and v possible.

First case μ and $v = t$

$\Gamma^r_{\mu v} = -\frac{(1-kr^2)}{2a^2}(g_{tr,t} + g_{tr,t} - g_{tt,r})$

So, derivatives give:

$g_{tr,t} = \frac{\partial g_{tr}}{\partial r} = 0$ because $g_{tr} = 0$

$g_{tt,r} = \frac{\partial c^2}{\partial r} = 0$ because c^2 independent of r

The g matrix is as follows:

$$\Gamma^{r}_{\mu\nu} = -\frac{(1-kr^2)}{2a^2}\begin{pmatrix} \times & & \\ & & \\ & & \end{pmatrix} => \Gamma^{r}_{\mu\nu} = -\frac{(1-kr^2)}{2a^2}\begin{pmatrix} 0 & & \\ & & \\ & & \end{pmatrix}$$

Second case $\mu = t$ and $\nu = r$

$$\Gamma^{r}_{\mu\nu} = -\frac{(1-kr^2)}{2a^2}(g_{\mu r,\nu} + g_{\nu r,\mu} - g_{\mu\nu,r}) => \Gamma^{r}_{tr} = -\frac{(1-kr^2)}{2a^2}(g_{tr,r} + g_{rr,t} - g_{tr,r})$$

The middle term is zero because derived from t.
$(g_{tr}) = 0 =>$ derivative $= 0$

$$g_{rr,t} = \frac{\partial g_{rr}}{\partial t} = \frac{-2a\dot{a}}{1-kr^2}$$

$$\Gamma^{r}_{tr} = -\frac{(1-kr^2)}{2a^2}(g_{tr,r} + g_{rr,t} - g_{tr,r}) = \Gamma^{r}_{tr} = -\frac{(1-kr^2)}{2a^2}\left(\frac{-2a\,\dot{a}}{(1-kr^2)}\right) = \frac{\dot{a}}{a} = \Gamma^{r}_{rt} = \frac{\dot{a}}{a} \tag{E.7}$$

$$\Gamma^{r}_{\mu\nu} = \begin{pmatrix} 0 & \times & \\ & & \\ & & \end{pmatrix} => \Gamma^{r}_{\mu\nu} = \begin{pmatrix} 0 & \frac{\dot{a}}{a} & \\ \frac{\dot{a}}{a} & & \\ & & \end{pmatrix}$$

Third case $\mu = r$ and $\nu = r$

$$\Gamma^{r}_{\mu\nu} = -\frac{(1-kr^2)}{2a^2}(g_{\mu r,\nu} + g_{\nu r,\mu} - g_{\mu\nu,r}) => \Gamma^{r}_{\mu\nu}$$
$$= -\frac{(1-kr^2)}{2a^2}(g_{rr,r} + g_{rr,r} - g_{rr,r}) => \Gamma^{r}_{rr} = -\frac{(1-kr^2)}{2a^2}(g_{rr,r})$$

$$g_{\mu\nu} = \begin{pmatrix} c^2 & 0 & 0 & 0 \\ 0 & -\frac{a^2}{1-kr^2} & 0 & 0 \\ 0 & 0 & -a^2r^2 & 0 \\ 0 & 0 & 0 & -a^2r^2\sin^2\theta \end{pmatrix} \begin{matrix} t \\ r \\ \theta \\ \varphi \end{matrix} => g_{\mu\nu} = \begin{pmatrix} g_{tt} & g_{tr} & g_{t\theta} & g_{t\varphi} \\ g_{rt} & g_{rr} & g_{r\theta} & g_{r\varphi} \\ g_{\theta t} & g_{\theta r} & g_{\theta\theta} & g_{\theta\varphi} \\ g_{\varphi t} & g_{\varphi r} & g_{\varphi\theta} & g_{\varphi\varphi} \end{pmatrix}$$

And $(g_{rr,r}) = \frac{\partial\left(\frac{-a^2}{1-kr^2}\right)}{\partial r}$

In link with r $\left(\frac{u}{v}\right)' = \frac{(u'v - v'u)}{v^2}$ $u = -a^2 \quad u' = 0$

$$v = \left(1 - kr^2\right) \quad v' = -2kr$$

And $(g_{rr,r}) = \frac{\partial\left(\frac{-a^2}{1-kr^2}\right)}{\partial r} = \left(\frac{-2a^2kr}{(1-kr^2)^2}\right)$

$$\Gamma^r_{rr} = -\frac{(1-kr^2)}{2a^2}(g_{rr,r}) = -\frac{(1-kr^2)}{2a^2}\left(-\frac{2a^2kr}{(1-kr^2)^2}\right) = \left(\frac{kr}{(1-kr^2)}\right) =>$$

$$\Gamma^r_{rr} = \frac{kr}{(1-kr^2)} \tag{E.8}$$

$$\Gamma^r_{\mu\nu} = \begin{pmatrix} 0 & \frac{\dot{a}}{a} & & \\ \frac{\dot{a}}{a} & \times & & \\ & & & \\ & & & \end{pmatrix} => \Gamma^r_{\mu\nu} = \begin{pmatrix} 0 & \frac{\dot{a}}{a} & & \\ \frac{\dot{a}}{a} & \frac{kr}{(1-kr^2)} & & \\ & & & \\ & & & \end{pmatrix}$$

So for not to be zero, it must be derivatives with respect to r, diagonal terms for which the indices are equal.

Fourth case $\mu = \theta$ and $\nu = \theta$

$$g_{\mu\nu} = \begin{pmatrix} c^2 & 0 & 0 & 0 \\ 0 & -\frac{a^2}{1-kr^2} & 0 & 0 \\ 0 & 0 & -a^2r^2 & 0 \\ 0 & 0 & 0 & -a^2r^2\sin^2\theta \end{pmatrix} \begin{matrix} t \\ r \\ \theta \\ \varphi \end{matrix} => g_{\mu\nu} = \begin{pmatrix} g_{tt} & g_{tr} & g_{t\theta} & g_{t\varphi} \\ g_{rt} & g_{rr} & g_{r\theta} & g_{r\varphi} \\ g_{\theta t} & g_{\theta r} & g_{\theta\theta} & g_{\theta\varphi} \\ g_{\varphi t} & g_{\varphi r} & g_{\varphi\theta} & g_{\varphi\varphi} \end{pmatrix}$$

$$\Gamma^r_{\mu\nu} = \frac{1}{2}g^{r\rho}(g_{\mu\rho,\nu} + g_{\nu\rho,\mu} - g_{\mu\nu,\rho})$$

$$\Gamma^r_{\vartheta\theta} = \frac{1}{2} g^{rr} (g_{\theta r,\theta} + g_{\theta r,\theta} - g_{\theta\theta,r}) => \Gamma^r_{\theta\theta} = -\frac{(1 - kr^2)}{2a^2} (g_{\theta r,\theta} + g_{\theta r,\theta} - g_{\theta\theta,r})$$

$g_{\theta r} = 0$ and $=> 0$.

So, it remains $g_{\theta\theta,r} = -\frac{\partial a^2 r^2}{\partial r} = -2a^2 r$

$$\Gamma^r_{\theta\theta} = -\frac{(1 - kr^2)}{2a^2} (2a^2 r) = -r(1 - kr^2) \quad \text{(E.9)}$$

$$=> \Gamma^r_{\mu v} = \begin{pmatrix} 0 & \frac{\dot{a}}{a} & & \\ \frac{\dot{a}}{a} & \frac{kr}{(1-kr^2)} & & \\ & & \times & \\ & & & \end{pmatrix} => \Gamma^r_{\mu v} = \begin{pmatrix} 0 & \frac{\dot{a}}{a} & & \\ \frac{\dot{a}}{a} & \frac{kr}{(1-kr^2)} & & \\ & & -r(1 - kr^2) & \\ & & & \end{pmatrix}$$

Ditto with $\boldsymbol{\mu = \varphi}$ **and** $\boldsymbol{\varphi = \varphi}$

$$g_{\mu v} = \begin{pmatrix} c^2 & 0 & 0 & 0 \\ 0 & -\frac{a^2}{1-kr^2} & 0 & 0 \\ 0 & 0 & -a^2 r^2 & 0 \\ 0 & 0 & 0 & -a^2 r^2 \sin^2\theta \end{pmatrix} \begin{matrix} t \\ r \\ \theta \\ \varphi \end{matrix} => g_{\mu v} = \begin{pmatrix} g_{tt} & g_{tr} & g_{t\theta} & g_{t\varphi} \\ g_{rt} & g_{rr} & g_{r\theta} & g_{r\varphi} \\ g_{\theta t} & g_{\theta r} & g_{\theta\theta} & g_{\theta\varphi} \\ g_{\varphi t} & g_{\varphi r} & g_{\varphi\theta} & g_{\varphi\varphi} \end{pmatrix}$$

$\Gamma^r_{\mu v} = \frac{1}{2} g^{r\rho} (g_{\mu\rho,v} + g_{v\rho,\mu} - g_{\mu v,\rho})$ is no different than 0 if $\rho = r$

$$\Gamma^r_{\varphi\varphi} = \frac{1}{2} g^{rr} (g_{\varphi r,\varphi} + g_{\varphi r,\varphi} - g_{\varphi\varphi,r}) => \Gamma^r_{\varphi\varphi} = -\frac{(1 - kr^2)}{2a^2} (g_{\varphi r,\varphi} + g_{\varphi r,\varphi} - g_{\varphi\varphi,r})$$

$g_{\varphi r} = 0$ and $=> 0$

Remain $g_{\varphi\varphi,r} = -\frac{\partial(-a^2 r^2 \sin^2\theta)}{\partial r} = 2ra^2 \sin^2\theta$

$$\Gamma^r_{\varphi\varphi} = -\frac{(1 - kr^2)}{2a^2} (2ra^2 \sin^2\theta) = \Gamma^r_{\varphi\varphi} = -(1 - kr^2)(r \sin^2\theta) \quad \text{(E.10)}$$

$$\begin{matrix} & t & r & \theta & \varphi \end{matrix}$$

$$\Gamma^r_{\mu\nu} = \begin{pmatrix} 0 & \frac{\dot{a}}{a} & & \\ \frac{\dot{a}}{a} & \frac{kr}{(1-kr^2)} & & \\ & & -r(1-kr^2) & \\ & & & \times \end{pmatrix} \Rightarrow \Gamma^r_{\mu\nu} = \begin{pmatrix} 0 & \frac{\dot{a}}{a} & & \\ \frac{\dot{a}}{a} & \frac{kr}{(1-kr^2)} & & \\ & & -r(1-kr^2) & \\ & & & -r(1-kr^2)\sin^2\theta \end{pmatrix}$$

$\mu = \theta$ **and** $\nu = t$

$$g_{\mu\nu} = \begin{pmatrix} c^2 & 0 & 0 & 0 \\ 0 & -\frac{a^2}{1-kr^2} & 0 & 0 \\ 0 & 0 & -a^2r^2 & 0 \\ 0 & 0 & 0 & -a^2r^2\sin^2\theta \end{pmatrix} \begin{matrix} t \\ r \\ \theta \\ \varphi \end{matrix} => g_{\mu\nu} = \begin{pmatrix} g_{tt} & g_{tr} & g_{t\theta} & g_{t\varphi} \\ g_{rt} & g_{rr} & g_{r\theta} & g_{r\varphi} \\ g_{\theta t} & g_{\theta r} & g_{\theta\theta} & g_{\theta\varphi} \\ g_{\varphi t} & g_{\varphi r} & g_{\varphi\theta} & g_{\varphi\varphi} \end{pmatrix}$$

$\Gamma^r_{\mu\nu} = \frac{1}{2} g^{r\rho}(g_{\mu\rho,\nu} + g_{\nu\rho,\mu} - g_{\mu\nu,\rho})$ is different 0 if $\rho = r$
$\Gamma^r_{\mu\nu} = \frac{1}{2} g^{rr}(g_{\theta r,t} + g_{tr,\theta} - g_{\theta t,r})$ is different 0 if $\rho = r$
As only the diagonal terms of g are different from 0 => all terms are zero.

Conclusion

All other terms will be null

So:

$$\Gamma^r_{\mu\nu} = \begin{pmatrix} 0 & \frac{\dot{a}}{a} & 0 & 0 \\ \frac{\dot{a}}{a} & \frac{kr}{(1-kr^2)} & 0 & 0 \\ 0 & 0 & -r(1-kr^2) & 0 \\ 0 & 0 & 0 & -r(1-kr^2)\sin^2\theta \end{pmatrix} \qquad \text{(E.11)}$$

$$\Gamma^r_{\varphi\varphi} = -r(1-kr^2)\sin^2\theta$$

$$\Gamma^r_{\theta\theta} = -r(1-kr^2)$$

$$\Gamma^r_{rr} = \frac{kr}{(1-kr^2)}$$

$$\Gamma^r_{tr} = \frac{\dot{a}}{a} = \Gamma^r_{rt} = \frac{\dot{a}}{a}$$

Study of the third term in θ

$\Gamma^\theta_{\mu\nu} = \frac{1}{2} g^{\theta\rho}(g_{\mu\rho,\nu} + g_{\nu\rho,\mu} - g_{\mu\nu,\rho})$ which will give a 4×4 matrix.

As the matrix $g^{\theta\rho}$ is diagonal, only the terms where $\theta = \rho$ are different of 0. So, you can replace it everywhere θ by ρ:

$$\Gamma^\theta_{\mu\nu} = \frac{1}{2} g^{\theta\theta}(g_{\mu\theta,\nu} + g_{\nu\theta,\mu} - g_{\mu\nu,\theta})$$

We can therefore replace $g^{\theta\theta}$ with its value (for memory it is the inverse matrix of g that must be read):

$$g_{\mu\nu} = \begin{pmatrix} c^2 & 0 & 0 & 0 \\ 0 & -\frac{a^2}{1-kr^2} & 0 & 0 \\ 0 & 0 & -a^2r^2 & 0 \\ 0 & 0 & 0 & -a^2r^2\sin^2\theta \end{pmatrix} \begin{matrix} t \\ r \\ \theta \\ \varphi \end{matrix} => g_{\mu\nu} = \begin{pmatrix} g_{tt} & g_{tr} & g_{t\theta} & g_{t\varphi} \\ g_{rt} & g_{rr} & g_{r\theta} & g_{r\varphi} \\ g_{\theta t} & g_{\theta r} & g_{\theta\theta} & g_{\theta\varphi} \\ g_{\varphi t} & g_{\varphi r} & g_{\varphi\theta} & g_{\varphi\varphi} \end{pmatrix}$$

$$=> g^{\theta\theta} = -\frac{1}{a^2r^2}$$

We get:

$$\Gamma^\theta_{\mu\nu} = -\frac{1}{2a^2r^2}(g_{\mu\theta,\nu} + g_{\nu\theta,\mu} - g_{\mu\nu,\theta})$$

First case μ and $\nu = t$

$\Gamma^\theta_{\mu\nu} = -\frac{1}{2a^2r^2}(g_{t\theta,t} + g_{t\theta,t} - g_{tt,\theta})$: The first two terms are null because the g non-diagonal $= 0$.

The g matrix is as follows:

$$g_{\mu\nu} = \begin{pmatrix} c^2 & 0 & 0 & 0 \\ 0 & -\frac{a^2}{1-kr^2} & 0 & 0 \\ 0 & 0 & -a^2r^2 & 0 \\ 0 & 0 & 0 & -a^2r^2\sin^2\theta \end{pmatrix} \begin{matrix} t \\ r \\ \theta \\ \varphi \end{matrix} => g_{\mu\nu} = \begin{pmatrix} g_{tt} & g_{tr} & g_{t\theta} & g_{t\varphi} \\ g_{rt} & g_{rr} & g_{r\theta} & g_{r\varphi} \\ g_{\theta t} & g_{\theta r} & g_{\theta\theta} & g_{\theta\varphi} \\ g_{\varphi t} & g_{\varphi r} & g_{\varphi\theta} & g_{\varphi\varphi} \end{pmatrix}$$

$$g_{tt,\theta} = \frac{\partial c^2}{\partial\theta} = 0$$

$$\Gamma^{\theta}_{\mu\nu} = -\frac{1}{2a^2r^2}\begin{pmatrix} \times & & & \\ & & & \\ & & & \\ & & & \end{pmatrix} => \Gamma^{\theta}_{\mu\nu} = -\frac{1}{2a^2r^2}\begin{pmatrix} 0 & & & \\ & & & \\ & & & \\ & & & \end{pmatrix}$$

Second case $\mu = t$ and $\nu = r$

$$\Gamma^{\theta}_{\mu\nu} = -\frac{1}{2a^2r^2}(g_{\mu\theta,\nu} + g_{\nu\theta,\mu} - g_{\mu\nu,\theta}) => \Gamma^{\theta}_{\mu\nu} = -\frac{1}{2a^2r^2}(g_{t\theta,r} + g_{r\theta,t} - g_{tr,\theta})$$

All the terms relate to the no diagonal terms of g which are zero $=> 0\ (g_{t\theta}) = 0$ and $(g_{tr}) = 0$

$$\Gamma^{\theta}_{\mu\nu} = -\frac{1}{2a^2r^2}\begin{pmatrix} 0 & \times & & \\ & & & \\ & & & \\ & & & \end{pmatrix} => \Gamma^{\theta}_{\mu\nu} = -\frac{1}{2a^2r^2}\begin{pmatrix} 0 & 0 & & \\ & & & \\ & & & \\ & & & \end{pmatrix}$$

By symmetry (because the $\Gamma^{t}_{\mu\nu}$) are symmetrical on $\mu\nu$:

$$\Gamma^{\theta}_{\mu\nu} = -\frac{1}{2a^2r^2}\begin{pmatrix} 0 & 0 & & \\ 0 & & & \\ & & & \\ & & & \end{pmatrix}$$

Third case $\mu = r$ and $\nu = r$

$$\Gamma^{\theta}_{\mu\nu} = -\frac{1}{2a^2r^2}(g_{\mu\theta,\nu} + g_{\nu\theta,\mu} - g_{\mu\nu,\theta}) => \Gamma^{\theta}_{\mu\nu} = -\frac{1}{2a^2r^2}(g_{r\theta,r} + g_{r\theta,r} - g_{rr,\theta})$$

$$g_{\mu\nu} = \begin{pmatrix} c^2 & 0 & 0 & 0 \\ 0 & -\frac{a^2}{1-kr^2} & 0 & 0 \\ 0 & 0 & -a^2r^2 & 0 \\ 0 & 0 & 0 & -a^2r^2\sin^2\theta \end{pmatrix} \begin{matrix} t \\ r \\ \theta \\ \varphi \end{matrix} => g_{\mu\nu} = \begin{pmatrix} g_{tt} & g_{tr} & g_{t\theta} & g_{t\varphi} \\ g_{rt} & g_{rr} & g_{r\theta} & g_{r\varphi} \\ g_{\theta t} & g_{\theta r} & g_{\theta\theta} & g_{\theta\varphi} \\ g_{\varphi t} & g_{\varphi r} & g_{\varphi\theta} & g_{\varphi\varphi} \end{pmatrix}$$

And $(g_{rr,\theta}) = 0$ because this term does not depend on θ.
For the other derivatives, they relate to no diagonal terms of g that are zero.

$$(g_{r\theta}) = 0$$

$$\Gamma^{\theta}_{\mu\nu} = -\frac{1}{2a^2r^2}\begin{pmatrix} 0 & 0 & & \\ 0 & \times & & \\ & & & \\ & & & \end{pmatrix} => \Gamma^{\theta}_{\mu\nu} = -\frac{1}{2a^2r^2}\begin{pmatrix} 0 & 0 & & \\ 0 & 0 & & \\ & & & \\ & & & \end{pmatrix}$$

So, in order not to be zero, it must be derivatives in relation to r, diagonal terms for which the indices are equal:

Fourth case $\mu = \theta$ and $v = \theta$

$$\Gamma^{\theta}_{\mu v} = -\frac{1}{2a^2r^2}(g_{\mu\theta,v} + g_{v\theta,\mu} - g_{\mu v,\theta}) => \Gamma^{\theta}_{\mu v} = -\frac{1}{2a^2r^2}(g_{\theta\theta,\theta} + g_{\theta\theta,\theta} - g_{\theta\theta,\theta}) => \Gamma^{\theta}_{\mu v} = -\frac{1}{2a^2r^2}(g_{\theta\theta,\theta})$$

Remain $g_{\theta\theta,\theta} = -\frac{\partial(-a^2r^2)}{\partial\theta} = 0$

$$\Gamma^{\theta}_{\mu v} = -\frac{1}{2a^2r^2}\begin{pmatrix} 0 & 0 & & \\ 0 & 0 & & \\ & & \times & \\ & & & \end{pmatrix} => \Gamma^{\theta}_{\mu v} = -\frac{1}{2a^2r^2}\begin{pmatrix} 0 & 0 & & \\ 0 & 0 & & \\ & & 0 & \\ & & & \end{pmatrix}$$

Ditto with $\mu = \varphi$ and $v = \varphi$

$$\Gamma^{\theta}_{\mu v} = -\frac{1}{2a^2r^2}(g_{\mu\theta,v} + g_{v\theta,\mu} - g_{\mu v,\theta}) => \Gamma^{\theta}_{\varphi\varphi} = -\frac{1}{2a^2r^2}(g_{\varphi\theta,\varphi} + g_{\varphi\theta,\varphi} - g_{\varphi\varphi,\theta})$$

$g_{\varphi\theta} = 0$ and => 0

Remain $g_{\varphi\varphi,\theta} = -\frac{\partial(-a^2r^2\sin^2\theta)}{\partial\theta} = 2a^2r^2\sin\theta \times \cos\theta$

We know that $\sin^2\theta = \sin\theta \times \sin\theta =>$ derivative of $(u \times v) = u'v + v'u => 2\sin\theta \times \cos\theta$

$\Gamma^{\theta}_{\varphi\varphi} = -\frac{1}{2a^2r^2}(2a^2r^2\sin\theta\cos\theta) = -\sin\theta\cos\theta$

$$\Gamma^{\theta}_{\mu v} = \begin{pmatrix} 0 & 0 & & \\ 0 & 0 & & \\ & & 0 & \\ & & & \times \end{pmatrix} => \Gamma^{\theta}_{\mu v} = \begin{pmatrix} 0 & 0 & & \\ 0 & 0 & & \\ & & 0 & \\ & & & -\sin\theta\cos\theta \end{pmatrix}$$

Fifth case $\mu = \theta$ and $v = r$

$$\Gamma^{\theta}_{\mu v} = -\frac{1}{2a^2r^2}(g_{\mu\theta,v} + g_{v\theta,\mu} - g_{\mu v,\theta}) => \Gamma^{\theta}_{\theta r} = -\frac{1}{2a^2r^2}(g_{\theta\theta,r} + g_{r\theta,\theta} - g_{\theta r,\theta})$$

$$g_{\mu v} = \begin{pmatrix} c^2 & 0 & 0 & 0 \\ 0 & -\frac{a^2}{1-kr^2} & 0 & 0 \\ 0 & 0 & -a^2r^2 & 0 \\ 0 & 0 & 0 & -a^2r^2\sin^2\theta \end{pmatrix} \begin{matrix} t \\ r \\ \theta \\ \varphi \end{matrix} => g_{\mu v} = \begin{pmatrix} g_{tt} & g_{tr} & g_{t\theta} & g_{t\varphi} \\ g_{rt} & g_{rr} & g_{r\theta} & g_{r\varphi} \\ g_{\theta t} & g_{\theta r} & g_{\theta\theta} & g_{\theta\varphi} \\ g_{\varphi t} & g_{\varphi r} & g_{\varphi\theta} & g_{\varphi\varphi} \end{pmatrix}$$

$g_{\theta r} = 0$ and $g_{r\theta} = 0$ and => 0 remains: $\Gamma^{\theta}_{\theta r} = -\frac{1}{2a^2r^2}(g_{\theta\theta,r})$

$$g_{\theta\theta,r} = \frac{\partial(-a^2r^2)}{\partial r} = -2a^2r$$

$$\Gamma^{\theta}_{\theta r} = -\frac{1}{2a^2r^2}(-2a^2r) = \frac{1}{r}$$

$$\Gamma^{\theta}_{\mu\nu} = \begin{pmatrix} 0 & 0 & & \\ 0 & 0 & \times & \\ & \times & 0 & \\ & & & -\sin\theta\cos\theta \end{pmatrix} => \Gamma^{\theta}_{\mu\nu} = \begin{pmatrix} 0 & 0 & & \\ 0 & 0 & \frac{1}{r} & \\ & \frac{1}{r} & 0 & \\ & & & -\sin\theta\cos\theta \end{pmatrix}$$

Sixth case $\mu = \theta$ and $\nu = t$

$\Gamma^{\theta}_{\mu\nu} = -\frac{1}{2a^2r^2}(g_{\mu\theta,\nu} + g_{\nu\theta,\mu} - g_{\mu\nu,\theta}) => \Gamma^{\theta}_{\theta t} = -\frac{1}{2a^2r^2}(g_{\theta\theta,t} + g_{t\theta,\theta} - g_{\theta t,\theta}) =>$ (terms of g in t and $\theta = 0$)

$$g_{\mu\nu} = \begin{pmatrix} c^2 & 0 & 0 & 0 \\ 0 & -\frac{a^2}{1-kr^2} & 0 & 0 \\ 0 & 0 & -a^2r^2 & 0 \\ 0 & 0 & 0 & -a^2r^2\sin^2\theta \end{pmatrix} \begin{matrix} t \\ r \\ \theta \\ \varphi \end{matrix} => g_{\mu\nu} = \begin{pmatrix} g_{tt} & g_{tr} & g_{t\theta} & g_{t\varphi} \\ g_{rt} & g_{rr} & g_{r\theta} & g_{r\varphi} \\ g_{\theta t} & g_{\theta r} & g_{\theta\theta} & g_{\theta\varphi} \\ g_{\varphi t} & g_{\varphi r} & g_{\varphi\theta} & g_{\varphi\varphi} \end{pmatrix}$$

$g_{\theta t} = 0$ and $g_{t\theta} = 0$ and $=> 0$ remains: $\Gamma^{\theta}_{\theta t} = -\frac{1}{2a^2r^2}(g_{\theta\theta,t})$

$$g_{\theta\theta,t} = \frac{\partial(-a^2r^2)}{\partial t} = -2a\,\dot{a}\,r^2$$

$$\Gamma^{\theta}_{\theta t} = -\frac{1}{2a^2r^2}(-2\,\dot{a}\,ar^2) = \frac{\dot{a}}{a} = \Gamma^{\theta}_{t\theta} = \frac{\dot{a}}{a}$$

$$\Gamma^{\theta}_{\mu\nu} = \begin{pmatrix} 0 & 0 & \times & \\ 0 & 0 & \frac{1}{r} & \\ \times & \frac{1}{r} & 0 & \\ & & & -\sin\theta\cos\theta \end{pmatrix} => \Gamma^{\theta}_{\mu\nu} = \begin{pmatrix} 0 & 0 & \frac{\dot{a}}{a} & \\ 0 & 0 & \frac{1}{r} & \\ \frac{\dot{a}}{a} & \frac{1}{r} & 0 & \\ & & & -\sin\theta\cos\theta \end{pmatrix}$$

Seventh case $\mu = \varphi$ and $\nu = t$

$\Gamma^{\theta}_{\mu\nu} = -\frac{1}{2a^2r^2}(g_{\mu\theta,\nu} + g_{\nu\theta,\mu} - g_{\mu\nu,\theta}) => \Gamma^{\theta}_{\varphi t} = -\frac{1}{2a^2r^2}(g_{\varphi\theta,t} + g_{t\theta,\varphi} - g_{\varphi t,\theta}) =>$ (diagonal terms of $g => 0$)

$$g_{\mu\nu} = \begin{pmatrix} c^2 & 0 & 0 & 0 \\ 0 & -\frac{a^2}{1-kr^2} & 0 & 0 \\ 0 & 0 & -a^2r^2 & 0 \\ 0 & 0 & 0 & -a^2r^2\sin^2\theta \end{pmatrix} \begin{matrix} t \\ r \\ \theta \\ \varphi \end{matrix} => g_{\mu\nu} = \begin{pmatrix} g_{tt} & g_{tr} & g_{t\theta} & g_{t\varphi} \\ g_{rt} & g_{rr} & g_{r\theta} & g_{r\varphi} \\ g_{\theta t} & g_{\theta r} & g_{\theta\theta} & g_{\theta\varphi} \\ g_{\varphi t} & g_{\varphi r} & g_{\varphi\theta} & g_{\varphi\varphi} \end{pmatrix}$$

$$\Gamma^{\theta}_{\mu v} = \begin{pmatrix} 0 & 0 & \frac{\dot{a}}{a} & \times \\ 0 & 0 & \frac{1}{r} & \\ \frac{\dot{a}}{a} & \frac{1}{r} & 0 & \\ \times & & & -\sin\theta\cos\theta \end{pmatrix} => \Gamma^{\theta}_{\mu v} = \begin{pmatrix} 0 & 0 & \frac{\dot{a}}{a} & 0 \\ 0 & 0 & \frac{1}{r} & \\ \frac{\dot{a}}{a} & \frac{1}{r} & 0 & \\ 0 & & & -\sin\theta\cos\theta \end{pmatrix}$$

The same will be true for φr and $\varphi\theta$:
Here all the other terms will be null.

$$\Gamma^{\theta}_{\mu v} = \begin{pmatrix} 0 & 0 & \frac{\dot{a}}{a} & 0 \\ 0 & 0 & \frac{1}{r} & 0 \\ \frac{\dot{a}}{a} & \frac{1}{r} & 0 & 0 \\ 0 & 0 & 0 & -\sin\theta\cos\theta \end{pmatrix} \tag{E.12}$$

$$\Gamma^{\theta}_{\theta t} = \frac{\dot{a}}{a} = \Gamma^{\theta}_{t\theta} = \frac{\dot{a}}{a}$$

$$\Gamma^{\theta}_{\theta r} = \frac{1}{r} = \Gamma^{\theta}_{r\theta} = \frac{1}{r}$$

$$\Gamma^{\theta}_{\varphi\varphi} = -\sin\theta\cos\theta$$

Study of the fourth term in φ

$\Gamma^{\varphi}_{\mu v} = \frac{1}{2} g^{\varphi\rho}(g_{\mu\rho,v} + g_{v\rho,\mu} - g_{\mu v,\rho})$ which will give a 4×4 matrix
As the matrix $g^{\varphi\rho}$ is diagonal, only the terms where $\varphi = \rho$ it is not zero.
So, we can replace everywhere φ by ρ.

$$\Gamma^{\varphi}_{\mu v} = \frac{1}{2} g^{\varphi\varphi}(g_{\mu\varphi,v} + g_{v\varphi,\mu} - g_{\mu v,\varphi})$$

We can therefore replace $g^{\varphi\varphi}$ with its value (for memory it is the inverse matrix of g that must be read).

$$\begin{array}{cccc} t & r & \theta & \varphi \end{array}$$

$$g_{\mu\nu} = \begin{pmatrix} c^2 & 0 & 0 & 0 \\ 0 & -\frac{a^2}{1-kr^2} & 0 & 0 \\ 0 & 0 & -a^2r^2 & 0 \\ 0 & 0 & 0 & -a^2r^2\sin^2\theta \end{pmatrix} \begin{matrix} t \\ r \\ \theta \\ \varphi \end{matrix} \quad => \quad g_{\mu\nu} = \begin{pmatrix} g_{tt} & g_{tr} & g_{t\theta} & g_{t\varphi} \\ g_{rt} & g_{rr} & g_{r\theta} & g_{r\varphi} \\ g_{\theta t} & g_{\theta r} & g_{\theta\theta} & g_{\theta\varphi} \\ g_{\varphi t} & g_{\varphi r} & g_{\varphi\theta} & g_{\varphi\varphi} \end{pmatrix}$$

$$=> g^{\varphi\varphi} = -\frac{1}{r^2 a^2 \sin^2\theta}$$

We get:

$$\Gamma^{\varphi}_{\mu\nu} = -\frac{1}{2r^2 a^2 \sin^2\theta}(g_{\mu\varphi,\nu} + g_{\nu\varphi,\mu} - g_{\mu\nu,\varphi})$$

So, we have to calculate this for all the components of μ and ν possible.

First case μ and $\nu = t$

$$\Gamma^{\varphi}_{\mu\nu} = -\frac{1}{2a^2r^2\sin^2\theta}(g_{\mu\varphi,\nu} + g_{\nu\varphi,\mu} - g_{\mu\nu,\varphi}) => \Gamma^{\varphi}_{\mu\nu}$$
$$= -\frac{1}{2a^2r^2\sin^2\theta}(g_{t\varphi,t} + g_{t\varphi,t} - g_{tt,\varphi})$$

Two first terms are zero because no diagonal terms of g remain the third.

$$g_{tt,\varphi} = \frac{\partial c^2}{\partial\varphi} = 0$$

$$\Gamma^{\varphi}_{\mu\nu} = -\frac{1}{2a^2r^2\sin^2\theta}\begin{pmatrix} \times \\ \\ \end{pmatrix} => \Gamma^{\varphi}_{\mu\nu} = -\frac{1}{2a^2r^2\sin^2\theta}\begin{pmatrix} 0 \\ \\ \end{pmatrix}$$

Second case $\mu = t$ and $\nu = r$

$$\Gamma^{\varphi}_{\mu\nu} = -\frac{1}{2a^2r^2\sin^2\theta}(g_{\mu\varphi,\nu} + g_{\nu\varphi,\mu} - g_{\mu\nu,\varphi}) => \Gamma^{\varphi}_{\mu\nu}$$
$$= -\frac{1}{2a^2r^2\sin^2\theta}(g_{t\varphi,r} + g_{r\varphi,t} - g_{tr,\varphi})$$

Concern only no diagonal term of $g => 0$

$$(g_{t\varphi}) = 0$$

and

$$(g_{r\varphi}) = 0\,(g_{tr}) = 0$$

$$\Gamma^{\varphi}_{\mu\nu} = -\frac{1}{2a^2r^2\sin^2\theta}\begin{pmatrix} 0 & \times & & \\ & & & \\ & & & \\ & & & \end{pmatrix} => \Gamma^{\varphi}_{\mu\nu} = -\frac{1}{2a^2r^2\sin^2\theta}\begin{pmatrix} 0 & 0 & & \\ & & & \\ & & & \\ & & & \end{pmatrix}$$

By symmetry (because the $\Gamma^{t}_{\mu\nu}$) are symmetrical on $\mu\nu$:

$$\Gamma^{\varphi}_{\mu\nu} = -\frac{1}{2a^2r^2\sin^2\theta}\begin{pmatrix} 0 & 0 & & \\ 0 & & & \\ & & & \\ & & & \end{pmatrix}$$

Third case $\mu = r$ and $\nu = r$

$$\Gamma^{\varphi}_{\mu\nu} = -\frac{1}{2a^2r^2\sin^2\theta}(g_{\mu\varphi,\nu} + g_{\nu\varphi,\mu} - g_{\mu\nu,\varphi}) => \Gamma^{\theta}_{\mu\nu}$$
$$= -\frac{1}{2a^2r^2\sin^2\theta}(g_{r\varphi,r} + g_{r\varphi,r} - g_{rr,\varphi})$$

And $(g_{r\varphi}) = 0$

$(g_{rr,\varphi}) = 0$ independent of (φ)

$$\Gamma^{\varphi}_{\mu\nu} = -\frac{1}{2a^2r^2\sin^2\theta}\begin{pmatrix} 0 & 0 & & \\ 0 & \times & & \\ & & & \\ & & & \end{pmatrix} => \Gamma^{\varphi}_{\mu\nu} = -\frac{1}{2a^2r^2\sin^2\theta}\begin{pmatrix} 0 & 0 & & \\ 0 & 0 & & \\ & & & \\ & & & \end{pmatrix}$$

So, in order not to be zero, it must be derivatives in relation to r, diagonal terms for which the indices are equal:

Fourth case $\mu = \theta$ and $\nu = \theta$

$$\Gamma^{\varphi}_{\mu\nu} = -\frac{1}{2a^2r^2\sin^2\theta}(g_{\mu\varphi,\nu} + g_{\nu\varphi,\mu} - g_{\mu\nu,\varphi}) => \Gamma^{\theta}_{\mu\nu}$$
$$= -\frac{1}{2a^2r^2\sin^2\theta}(g_{\theta\varphi,\theta} + g_{\theta\varphi,\theta} - g_{\theta\theta,\varphi})$$

$g_{\theta\varphi} = 0$ and $g_{\theta\theta,\varphi} => 0$

$$\Gamma^{\varphi}_{\mu\nu} = -\frac{1}{2a^2r^2\sin^2\theta}\begin{pmatrix} 0 & 0 & & \\ 0 & 0 & & \\ & & \times & \\ & & & \end{pmatrix} => \Gamma^{\varphi}_{\mu\nu} = -\frac{1}{2a^2r^2\sin^2\theta}\begin{pmatrix} 0 & 0 & & \\ 0 & 0 & & \\ & & 0 & \\ & & & \end{pmatrix}$$

Ditto with $\boldsymbol{\mu = \varphi}$ **and** $\boldsymbol{\nu = \varphi}$

$$\Gamma^{\varphi}_{\mu\nu} = -\frac{1}{2a^2r^2\sin^2\theta}(g_{\mu\varphi,\nu} + g_{\nu\varphi,\mu} - g_{\mu\nu,\varphi}) => \Gamma^{\varphi}_{\mu\nu}$$
$$= -\frac{1}{2a^2r^2\sin^2\theta}(g_{\varphi\varphi,\varphi} + g_{\varphi\varphi,\varphi} - g_{\varphi\varphi,\varphi})$$

No term depends on φ:

$$\Gamma^{\varphi}_{\mu v} = -\frac{1}{2a^2r^2\sin^2\theta}\begin{pmatrix} 0 & 0 & & \\ 0 & 0 & & \\ & & 0 & \\ & & & \times \end{pmatrix} => \Gamma^{\varphi}_{\mu v} = -\frac{1}{2a^2r^2\sin^2\theta}\begin{pmatrix} 0 & 0 & & \\ 0 & 0 & & \\ & & 0 & \\ & & & 0 \end{pmatrix}$$

Fifth case $\mu = \varphi$ and $v = r$

$$\Gamma^{\varphi}_{\mu v} = -\frac{1}{2a^2r^2\sin^2\theta}(g_{\mu\varphi,v} + g_{v\varphi,\mu} - g_{\mu v,\varphi}) =>$$
$$\Gamma^{\varphi}_{\mu v} = -\frac{1}{2a^2r^2\sin^2\theta}(g_{\varphi\varphi,r} + g_{r\varphi,\varphi} - g_{\varphi r,\varphi}) =>$$

Derivatives in respect to φ are null:
Remain:

$$g_{\varphi\varphi,r} = -\frac{\partial a^2r^2\sin^2\theta}{\partial r} = -2a^2r\sin^2\theta$$

$$\Gamma^{\varphi}_{\varphi r} = -\frac{1}{2a^2r^2\sin^2\theta}(g_{\varphi\varphi,r}) = -\frac{1}{2a^2r^2\sin^2\theta}(-2a^2r\sin^2\theta) = \frac{1}{r} = \Gamma^{\varphi}_{r\varphi}$$

$$\Gamma^{\varphi}_{\mu v} = \begin{pmatrix} 0 & 0 & & \\ 0 & 0 & & \times \\ & & 0 & \\ & & & 0 \end{pmatrix} => \Gamma^{\varphi}_{\mu v} = \begin{pmatrix} 0 & 0 & & \\ 0 & 0 & & \frac{1}{r} \\ & & 0 & \\ & \frac{1}{r} & & 0 \end{pmatrix}$$

Sixth case $\mu = \varphi$ and $v = \theta$

$$\Gamma^{\varphi}_{\mu v} = -\frac{1}{2a^2r^2\sin^2\theta}(g_{\mu\varphi,v} + g_{v\varphi,\mu} - g_{\mu v,\varphi}) =>$$
$$\Gamma^{\varphi}_{\mu v} = -\frac{1}{2a^2r^2\sin^2\theta}(g_{\varphi\varphi,\theta} + g_{\theta\varphi,\varphi} - g_{\varphi\theta,\varphi}) =>$$

The terms of g no diagonals are zero remain $g_{\varphi\varphi,\theta} = -\frac{\partial r^2a^2\sin^2\theta}{\partial\theta} = -2r^2a^2\sin\theta\cos\theta$

$$\Gamma^{\varphi}_{\varphi\theta} = -\frac{1}{2a^2r^2\sin^2\theta}(g_{\varphi\varphi,\theta}) = \Gamma^{\varphi}_{\varphi\theta} = -\frac{1}{2a^2r^2\sin^2\theta}(-2r^2a^2\sin\theta\cos\theta) = \frac{\cos\theta}{\sin\theta}$$
$$= \frac{1}{tg\theta}\Gamma^{\varphi}_{\varphi\theta}$$

$$\Gamma^{\varphi}_{\mu v} = \begin{pmatrix} 0 & 0 & & \\ 0 & 0 & & \frac{1}{r} \\ & & 0 & \\ & \frac{1}{r} & \times & 0 \end{pmatrix} => \Gamma^{\varphi}_{\mu v} = \begin{pmatrix} 0 & 0 & & \\ 0 & 0 & & \frac{1}{r} \\ & & 0 & \frac{1}{tg\theta} \\ & \frac{1}{r} & \frac{1}{tg\theta} & 0 \end{pmatrix}$$

Seventh case $\mu = \varphi$ and $v = t$

$$\Gamma^{\varphi}_{\mu v} = -\frac{1}{2a^2r^2\sin^2\theta}(g_{\mu\varphi,v} + g_{v\varphi,\mu} - g_{\mu v,\varphi}) => \Gamma^{\varphi}_{\mu v}$$
$$= -\frac{1}{2a^2r^2\sin^2\theta}(g_{\varphi\varphi,t} + g_{t\varphi,\varphi} - g_{\varphi t,\varphi})$$

The terms out of diagonals are zero remain $g_{\varphi\varphi,t} = -\frac{\partial(r^2a^2\sin^2\theta)}{\partial t} = -2r^2 a \overset{\bullet}{a} \sin^2\theta$

$$\Gamma^{\varphi}_{\varphi t} = -\frac{1}{2a^2r^2\sin^2\theta}(g_{\varphi\varphi,t}) = \Gamma^{\varphi}_{\varphi t} = -\frac{1}{2a^2r^2\sin^2\theta}\left(-2r^2 a \overset{\bullet}{a} \sin^2\theta\right) = \frac{\overset{\bullet}{a}}{a}\Gamma^{\varphi}_{t\varphi}$$

$$=> \Gamma^{\varphi}_{\mu v} = \begin{pmatrix} 0 & 0 & & \times \\ 0 & 0 & & \frac{1}{r} \\ & & 0 & \frac{1}{tg\theta} \\ \times & \frac{1}{r} & \frac{1}{tg\theta} & 0 \end{pmatrix} => \Gamma^{\varphi}_{\mu v} = \begin{pmatrix} 0 & 0 & & \frac{\overset{\bullet}{a}}{a} \\ 0 & 0 & & \frac{1}{r} \\ & & 0 & \frac{1}{tg\theta} \\ \frac{\overset{\bullet}{a}}{a} & \frac{1}{r} & \frac{1}{tg\theta} & 0 \end{pmatrix}$$

Eighth case $\mu = \theta$ and $v = t$ and for $\mu = \theta$ and $v = r$

$$\Gamma^{\varphi}_{\mu v} = -\frac{1}{2r^2a^2\sin^2\theta}(g_{\mu\varphi,v} + g_{v\varphi,\mu} - g_{\mu v,\varphi})$$

$\Gamma^{\varphi}_{\mu v} = -\frac{1}{2a^2r^2\sin^2\theta}(g_{\theta\varphi,t} + g_{t\varphi,\theta} - g_{\theta\varphi,\varphi})$ with no diagonal g terms $= 0$
$\Gamma^{\varphi}_{\mu v} = -\frac{1}{2a^2r^2\sin^2\theta}(g_{\theta\varphi,r} + g_{r\varphi,\theta} - g_{\theta r,\varphi})$ with no diagonal g terms $= 0$

All other terms will be null

So:

$$\Gamma^{\varphi}_{\mu v} = \begin{pmatrix} 0 & 0 & 0 & \frac{\overset{\bullet}{a}}{a} \\ 0 & 0 & 0 & \frac{1}{r} \\ 0 & 0 & 0 & \frac{1}{tg\theta} \\ \frac{\overset{\bullet}{a}}{a} & \frac{1}{r} & \frac{1}{tg\theta} & 0 \end{pmatrix} \tag{E.13}$$

$$=> \Gamma^{\varphi}_{\varphi t} = \frac{\overset{\bullet}{a}}{a} = \Gamma^{\varphi}_{t\varphi}$$

$$=> \Gamma^{\varphi}_{\varphi\theta} = \frac{\cos\theta}{\sin\theta} = \frac{1}{tg\theta} = \Gamma^{\varphi}_{\varphi\theta}$$

$$\Gamma^{\varphi}_{\varphi r} = \frac{1}{r} = \Gamma^{\varphi}_{r\varphi}$$

$$
\Gamma^t_{\mu\nu} = \begin{pmatrix} 0 & 0 & 0 & 0 \\ 0 & \frac{a\,\dot{a}}{c^2(1-kr^2)} & 0 & 0 \\ 0 & 0 & \frac{a\,\dot{a}\,r^2}{c^2} & 0 \\ 0 & 0 & 0 & \frac{a\,\dot{a}\,r^2\sin^2\theta}{c^2} \end{pmatrix} \Gamma^r_{\mu\nu} = \begin{pmatrix} \Gamma^t_{tt} & \Gamma^t_{tr} & \Gamma^t_{t\theta} & \Gamma^t_{t\varphi} \\ \Gamma^t_{rt} & \Gamma^t_{rr} & \Gamma^t_{r\theta} & \Gamma^t_{r\varphi} \\ \Gamma^t_{\theta t} & \Gamma^t_{\theta r} & \Gamma^t_{\theta\theta} & \Gamma^t_{\theta\varphi} \\ \Gamma^t_{\varphi t} & \Gamma^t_{\varphi r} & \Gamma^t_{\varphi\theta} & \Gamma^t_{\varphi\varphi} \end{pmatrix} \quad \text{(E.14)}
$$

$$\Gamma^t_{tt} = 0$$

$$\Gamma^t_{rr} = \frac{a\,\dot{a}}{c^2(1-kr^2)}$$

$$\Gamma^t_{\theta\theta} = \frac{a\,\dot{a}\,r^2}{c^2}$$

$$\Gamma^t_{\varphi\varphi} = \frac{a\,\dot{a}\,r^2\sin^2\theta}{c^2}$$

$$
\Gamma^r_{\mu\nu} = \begin{pmatrix} 0 & \frac{\dot{a}}{a} & 0 & 0 \\ \frac{\dot{a}}{a} & \frac{kr}{(1-kr^2)} & 0 & 0 \\ 0 & 0 & -r(1-kr^2) & 0 \\ 0 & 0 & 0 & -r(1-kr^2)\sin^2\theta \end{pmatrix} \quad \text{(E.15)}
$$

$$
\Gamma^r_{\mu\nu} = \begin{pmatrix} \Gamma^r_{tt} & \Gamma^r_{tr} & \Gamma^r_{t\theta} & \Gamma^r_{t\varphi} \\ \Gamma^r_{rt} & \Gamma^r_{rr} & \Gamma^r_{r\theta} & \Gamma^r_{r\varphi} \\ \Gamma^r_{\theta t} & \Gamma^r_{\theta r} & \Gamma^r_{\theta\theta} & \Gamma^r_{\theta\varphi} \\ \Gamma^r_{\varphi t} & \Gamma^r_{\varphi r} & \Gamma^r_{\varphi\theta} & \Gamma^r_{\varphi\varphi} \end{pmatrix}
$$

$$\Gamma^r_{\varphi\varphi} = -r\left(1-kr^2\right)\sin^2\theta$$

$$\Gamma^r_{\theta\theta} = -r\left(1-kr^2\right)$$

$$\Gamma^r_{rr} = \frac{kr}{(1-kr^2)}$$

$$\Gamma^r_{tr} = \frac{a}{a} = \Gamma^r_{tr} = \frac{a}{a}$$

$$\Gamma^{\theta}_{\mu\nu} = \begin{pmatrix} 0 & 0 & \frac{\dot{a}}{a} & 0 \\ 0 & 0 & \frac{1}{r} & 0 \\ \frac{\dot{a}}{a} & \frac{1}{r} & 0 & 0 \\ 0 & 0 & 0 & -\sin\theta\cos\theta \end{pmatrix} \Gamma^{\theta}_{\mu\nu} = \begin{pmatrix} \Gamma^{\theta}_{tt} & \Gamma^{\theta}_{tr} & \Gamma^{\theta}_{t\theta} & \Gamma^{\theta}_{t\varphi} \\ \Gamma^{\theta}_{rt} & \Gamma^{\theta}_{rr} & \Gamma^{\theta}_{r\theta} & \Gamma^{\theta}_{r\varphi} \\ \Gamma^{\theta}_{\theta t} & \Gamma^{\theta}_{\theta r} & \Gamma^{\theta}_{\theta\theta} & \Gamma^{\theta}_{\theta\varphi} \\ \Gamma^{\theta}_{\varphi t} & \Gamma^{\theta}_{\varphi r} & \Gamma^{\theta}_{\varphi\theta} & \Gamma^{\theta}_{\varphi\varphi} \end{pmatrix} \tag{E.16}$$

$$\Gamma^{\theta}_{\theta t} = \frac{\dot{a}}{a} = \Gamma^{\theta}_{t\theta} = \frac{\dot{a}}{a}$$

$$\Gamma^{\theta}_{\theta r} = \frac{1}{r} = \Gamma^{\theta}_{r\theta} = \frac{1}{r}$$

$$\Gamma^{\theta}_{\varphi\varphi} = -\sin\theta\cos\theta$$

$$\Gamma^{\varphi}_{\mu\nu} = \begin{pmatrix} 0 & 0 & 0 & \frac{\dot{a}}{a} \\ 0 & 0 & 0 & \frac{1}{r} \\ 0 & 0 & 0 & \frac{1}{tg\theta} \\ \frac{\dot{a}}{a} & \frac{1}{r} & \frac{1}{tg\theta} & 0 \end{pmatrix} \Gamma^{\varphi}_{\mu\nu} = \begin{pmatrix} \Gamma^{\varphi}_{tt} & \Gamma^{\varphi}_{tr} & \Gamma^{\varphi}_{t\theta} & \Gamma^{\varphi}_{t\varphi} \\ \Gamma^{\varphi}_{rt} & \Gamma^{\varphi}_{rr} & \Gamma^{\varphi}_{r\theta} & \Gamma^{\varphi}_{r\varphi} \\ \Gamma^{\varphi}_{\theta t} & \Gamma^{\varphi}_{\theta r} & \Gamma^{\varphi}_{\theta\theta} & \Gamma^{\varphi}_{\theta\varphi} \\ \Gamma^{\varphi}_{\varphi t} & \Gamma^{\varphi}_{\varphi r} & \Gamma^{\varphi}_{\varphi\theta} & \Gamma^{\varphi}_{\varphi\varphi} \end{pmatrix} \tag{E.17}$$

$$=> \Gamma^{\varphi}_{\varphi t} = \frac{\dot{a}}{a} = \Gamma^{\varphi}_{t\varphi}$$

$$=> \Gamma^{\varphi}_{\varphi\theta} = \frac{\cos\theta}{\sin\theta} = \frac{1}{tg\theta} = \Gamma^{\varphi}_{\varphi\theta}$$

$$\Gamma^{\varphi}_{\varphi r} = \frac{1}{r} = \Gamma^{\varphi}_{r\varphi}$$

E.3 Step 2

We will now calculate the components of curvature tensor and Ricci's tensor
According to [52] and [53]

$$R^{\lambda}_{\mu\nu\alpha} = \Gamma^{\lambda}_{\mu\alpha,\nu} - \Gamma^{\lambda}_{\mu\nu,\alpha} + \Gamma^{\lambda}_{\mu\eta}\Gamma^{\eta}_{\nu\alpha} - \Gamma^{\lambda}_{\alpha\eta}\Gamma^{\eta}_{\mu\nu}$$

For memory, Ricci's tensor is written:
$R_{\mu\nu} = R^{\lambda}_{\mu\nu\lambda}$ with a contraction on λ:
What we need in the end is contraction ($\alpha => \lambda$):

$$R^{\lambda}_{\mu\nu\lambda} = \Gamma^{\lambda}_{\mu\lambda,\nu} - \Gamma^{\lambda}_{\mu\nu,\lambda} + \Gamma^{\lambda}_{\nu\eta}\Gamma^{\eta}_{\mu\lambda} - \Gamma^{\lambda}_{\lambda\eta}\Gamma^{\eta}_{\mu\nu}$$

Part One: Calculating R_{rr}

The case is being looked at $\mu = \nu = r$

$$R^{\lambda}_{rr\lambda} = R_{rr} = \Gamma^{\lambda}_{r\lambda,r} - \Gamma^{\lambda}_{rr,\lambda} + \Gamma^{\lambda}_{r\eta}\Gamma^{\eta}_{r\lambda} - \Gamma^{\lambda}_{\lambda\eta}\Gamma^{\eta}_{rr} \qquad \text{(E.19)}$$

T1 T2 T3 T4

T1 calculation:

$$\Gamma^{\lambda}_{r\lambda,r} = \Gamma^{t}_{rt,r} + \Gamma^{r}_{rr,r} + \Gamma^{\theta}_{r\theta,r} + \Gamma^{\varphi}_{r\varphi,r}$$

We will then look for each of the terms in the matrix calculated at the previous stage.

$$\Gamma^{\lambda}_{r\lambda,r} = \frac{\partial \Gamma^{t}_{rt}}{\partial r} + \frac{\partial \Gamma^{r}_{rr}}{\partial r} + \frac{\partial \Gamma^{\theta}_{r\theta}}{\partial r} + \frac{\partial^{\varphi}_{r\varphi}}{\partial r}$$

$$\Gamma^{\lambda}_{r\lambda,r} = \frac{\partial(0)}{\partial r} + \frac{\partial\left(\frac{kr}{1-kr^2}\right)}{\partial r} + \frac{\partial\left(\frac{1}{r}\right)}{\partial r} + \frac{\partial\left(\frac{1}{r}\right)}{\partial r}$$

$$\Gamma^{\lambda}_{r\lambda,r} = \left(\frac{k(1+kr^2)}{(1-kr^2)^2}\right)' - \frac{2}{r^2}$$

$$u = kr \quad u' = k$$

$$v = (1-kr^2) \quad v' = -2kr$$

$$\frac{k\,(1\text{-}kr^2) + 2k^2r^2}{(1\text{-}kr^2)^2}$$

$$\frac{(k\text{-}k^2r^2) + 2k^2r^2}{(1\text{-}kr^2)^2}$$

$$\frac{k(1 + kr^2)}{(1\text{-}kr^2)^2}$$

Calculating T2:

$$\Gamma^{\lambda}_{r\lambda,r} = \Gamma^{t}_{rt,r} + \Gamma^{r}_{rr,r} + \Gamma^{\theta}_{r\theta,r} + \Gamma^{\varphi}_{r\varphi,r}$$

$$\Gamma^{\lambda}_{rr,\lambda} = \Gamma^{t}_{rr,t} + \Gamma^{r}_{rr,r} + \Gamma^{\theta}_{rr,\theta} + \Gamma^{\varphi}_{rr,\varphi}$$

No term depends on t or φ first part $= 0$.

$$\Gamma^{\lambda}_{r\lambda,r} = \frac{\partial \Gamma^{t}_{rr}}{\partial t} + \frac{\partial \Gamma^{r}_{rr}}{\partial r} + \frac{\partial \Gamma^{\theta}_{rr}}{\partial \theta} + \frac{\partial \Gamma^{\varphi}_{rr}}{\partial \varphi}$$

$$\Gamma^{\lambda}_{rr,\lambda} = \frac{\partial\left(\frac{\dot{a}\,a}{c^2(1-kr^2)}\right)}{\partial t} + \frac{\partial \frac{kr}{1-kr^2}}{\partial r} + \frac{\partial(0)}{\partial \theta} + \frac{\partial(0)}{\partial \varphi}$$

$$\boxed{\Gamma^{\lambda}_{rr,\lambda} = \frac{\ddot{a}\,a + (\dot{a})^2}{c^2(1-kr^2)} + \left(\frac{k(1+kr^2)}{(1-kr^2)^2}\right)'} \tag{E.20}$$

T3 calculation:

Attention we have two sums entangled, one on λ and one on η.

$$\begin{aligned}
\Gamma^{\lambda}_{r\eta}\Gamma^{\eta}_{r\lambda} &= \Gamma^{t}_{r\eta}\Gamma^{\eta}_{rt} + \Gamma^{r}_{r\eta}\Gamma^{\eta}_{rr} + \Gamma^{\theta}_{r\eta}\Gamma^{\eta}_{r\theta} + \Gamma^{\varphi}_{r\eta}\Gamma^{\eta}_{r\varphi} \\
&= \Gamma^{t}_{rt}\Gamma^{t}_{rt} + \Gamma^{t}_{rr}\Gamma^{r}_{rt} + \Gamma^{t}_{r\theta}\Gamma^{\theta}_{rt} + \Gamma^{t}_{r\varphi}\Gamma^{\varphi}_{rt} + \Gamma^{r}_{rt}\Gamma^{t}_{rr} + \Gamma^{r}_{rr}\Gamma^{r}_{rr} + \Gamma^{r}_{r\theta}\Gamma^{\theta}_{rr} + \Gamma^{r}_{r\varphi}\Gamma^{\varphi}_{rr} \\
&\quad + \Gamma^{\theta}_{rt}\Gamma^{t}_{r\theta} + \Gamma^{\theta}_{rr}\Gamma^{r}_{r\theta} + \Gamma^{\theta}_{r\theta}\Gamma^{\theta}_{r\theta} + \Gamma^{\theta}_{r\varphi}\Gamma^{\varphi}_{r\theta} + \Gamma^{\varphi}_{rt}\Gamma^{t}_{r\varphi} + \Gamma^{\varphi}_{rr}\Gamma^{r}_{r\varphi} + \Gamma^{\varphi}_{r\theta}\Gamma^{\theta}_{r\varphi} + \Gamma^{\varphi}_{r\varphi}\Gamma^{\varphi}_{r\varphi} \\
&= \left((0)^2\left(\frac{\dot{a}\,a}{c^2(1-kr^2)}\right)\left(\frac{\dot{a}}{a}\right) + 0 + 0\right) \\
&\quad + \left(\left(\frac{\dot{a}}{a}\right)\left(\frac{\dot{a}\,a}{c^2(1-kr^2)}\right) + \left(\frac{k^2r^2}{(1-kr^2)^2}\right) + 0 + 0\right) \\
&\quad + \left(0 + 0 + \left(\frac{1}{r^2}\right) + 0\right) + \left(0 + 0 + 0 + \left(\frac{1}{r^2}\right)\right)
\end{aligned}$$

$$\Gamma^{\lambda}_{r\eta}\Gamma^{\eta}_{r\lambda} = +2\left(\frac{\dot{a}\,a}{c^2(1-kr^2)}\right)\left(\frac{\dot{a}}{a}\right) + \frac{2}{r^2} + \left(\frac{k^2r^2}{(1-kr^2)^2}\right) \tag{E.21}$$

T4 calculation:

Attention we have two sums entangled, one on λ and one on η:

$$\Gamma^{\lambda}_{\lambda\eta}\Gamma^{\eta}_{rr} = \Gamma^{t}_{t\eta}\Gamma^{\eta}_{rr} + \Gamma^{r}_{r\eta}\Gamma^{\eta}_{rr} + \Gamma^{\theta}_{\theta\eta}\Gamma^{\eta}_{rr} + \Gamma^{\varphi}_{\varphi\eta}\Gamma^{\eta}_{rr}$$

$$\begin{aligned}\Gamma^{\lambda}_{\lambda\eta}\Gamma^{\eta}_{rr} = {} & \Gamma^{t}_{tt}\Gamma^{t}_{rr} + \Gamma^{t}_{tr}\Gamma^{r}_{rr} + \Gamma^{t}_{t\theta}\Gamma^{\theta}_{rr} + \Gamma^{t}_{t\varphi}\Gamma^{\varphi}_{rr} + \Gamma^{r}_{rt}\Gamma^{t}_{rr} + \Gamma^{r}_{rr}\Gamma^{r}_{rr} + \Gamma^{r}_{r\theta}\Gamma^{\theta}_{rr} + \Gamma^{r}_{r\varphi}\Gamma^{\varphi}_{rr} \\ & + \Gamma^{\theta}_{\theta t}\Gamma^{t}_{rr} + \Gamma^{\theta}_{\theta r}\Gamma^{r}_{rr} + \Gamma^{\theta}_{\theta\theta}\Gamma^{\theta}_{rr} + \Gamma^{\theta}_{\theta\varphi}\Gamma^{\varphi}_{rr} + \Gamma^{\varphi}_{\varphi t}\Gamma^{t}_{rr} + \Gamma^{\varphi}_{\varphi r}\Gamma^{r}_{rr} + \Gamma^{\varphi}_{\varphi\theta}\Gamma^{\theta}_{rr} + \Gamma^{\varphi}_{\varphi\varphi}\Gamma^{\varphi}_{rr}\end{aligned}$$

$$\begin{aligned}& (0+0+0+0) + \left(\left(\frac{\dot{a}}{a}\right)\left(\frac{\dot{a}\,a}{c^2(1-kr^2)}\right) + \left(\frac{k^2r^2}{(1-kr^2)^2}\right) + 0 + 0\right) \\ & + \left(\left(\frac{\dot{a}}{a}\right)\left(\frac{\dot{a}\,a}{c^2(1-kr^2)}\right) + \left(\frac{1}{r}\right)\left(\frac{kr}{(1-kr^2)}\right) + 0 + 0\right) \\ & + \left(\frac{\dot{a}}{a}\right)\left(\frac{\dot{a}\,a}{c^2(1-kr^2)}\right) + \left(\left(\frac{1}{r}\right)\left(\frac{kr}{(1-kr^2)}\right) + 0 + 0\right)\end{aligned}$$

$$\Gamma^{\lambda}_{\lambda\eta}\Gamma^{\eta}_{rr} = 3\left(\frac{\dot{a}}{a}\right)\left(\frac{\dot{a}\,a}{c^2(1-kr^2)}\right) + \left(\frac{k^2r^2}{(1-kr^2)^2}\right) + \left(\frac{1}{r}\right)\left(\frac{2kr}{(1-kr^2)}\right) \tag{E.22}$$

Summarize:

The case is being looked at $\mu = v = r$

$$R^{\lambda}_{rr\lambda} = R_{rr} = \Gamma^{\lambda}_{r\lambda,r} - \Gamma^{\lambda}_{rr,\lambda} + \Gamma^{\lambda}_{r\eta}\Gamma^{\eta}_{r\lambda} - \Gamma^{\lambda}_{\lambda\eta}\Gamma^{\eta}_{rr}$$

T1 T2 T3 T4

So, by doing the sum there remains:

$$R^{\lambda}_{rr\lambda} = R_{rr} = \Gamma^{\lambda}_{r\lambda,r} - \Gamma^{\lambda}_{rr,\lambda} + \Gamma^{\lambda}_{r\eta}\Gamma^{\eta}_{r\lambda} - \Gamma^{\lambda}_{\lambda\eta}\Gamma^{\eta}_{rr}$$

$$R_{rr} =$$

$$\Gamma^{\lambda}_{r\lambda,r} = \left(\frac{k(1+kr^2)}{(1-kr^2)^2}\right)' - \frac{2}{r^2}$$

$$-\Gamma^{\lambda}_{rr,\lambda} = -\frac{\ddot{a}\,a + \left(\dot{a}\right)^2}{c^2(1-kr^2)} - \left(\frac{k(1+kr^2)}{(1-kr^2)^2}\right)'$$

$$\Gamma^{\lambda}_{r\eta}\Gamma^{\eta}_{r\lambda} = +2\left(\frac{\dot{a}\,a}{c^2(1-kr^2)}\right)\left(\frac{\dot{a}}{a}\right) + \frac{2}{r^2} + \left(\frac{k^2r^2}{(1-kr^2)^2}\right)$$

$$\Gamma^{\lambda}_{\lambda\eta}\Gamma^{\eta}_{rr} = -3\left(\frac{\dot{a}}{a}\right)\left(\frac{\dot{a}\,a}{c^2(1-kr^2)}\right) - \left(\frac{k^2r^2}{(1-kr^2)^2}\right) - \left(\frac{1}{r}\right)\left(\frac{2kr}{(1-kr^2)}\right)$$

$$R_{rr} = -\left(\frac{\dot{a}}{a}\right)\left(\frac{\dot{a}\,a}{c^2(1-kr^2)}\right) - \left(\frac{1}{r}\right)\left(\frac{2kr}{(1-kr^2)}\right) - \frac{\ddot{a}\,a + \left(\dot{a}\right)^2}{c^2(1-kr^2)}$$

$$R_{rr} = -\left(\frac{\left(\dot{a}\right)^2}{c^2(1-kr^2)}\right) - \left(\frac{2k}{(1-kr^2)}\right) - \frac{\ddot{a}\,a + \left(\dot{a}\right)^2}{c^2(1-kr^2)}$$

$$R_{rr} = -\left(\frac{2k}{(1-kr^2)}\right) - \frac{\ddot{a}\,a}{c^2(1-kr^2)} - \frac{2\left(\dot{a}\right)^2}{c^2(1-kr^2)} \qquad \text{(E.23)}$$

Part Two: Calculating R_{tt}

$$R^{\lambda}_{t t\lambda} = R_{tt} = \Gamma^{\lambda}_{t\lambda,t} - \Gamma^{\lambda}_{tt,\lambda} + \Gamma^{\lambda}_{t\eta}\Gamma^{\eta}_{t\lambda} - \Gamma^{\lambda}_{\lambda\eta}\Gamma^{\eta}_{tt}$$

The case is being looked at $\mu = \nu = t$

T1 T2 T3 T4

T1 calculation:
$$\Gamma^{\lambda}_{t\lambda,t} = \Gamma^{t}_{tt,t} + \Gamma^{r}_{tr,t} + \Gamma^{\theta}_{t\theta,t} + \Gamma^{\varphi}_{t\varphi,t}$$

On will then look for each of the terms in the matrix calculated at the previous stage.

$$\Gamma^{\lambda}_{t\lambda,t} = \frac{\partial \Gamma^{t}_{tt}}{\partial t} + \frac{\partial \Gamma^{r}_{tr}}{\partial t} + \frac{\partial \Gamma^{\theta}_{t\theta}}{\partial t} + \frac{\partial^{\varphi}_{t\varphi}}{\partial t}$$

$$\Gamma^{\lambda}_{t\lambda,t} = \frac{\partial(0)}{\partial t} + \frac{\partial\left(\frac{\dot{a}}{a}\right)}{\partial t} + \frac{\partial\left(\frac{\dot{a}}{a}\right)}{\partial t} + \frac{\partial\left(\frac{\dot{a}}{a}\right)}{\partial t}$$

$$u = \dot{a} \quad u' = \ddot{a}$$

$$v = a \quad v' = \dot{a}$$

$$\Gamma^{\lambda}_{t\lambda,t} = 3\left(\frac{\ddot{a}\,a - \left(\dot{a}\right)^2}{a^2}\right)$$

Calculating T2:
$$\Gamma^{\lambda}_{tt,\lambda} = \Gamma^{t}_{tt,t} + \Gamma^{r}_{tt,r} + \Gamma^{\theta}_{tt,\theta} + \Gamma^{\varphi}_{tt,\varphi}$$

No term depends on t or φ => first part = 0

$$\Gamma^{\lambda}_{t\lambda,t} = \frac{\partial(0)}{\partial t} + \frac{\partial(0)}{\partial r} + \frac{\partial(0)}{\partial \theta} + \frac{\partial(0)}{\partial \varphi} = 0$$

$$\boxed{\Gamma^{\lambda}_{tt,t} = 0}$$

T3 calculation:

Attention we have two sums entangled, one on λ and one on η.

$$\begin{aligned}\Gamma^{\lambda}_{t\eta}\Gamma^{\eta}_{t\lambda} &= \Gamma^{t}_{t\eta}\Gamma^{\eta}_{tt} + \Gamma^{r}_{t\eta}\Gamma^{\eta}_{tr} + \Gamma^{\theta}_{t\eta}\Gamma^{\eta}_{t\theta} + \Gamma^{\varphi}_{t\eta}\Gamma^{\eta}_{t\varphi} \\ &= \Gamma^{t}_{tt}\Gamma^{t}_{tt} + \Gamma^{t}_{tr}\Gamma^{r}_{tt} + \Gamma^{t}_{t\theta}\Gamma^{\theta}_{tt} + \Gamma^{t}_{t\varphi}\Gamma^{\varphi}_{tt} + \Gamma^{r}_{tt}\Gamma^{t}_{tr} + \Gamma^{r}_{tr}\Gamma^{r}_{tr} + \Gamma^{r}_{t\theta}\Gamma^{\theta}_{tr} \\ &\quad + \Gamma^{r}_{t\varphi}\Gamma^{\varphi}_{tr} + \Gamma^{\theta}_{tt}\Gamma^{t}_{t\theta} + \Gamma^{\theta}_{tr}\Gamma^{r}_{t\theta} + \Gamma^{\theta}_{t\theta}\Gamma^{\theta}_{t\theta} \\ &\quad + \Gamma^{\theta}_{t\varphi}\Gamma^{\varphi}_{t\theta} + \Gamma^{\varphi}_{tt}\Gamma^{t}_{t\varphi} + \Gamma^{\varphi}_{tr}\Gamma^{r}_{t\varphi} + \Gamma^{\varphi}_{t\theta}\Gamma^{\theta}_{t\varphi} + \Gamma^{\varphi}_{t\varphi}\Gamma^{\varphi}_{t\varphi}\end{aligned}$$

$$= (0+0+0+0) + \left(0 + \left(\frac{\dot{a}}{a}\right)^2 + 0 + 0\right) + \left(0 + 0 + \left(\frac{\dot{a}}{a}\right)^2 + 0\right) + \left(0 + 0 + 0\left(\frac{\dot{a}}{a}\right)^2\right)$$

$$\boxed{\Gamma^{\lambda}_{t\eta}\Gamma^{\eta}_{t\lambda} = 3\left(\frac{\dot{a}}{a}\right)^2} \tag{E.24}$$

T4 calculation:

Attention we have two sums entangled, one on λ and one on η.

$$\Gamma^{\lambda}_{\lambda\eta}\Gamma^{\eta}_{tt} = \Gamma^{t}_{t\eta}\Gamma^{\eta}_{tt} + \Gamma^{r}_{r\eta}\Gamma^{\eta}_{tt} + \Gamma^{\theta}_{\theta\eta}\Gamma^{\eta}_{tt} + \Gamma^{\varphi}_{\varphi\eta}\Gamma^{\eta}_{tt}$$

$$\begin{aligned}\Gamma^{\lambda}_{\lambda\eta}\Gamma^{\eta}_{tt} &= \Gamma^{t}_{tt}\Gamma^{t}_{tt} + \Gamma^{t}_{tr}\Gamma^{r}_{tt} + \Gamma^{t}_{t\theta}\Gamma^{\theta}_{tt} + \Gamma^{t}_{t\varphi}\Gamma^{\varphi}_{tt} + \Gamma^{r}_{rt}\Gamma^{t}_{tt} + \Gamma^{r}_{rr}\Gamma^{r}_{tt} \\ &\quad + \Gamma^{r}_{r\theta}\Gamma^{\theta}_{tt} + \Gamma^{r}_{r\varphi}\Gamma^{\varphi}_{tt} + \Gamma^{\theta}_{\theta t}\Gamma^{t}_{tt} + \Gamma^{\theta}_{\theta r}\Gamma^{r}_{tt} + \Gamma^{\theta}_{\theta\theta}\Gamma^{\theta}_{tt} \\ &\quad + \Gamma^{\theta}_{\theta\varphi}\Gamma^{\varphi}_{tt} + \Gamma^{\varphi}_{\varphi t}\Gamma^{t}_{tt} + \Gamma^{\varphi}_{\varphi r}\Gamma^{r}_{tt} + \Gamma^{\varphi}_{\varphi\theta}\Gamma^{\theta}_{tt} + \Gamma^{\varphi}_{\varphi\varphi}\Gamma^{\varphi}_{tt}\end{aligned}$$

$$(0+0+0+0) + (0+0+0+0) + (0+0+0+0) + (0+0+0+0)$$

$$\boxed{\Gamma^{\lambda}_{\lambda\eta}\Gamma^{\eta}_{tt} = 0}$$

Summary

The case is being looked at $\mu = \nu = t$

$$\boxed{R^{\lambda}_{t t\lambda} = R_{tt} = \Gamma^{\lambda}_{t\lambda,t} - \Gamma^{\lambda}_{tt,\lambda} + \Gamma^{\lambda}_{t\eta}\Gamma^{\eta}_{t\lambda} - \Gamma^{\lambda}_{\lambda\eta}\Gamma^{\eta}_{tt}}$$

T1 T2 T3 T4

The calculation on $\mu = \nu = t$ therefore gives:

$$R_{tt} =$$

$$\Gamma^{\lambda}_{t\lambda,t} = 3\left(\frac{\ddot{a}\,a - \left(\dot{a}\right)^2}{a^2}\right)$$

$$\Gamma^{\lambda}_{tt,t} = 0$$

$$\Gamma^{\lambda}_{t\eta}\Gamma^{\eta}_{t\lambda} = 3\left(\frac{\dot{a}}{a}\right)^2$$

$$\Gamma^{\lambda}_{\lambda\eta}\Gamma^{\eta}_{tt} = 0$$

$$R_{tt} = 3\left(\frac{\ddot{a}}{a}\right) \tag{E.25}$$

Part Three: Calculating $R_{\theta\theta}$.
The case is being looked at $\mu - \nu - \theta$

$$R^{\lambda}_{\mu\nu\alpha} = \Gamma^{\lambda}_{\mu\alpha,\nu} - \Gamma^{\lambda}_{\mu\nu,\alpha} + \Gamma^{\lambda}_{\mu\eta}\Gamma^{\eta}_{\nu\alpha} - \Gamma^{\lambda}_{\alpha,\eta}\Gamma^{\eta}_{\mu\nu}$$

$$R^{\lambda}_{\theta\theta\lambda} = R_{\theta\theta} = \Gamma^{\lambda}_{\theta\lambda,\theta} - \Gamma^{\lambda}_{\theta\theta,\lambda} + \Gamma^{\lambda}_{\theta\eta}\Gamma^{\eta}_{\theta\lambda} - \Gamma^{\lambda}_{\lambda\eta}\Gamma^{\eta}_{\theta\theta}$$

T1 T2 T3 T4

T1 calculation:

$$\Gamma^{\lambda}_{\theta\lambda,\theta} = \Gamma^{t}_{\theta t,\theta} + \Gamma^{r}_{\theta r,\theta} + \Gamma^{\theta}_{\theta\theta,\theta} + \Gamma^{\varphi}_{\theta\varphi,\theta}$$

On will then look for each of the terms in the matrix calculated at the previous stage.

$$\Gamma^{\lambda}_{\theta\lambda,\theta} = \frac{\partial\Gamma^{t}_{\theta t}}{\partial\theta} + \frac{\partial\Gamma^{r}_{\theta r}}{\partial\theta} + \frac{\partial\Gamma^{\theta}_{\theta\theta}}{\partial\theta} + \frac{\partial^{\varphi}_{\theta\varphi}}{\partial\theta}$$

$$\Gamma^{\lambda}_{\theta\lambda,\theta} = \frac{\partial(0)}{\partial\theta} + \frac{\partial(0)}{\partial\theta} + \frac{\partial(0)}{\partial\theta} + \frac{\partial\left(\frac{\cos\theta}{\sin\theta}\right)}{\partial\theta}$$

$$u = \cos\theta \quad u' = -\sin\theta$$

$$v = \sin\theta \quad v' = \cos\theta$$

$$\boxed{\Gamma^{\theta}_{\theta\lambda,\theta} = \left(\frac{-\sin^2\theta - \cos^2\theta}{\sin^2\theta}\right)' = -\frac{1}{\sin^2\theta}} \tag{E.26}$$

Calculating T2:

$$\Gamma^{\lambda}_{\theta\lambda,\theta} = \Gamma^{t}_{\theta t,\theta} + \Gamma^{r}_{\theta r,\theta} + \Gamma^{\theta}_{\theta\theta,\theta} + \Gamma^{\varphi}_{\theta\varphi,\theta}$$

$$\Gamma^{\lambda}_{\theta\theta,\lambda} = \Gamma^{t}_{\theta\theta,t} + \Gamma^{r}_{\theta\theta,r} + \Gamma^{\theta}_{\theta\theta,\theta} + \Gamma^{\varphi}_{\theta\theta,\varphi}$$

No term depends on t or $= \theta =>$ part one $= 0$

$$\Gamma^{\lambda}_{\theta\lambda,\theta} = \frac{\partial\Gamma^{t}_{\theta\theta}}{\partial t} + \frac{\partial\Gamma^{r}_{\theta\theta}}{\partial r} + \frac{\partial\Gamma^{\theta}_{\theta\theta}}{\partial\theta} + \frac{\partial\Gamma^{\varphi}_{\theta\theta}}{\partial\varphi}$$

$$\Gamma^{\lambda}_{\theta\theta,\lambda} = \frac{\partial\left(\frac{a\dot{a}r^2}{c^2}\right)}{\partial t} + \frac{\partial(-r(1-kr^2))}{\partial r} + \frac{\partial(0)}{\partial\theta} + \frac{\partial(0)}{\partial\varphi}$$

$$\boxed{\Gamma^{\lambda}_{\theta\theta,\lambda} = \frac{r^2}{c^2}((\dot{a})^2 + a\ddot{a}) - 1 + 3kr^2} \tag{E.27}$$

T3 calculation:

Attention we have two sums entangled, one on λ and one on η.

$$\begin{aligned}
\Gamma^{\lambda}_{\theta\eta}\Gamma^{\eta}_{\theta\lambda} &= \Gamma^{t}_{\theta\eta}\Gamma^{\eta}_{\theta t} + \Gamma^{r}_{\theta\eta}\Gamma^{\eta}_{\theta r} + \Gamma^{\theta}_{\theta\eta}\Gamma^{\eta}_{\theta\theta} + \Gamma^{\varphi}_{\theta\eta}\Gamma^{\eta}_{\theta\varphi} \\
&= \Gamma^{t}_{\theta t}\Gamma^{t}_{\theta t} + \Gamma^{t}_{\theta r}\Gamma^{r}_{\theta t} + \Gamma^{t}_{\theta\theta}\Gamma^{\theta}_{\theta t} + \Gamma^{t}_{\theta\varphi}\Gamma^{\varphi}_{\theta t} + \Gamma^{r}_{\theta t}\Gamma^{t}_{\theta r} + \Gamma^{r}_{\theta r}\Gamma^{r}_{\theta r} \\
&\quad + \Gamma^{r}_{\theta\theta}\Gamma^{\theta}_{\theta r} + \Gamma^{r}_{\theta\varphi}\Gamma^{\varphi}_{\theta r} + \Gamma^{\theta}_{\theta t}\Gamma^{t}_{\theta\theta} + \Gamma^{\theta}_{\theta r}\Gamma^{r}_{\theta\theta} + \Gamma^{\theta}_{\theta\theta}\Gamma^{\theta}_{\theta\theta} \\
&\quad + \Gamma^{\theta}_{\theta\varphi}\Gamma^{\varphi}_{\theta\theta} + \Gamma^{\varphi}_{\theta t}\Gamma^{t}_{\theta\varphi} + \Gamma^{\varphi}_{\theta r}\Gamma^{r}_{\theta\varphi} + \Gamma^{\varphi}_{\theta\theta}\Gamma^{\theta}_{\theta\varphi} + \Gamma^{\varphi}_{\theta\varphi}\Gamma^{\varphi}_{\theta\varphi}
\end{aligned}$$

$$\begin{aligned}
&= \left(0 + 0 + \left(\frac{a\dot{a}r^2}{c^2}\right)\left(\frac{\dot{a}}{a}\right) + 0\right) \\
&\quad + \left(0 + 0 + (-r(1-kr^2))\left(\frac{1}{r}\right) + 0\right) + \left(\frac{\dot{a}}{a}\right)\left(\frac{a\dot{a}r^2}{c^2}\right) \\
&\quad + \left(\frac{1}{r}\right)(-r(1-kr^2)) + 0 + 0) + \left(0 + 0 + 0 + \left(\frac{\cos^2\theta}{\sin^2\theta}\right)\right)
\end{aligned}$$

$$\Gamma^{\lambda}_{\theta\eta}\Gamma^{\eta}_{\theta\lambda} = 2\left(\frac{(\overset{\bullet}{a})^2}{c^2}\right) r^2 - 2(1 - kr^2) + \left(\frac{\cos^2\theta}{\sin^2\theta}\right) \tag{E.28}$$

T4 calculation:

Attention we have two sums entangled, one on λ and one on η.

$$\Gamma^{\lambda}_{\lambda\eta}\Gamma^{\eta}_{\theta\theta} = \Gamma^{t}_{t\eta}\Gamma^{\eta}_{\theta\theta} + \Gamma^{r}_{r\eta}\Gamma^{\eta}_{\theta\theta} + \Gamma^{\theta}_{\theta\eta}\Gamma^{\eta}_{\theta\theta} + \Gamma^{\varphi}_{\varphi\eta}\Gamma^{\eta}_{\theta\theta}$$

$$\begin{aligned}\Gamma^{\lambda}_{\lambda\eta}\Gamma^{\eta}_{\theta\theta} = {} & \Gamma^{t}_{tt}\Gamma^{t}_{\theta\theta} + \Gamma^{t}_{tr}\Gamma^{r}_{\theta\theta} + \Gamma^{t}_{t\theta}\Gamma^{\theta}_{\theta\theta} + \Gamma^{t}_{t\varphi}\Gamma^{\varphi}_{\theta\theta} + \Gamma^{r}_{rt}\Gamma^{t}_{\theta\theta} \\ & + \Gamma^{r}_{rr}\Gamma^{r}_{\theta\theta} + \Gamma^{r}_{r\theta}\Gamma^{\theta}_{\theta\theta} + \Gamma^{r}_{r\varphi}\Gamma^{\varphi}_{\theta\theta} + \Gamma^{\theta}_{\theta t}\Gamma^{t}_{\theta\theta} + \Gamma^{\theta}_{\theta r}\Gamma^{r}_{\theta\theta} \\ & + \Gamma^{\theta}_{\theta\theta}\Gamma^{\theta}_{\theta\theta} + \Gamma^{\theta}_{\theta\varphi}\Gamma^{\varphi}_{\theta\theta} + \Gamma^{\varphi}_{\varphi t}\Gamma^{t}_{\theta\theta} + \Gamma^{\varphi}_{\varphi r}\Gamma^{r}_{\theta\theta} + \Gamma^{\varphi}_{\varphi\theta}\Gamma^{\theta}_{\theta\theta} + \Gamma^{\varphi}_{\varphi\varphi}\Gamma^{\varphi}_{\theta\theta}\end{aligned}$$

$$\begin{aligned}(0 + 0 + 0 + 0) & + \left(\left(\frac{\overset{\bullet}{a}}{a}\right)\left(\frac{a\,\overset{\bullet}{a}\,r^2}{c^2}\right) + \left(\frac{kr}{(1-kr^2)}\right)(-r(1-kr^2))\,0 + 0\right) \\ & + \left(\left(\frac{\overset{\bullet}{a}}{a}\right)\left(\frac{a\,\overset{\bullet}{a}\,r^2}{c^2}\right) + \left(\frac{1}{r}\right)(-r(1-kr^2)) + 0 + 0\right) \\ & + \left(\left(\frac{\overset{\bullet}{a}}{a}\right)\left(\frac{a\,\overset{\bullet}{a}\,r^2}{c^2}\right) + \left(\frac{1}{r}\right)(-r(1-kr^2)) + 0 + 0\right)\end{aligned}$$

$$\Gamma^{\lambda}_{\lambda\eta}\Gamma^{\eta}_{\theta\theta} = 3r^2\left(\frac{(\overset{\bullet}{a^2})}{c^2}\right) - kr^2 + -2(1-kr^2) \tag{E.29}$$

In summary:

$$R^{\lambda}_{\theta\theta\lambda} = R_{\theta\theta} = \Gamma^{\lambda}_{\theta\lambda,\theta} - \Gamma^{\lambda}_{\theta\theta,\lambda} + \Gamma^{\lambda}_{\theta\eta}\Gamma^{\eta}_{\theta\lambda} - \Gamma^{\lambda}_{\lambda\eta}\Gamma^{\eta}_{\theta\theta}$$

$$R_{\theta\theta} =$$

$$\Gamma^{\theta}_{\theta\lambda,\theta} = -\frac{1}{\sin^2\theta}$$

$$\Gamma^{\lambda}_{\theta\theta,\lambda} = -\frac{r^2}{c^2}\left((\dot{a})^2 + a\ddot{a}\right) + 1 - 3kr^2$$

$$\Gamma^{\lambda}_{\theta\eta}\Gamma^{\eta}_{\theta\lambda} = 2\left(\frac{(\dot{a})^2}{c^2}\right)r^2 - 2(1-kr^2) + \left(\frac{\cos^2\theta}{\sin^2\theta}\right)$$ for memory $\cos^2\theta + \sin^2\theta = 1$

$$\Gamma^{\lambda}_{\lambda\eta}\Gamma^{\eta}_{\theta\theta} = -3r^2\left(\frac{(\dot{a})^2}{c^2}\right) + kr^2 + 2(1-kr^2)$$

$$R_{\theta\theta} = -2\left(\frac{r^2}{c^2}\right)\left(\dot{a}\right)^2 - 2kr^2 - \left(\frac{r^2}{c^2}\right) a\ \ddot{a} \tag{E.30}$$

Part Four: Calculating $R_{\varphi\varphi}$.
The case is being looked at $\mu = \nu = \varphi$

$$R^{\lambda}_{\mu\nu\alpha} = \Gamma^{\lambda}_{\mu\alpha,\nu} - \Gamma^{\lambda}_{\mu\nu,\alpha} + \Gamma^{\lambda}_{\mu\eta}\Gamma^{\eta}_{\nu\alpha} - \Gamma^{\lambda}_{\alpha\eta}\Gamma^{\eta}_{\mu\nu}$$

$$R^{\lambda}_{\varphi\varphi\lambda} = R_{\varphi\varphi} = \Gamma^{\lambda}_{\varphi\lambda,\varphi} - \Gamma^{\lambda}_{\varphi\varphi,\lambda} + \Gamma^{\lambda}_{\varphi\eta}\Gamma^{\eta}_{\varphi\lambda} - \Gamma^{\lambda}_{\lambda\eta}\Gamma^{\eta}_{\varphi\varphi}$$

T1 T2 T3 T4

T1 calculation:

$$\Gamma^{\lambda}_{\varphi\lambda,\varphi} = \Gamma^{t}_{\varphi t,\varphi} + \Gamma^{r}_{\varphi r,\varphi} + \Gamma^{\theta}_{\varphi\theta,\varphi} + \Gamma^{\varphi}_{\varphi\varphi,\varphi}$$

One will then look for each of the terms in the matrix calculated at the previous stage.

$$\Gamma^{\lambda}_{\varphi\lambda,\varphi} = \frac{\partial \Gamma^{t}_{\varphi t}}{\partial \varphi} + \frac{\partial \Gamma^{r}_{\varphi r}}{\partial \varphi} + \frac{\partial \Gamma^{\theta}_{\varphi \theta}}{\partial \varphi} + \frac{\partial^{\varphi}_{\varphi\varphi}}{\partial \varphi}$$

$$\Gamma^{\lambda}_{\varphi\lambda,\kappa} = \frac{\partial(0)}{\partial \varphi} + \frac{\partial(0)}{\partial \varphi} + \frac{\partial(0)}{\partial \varphi} + \frac{\partial(0)}{\partial \varphi}$$

$$\boxed{\Gamma^{\theta}_{\varphi\lambda,\varphi} = 0}$$

Calculating T2:

$$\begin{aligned}\Gamma^{\lambda}_{\varphi\lambda,\varphi} &= \Gamma^{t}_{\varphi t,\varphi} + \Gamma^{r}_{\varphi r,\varphi} + \Gamma^{\theta}_{\varphi\theta,\varphi} + \Gamma^{\varphi}_{\varphi\varphi,\varphi}\\ \Gamma^{\lambda}_{\varphi\varphi,\lambda} &= \Gamma^{t}_{\varphi\varphi,t} + \Gamma^{r}_{\varphi\varphi,r} + \Gamma^{\theta}_{\varphi\varphi,\theta} + \Gamma^{\varphi}_{\varphi\varphi,\varphi}\end{aligned}$$

No term depends on φ => first part = 0

$$\Gamma^{\lambda}_{\varphi\lambda,\varphi} = \frac{\partial \Gamma^{t}_{\varphi\varphi}}{\partial t} + \frac{\partial \Gamma^{r}_{\varphi\varphi}}{\partial r} + \frac{\partial \Gamma^{\theta}_{\varphi\varphi}}{\partial \theta} + \frac{\partial \Gamma^{\varphi}_{\varphi\varphi}}{\partial \varphi}$$

$$\Gamma^{\lambda}_{\varphi\varphi,\lambda} = \frac{\partial\left(\frac{\dot{a}\; ar^2 \sin^2\theta}{c^2}\right)}{\partial t} + \frac{\partial(-r(1-kr^2)\sin^2\theta)}{\partial r} + \frac{\partial(-\sin\theta\,\cos\theta)}{\partial\theta} + \frac{\partial(0)}{\partial\varphi}$$

$$\boxed{\Gamma^{\lambda}_{\varphi\varphi,\lambda} = \frac{(\ddot{a}\,a + (\dot{a})^2) r^2 \sin^2\theta}{c^2} + (-\sin^2\theta + 3kr^2\sin^2\theta) + (-\cos^2\theta + \sin^2\theta)} \qquad \text{(E.31)}$$

T3 calculation:

Attention we have two sums entangled, one on λ and one on η.

$$\begin{aligned}\Gamma^{\lambda}_{\varphi\eta}\Gamma^{\eta}_{\varphi\lambda} &= \Gamma^{t}_{\varphi\eta}\Gamma^{\eta}_{\varphi t} + \Gamma^{r}_{\varphi\eta}\Gamma^{\eta}_{\varphi r} + \Gamma^{\theta}_{\varphi\eta}\Gamma^{\eta}_{\varphi\theta} + \Gamma^{\varphi}_{\varphi\eta}\Gamma^{\eta}_{\varphi\varphi}\\ &= \Gamma^{t}_{\varphi t}\Gamma^{t}_{\varphi t} + \Gamma^{t}_{\varphi r}\Gamma^{r}_{\varphi t} + \Gamma^{t}_{\varphi\theta}\Gamma^{\theta}_{\varphi t} + \Gamma^{t}_{\varphi\varphi}\Gamma^{\varphi}_{\varphi t} + \Gamma^{r}_{\varphi t}\Gamma^{t}_{\varphi r} + \Gamma^{r}_{\varphi r}\Gamma^{r}_{\varphi r} + \Gamma^{r}_{\varphi\theta}\Gamma^{\theta}_{\varphi r} + \Gamma^{r}_{\varphi\varphi}\Gamma^{\varphi}_{\varphi r}\\ &\quad + \Gamma^{\theta}_{\varphi t}\Gamma^{t}_{\varphi\theta} + \Gamma^{\theta}_{\varphi r}\Gamma^{r}_{\varphi\theta} + \Gamma^{\theta}_{\varphi\theta}\Gamma^{\theta}_{\varphi\theta} + \Gamma^{\theta}_{\varphi\varphi}\Gamma^{\varphi}_{\varphi\theta} + \Gamma^{\varphi}_{\varphi t}\Gamma^{t}_{\varphi\varphi} + \Gamma^{\varphi}_{\varphi r}\Gamma^{r}_{\varphi\varphi} + \Gamma^{\varphi}_{\varphi\theta}\Gamma^{\theta}_{\varphi\varphi} + \Gamma^{\varphi}_{\varphi\varphi}\Gamma^{\varphi}_{\varphi\varphi}\end{aligned}$$

$$\begin{aligned}&= \left(0 + 0 + 0 + \left(\frac{\dot{a}\; ar^2\sin^2\theta}{c^2}\right)\left(\frac{\dot{a}}{a}\right)\right)\\ &\quad + (0+0+0+0) + \left((1-kr)\sin\theta(-r)\left(\frac{1}{r}\right) + 0 + 0 \quad + (-\sin\theta\cos\theta)\frac{\cos\theta}{\sin\theta}\right)\\ &\quad + \left(\left(\frac{\dot{a}\; ar^2\sin^2\theta}{c^2}\right)\left(\frac{\dot{a}}{a}\right) + (1-kr^2)\sin^2\theta(-r)\left(\frac{1}{r}\right) + (-\sin\theta\cos\theta)\frac{\cos\theta}{\sin\theta} + 0\right)\end{aligned}$$

$$\Gamma^{\lambda}_{\varphi\eta}\Gamma^{\eta}_{\varphi\lambda} = 2\left(\frac{(\dot{a})^2\, r^2 \sin^2\theta}{c^2}\right) - 2(1-kr^2)\sin^2\theta - 2\cos^2\theta \qquad \text{(E.32)}$$

T4 calculation:

Attention we have two sums entangled, one on λ and one on η.

$$\Gamma^{\lambda}_{\lambda\eta}\Gamma^{\eta}_{\varphi\varphi} = \Gamma^{t}_{t\eta}\Gamma^{\eta}_{\varphi\varphi} + \Gamma^{r}_{r\eta}\Gamma^{\eta}_{\varphi\varphi} + \Gamma^{\theta}_{\theta\eta}\Gamma^{\eta}_{\varphi\varphi} + \Gamma^{\varphi}_{\varphi\eta}\Gamma^{\eta}_{\varphi\varphi}$$

$$\begin{aligned}\Gamma^{\lambda}_{\lambda\eta}\Gamma^{\eta}_{\varphi\varphi} = {}& \Gamma^{t}_{tt}\Gamma^{t}_{\varphi\varphi} + \Gamma^{t}_{tr}\Gamma^{r}_{\varphi\varphi} + \Gamma^{t}_{t\theta}\Gamma^{\theta}_{\varphi\varphi} + \Gamma^{t}_{t\varphi}\Gamma^{\varphi}_{\varphi\varphi} + \Gamma^{r}_{rt}\Gamma^{t}_{\varphi\varphi}\\ &+ \Gamma^{r}_{rr}\Gamma^{r}_{\varphi\varphi} + \Gamma^{r}_{r\theta}\Gamma^{\theta}_{\varphi\varphi} + \Gamma^{r}_{r\varphi}\Gamma^{\varphi}_{\varphi\varphi} + \Gamma^{\theta}_{\theta t}\Gamma^{t}_{\varphi\varphi} + \Gamma^{\theta}_{\theta r}\Gamma^{r}_{\varphi\varphi}\\ &+ \Gamma^{\theta}_{\theta\theta}\Gamma^{\theta}_{\varphi\varphi} + \Gamma^{\theta}_{\theta\varphi}\Gamma^{\varphi}_{\varphi\varphi} + \Gamma^{\varphi}_{\varphi t}\Gamma^{t}_{\varphi\varphi} + \Gamma^{\varphi}_{\varphi r}\Gamma^{r}_{\varphi\varphi} + \Gamma^{\varphi}_{\varphi\theta}\Gamma^{\theta}_{\varphi\varphi} + \Gamma^{\varphi}_{\varphi\varphi}\Gamma^{\varphi}_{\varphi\varphi}\end{aligned}$$

$$(0+0+0+0) + \left(\frac{\dot{a}}{a}\right)\left(\frac{\dot{a}\,a r^2\sin^2\theta}{c^2}\right) + \left(\frac{kr}{(1-kr^2)}\right)(-r(1-kr^2)\sin^2\theta) + 0 + 0) + \left(\frac{\dot{a}}{a}\right)\left(\frac{\dot{a}\,a r^2\sin^2\theta}{c^2}\right)$$

$$+\left(\frac{1}{r}\right)(-r(1-kr^2)\sin^2\theta) + 0 + 0) + \left(\frac{\dot{a}}{a}\right)\left(\frac{\dot{a}\,a r^2\sin^2\theta}{c^2}\right) + \left(\frac{1}{r}\right)(-r(1-kr^2)\sin^2\theta) +$$

$$(-\sin\theta\cos\theta)\frac{\cos\theta}{\sin\theta} + 0)$$

$$\Gamma^{\lambda}_{\lambda\eta}\Gamma^{\eta}_{\varphi\varphi} = 3\left(\frac{(\dot{a})^2\, r^2 \sin^2\theta}{c^2}\right) - 2\sin^2\theta - \cos^2\theta + kr^2(\sin^2\theta) \qquad \text{(E.33)}$$

In summary:

$$R^{\lambda}_{\theta\theta\lambda} = R_{\theta\theta} = \Gamma^{\lambda}_{\theta\lambda,\theta} - \Gamma^{\lambda}_{\theta\theta,\lambda} + \Gamma^{\lambda}_{\theta\eta}\Gamma^{\eta}_{\theta\lambda} - \Gamma^{\lambda}_{\lambda\eta}\Gamma^{\eta}_{\theta\theta}$$

$$R_{\theta\theta} =$$

$$\Gamma^{\theta}_{\varphi\lambda,\varphi} = 0$$

$$\Gamma^{\lambda}_{\varphi\varphi,\lambda} = -\frac{(\ddot{a}a + (\dot{a})^2)r^2\sin^2\theta}{c^2} - (-\sin^2\theta + 3kr^2\sin^2\theta)' - (-\cos^2\theta + \sin^2\theta)$$

$$\Gamma^{\lambda}_{\varphi\eta}\Gamma^{\eta}_{\varphi\lambda} = 2\left(\frac{(\dot{a})^2 r^2\sin^2\theta}{c^2}\right) - 2(1 - kr^2)\sin^2\theta - 2\cos^2\theta$$

$$\Gamma^{\lambda}_{\lambda\eta}\Gamma^{\eta}_{\varphi\varphi} = -3\left(\frac{(\dot{a})^2 r^2\sin^2\theta}{c^2}\right) + 2\sin^2\theta + \cos^2\theta - kr^2(\sin^2\theta)$$

$$R_{\varphi\varphi} = -2\left(\frac{(\dot{a})^2\, r^2\sin^2\theta}{c^2}\right) - 2kr^2(\sin^2\theta) - \frac{(\ddot{a}\,a)r^2\sin^2\theta}{c^2} \tag{E.34}$$

Final expression of Ricci's tensor.

$$R_{\mu\nu} = \begin{pmatrix} R_{tt} & R_{tr} & R_{t\theta} & R_{t\varphi} \\ R_{rt} & R_{rr} & R_{r\theta} & R_{r\varphi} \\ R_{\theta t} & R_{\theta r} & R_{\theta\theta} & R_{\theta\varphi} \\ R_{\varphi t} & R_{\varphi r} & R_{\varphi\theta} & R_{\varphi\varphi} \end{pmatrix} = R_{\mu\nu} = \begin{pmatrix} R_{tt} & 0 & 0 & 0 \\ 0 & R_{rr} & 0 & 0 \\ 0 & 0 & R_{\theta\theta} & 0 \\ 0 & 0 & 0 & R_{\varphi\varphi} \end{pmatrix}$$

$$R_{\mu\nu} = \begin{pmatrix} 3\left(\frac{\ddot{a}}{a}\right) & 0 & 0 & 0 \\ 0 & -\frac{2k}{1-kr^2} - \frac{\ddot{a}\,a}{c^2(1-kr^2)} - \frac{2\,\dot{a}^2}{c^2(1-kr^2)} & 0 & 0 \\ 0 & 0 & -\frac{2r^2\left(\dot{a}\right)^2}{c^2} - 2kr^2 - \frac{r^2 a\,\ddot{a}}{c^2} & 0 \\ 0 & 0 & 0 & -2\left(\frac{(\dot{a})^2 r^2\sin^2\theta}{c^2}\right) - 2kr^2\sin^2\theta - \frac{\left(\ddot{a}\,ar^2\sin^2\theta\right)}{c^2} \end{pmatrix} \tag{E.35}$$

$$R_{tt} = 3\left(\frac{\ddot{a}}{a}\right)$$

$$R_{rr} = -\left(\frac{2k}{(1-kr^2)}\right) - \frac{\ddot{a}\,a}{c^2(1-kr^2)} - \frac{2\left(\dot{a}\right)^2}{c^2(1-kr^2)}$$

$$R_{\theta\theta} = -2\left(\frac{r^2}{c^2}\right)\left(\dot{a}\right)^2 - 2kr^2 - \left(\frac{r^2}{c^2}\right)a\,\ddot{a}$$

$$R_{\varphi\varphi} = -2\left(\frac{(\dot{a})^2\, r^2 \sin^2\theta}{c^2}\right) - 2kr^2\left(\sin^2\theta\right) - \frac{(\ddot{a}\,a)r^2\sin^2\theta}{c^2}$$

$$g_{\mu\nu} = \begin{pmatrix} c^2 & 0 & 0 & 0 \\ 0 & -\frac{a}{1-kr^2} & 0 & 0 \\ 0 & 0 & -a^2r^2 & 0 \\ 0 & 0 & 0 & -a^2r^2\sin^2\theta \end{pmatrix} \begin{matrix} t \\ r \\ \theta \\ \varphi \end{matrix} \quad => $$

$$g_{\mu\nu} = \begin{pmatrix} g_{tt} & g_{tr} & g_{t\theta} & g_{t\varphi} \\ g_{rt} & g_{rr} & g_{r\theta} & g_{r\varphi} \\ g_{\theta t} & g_{\theta r} & g_{\theta\theta} & g_{\theta\varphi} \\ g_{\varphi t} & g_{\varphi r} & g_{\varphi\theta} & g_{\varphi\varphi} \end{pmatrix}$$

We notice that the space components Ricci's tensor are:

$$R_{tt} = 3\left(\frac{\ddot{a}}{a}\right) \qquad \text{(E.36)}$$

$$R_{ii} = \frac{g_{ii}}{a^2}\left(\frac{\ddot{a}\,a}{c^2} + \frac{2(\dot{a})^2}{c^2} + 2k\right) \text{ and 0 for all other components} \qquad \text{(E.37)}$$

$$R_{rr} = -\left(\frac{2k}{(1-kr^2)}\right) - \frac{\ddot{a}\,a}{c^2(1-kr^2)} - \frac{2\left(\dot{a}\right)^2}{c^2(1-kr^2)} \text{ and } g_{rr} = \frac{-a^2}{1-kr^2}$$

$$R_{\theta\theta} = -2\left(\frac{r^2}{c^2}\right)\left(\dot{a}\right)^2 - 2kr^2 - \left(\frac{r^2}{c^2}\right) a\,\ddot{a} \text{ and } g_{\theta\theta} = -a^2 r^2$$

$$R_{\varphi\varphi} = -2\left(\frac{(\dot{a})^2\, r^2 \sin^2\theta}{c^2}\right) - 2kr^2\left(\sin^2\theta\right) - \frac{(\ddot{a}\,a) r^2 \sin^2\theta}{c^2} \text{ and } g_{\varphi\varphi} = -a^2 r^2 \sin^2\theta$$

E.4 Steps 3 Determination of Scalar R

According to [51] and [52]

$$R = g^{\mu\nu} R_{\mu\nu} = g^{tt} R_{tt} + g^{rr} R_{rr} + g^{\theta\theta} R_{\theta\theta} + g^{\varphi\varphi} R_{\varphi\varphi}$$

$$R_{ii} = \frac{g_{ii}}{a^2}\left(\frac{\ddot{a}\,a}{c^2} + \frac{2(\dot{a})^2}{c^2} + 2k\right)$$

With:

$$g^{\mu\nu} = \begin{pmatrix} \frac{1}{c^2} & 0 & 0 & 0 \\ 0 & -\frac{(1-kr^2)}{a^2} & 0 & 0 \\ 0 & 0 & -\frac{1}{a^2 r^2} & 0 \\ 0 & 0 & 0 & -\frac{1}{a^2 r^2 \sin^2\theta} \end{pmatrix}$$

$$g_{\mu\nu} = \begin{pmatrix} c^2 & 0 & 0 & 0 \\ 0 & -\frac{a^2}{1-kr^2} & 0 & 0 \\ 0 & 0 & -a^2 r^2 & 0 \\ 0 & 0 & 0 & -a^2 r^2 \sin^2\theta \end{pmatrix}$$

$$\begin{aligned} R = {} & 3\left(\frac{\ddot{a}}{a}\right) \times \left(\frac{1}{c^2}\right) + -\frac{(1-kr^2)}{a^2}\frac{-a^2}{a^2(1-kr^2)}\left(\frac{\ddot{a}\,a}{c^2} + \frac{2(\dot{a})^2}{c^2} + 2k\right) + \\ & -\frac{1}{a^2 r^2}\left(-\frac{a^2 r^2}{a^2}\right)\left(\frac{\ddot{a}\,a}{c^2} + \frac{2(\dot{a})^2}{c^2} + 2k\right) \\ & -\frac{1}{a^2 r^2 \sin^2\theta}\left(-\frac{a^2 r^2 \sin^2\theta}{a^2}\right)\left(\frac{\ddot{a}\,a}{c^2} + \frac{2(\dot{a})^2}{c^2} + 2k\right) \end{aligned}$$

$$R = 3\left(\frac{\ddot{a}}{a}\right) \times \left(\frac{1}{c^2}\right) + \left(\frac{\ddot{a}}{ac^2} + \frac{2(\dot{a})^2}{a^2c^2} + \frac{2k}{a^2}\right) + \left(\frac{\ddot{a}}{ac^2} + \frac{2(\dot{a})^2}{a^2c^2} + \frac{2k}{a^2}\right) + \left(\frac{\ddot{a}}{ac^2} + \frac{2(\dot{a})^2}{a^2c^2} + \frac{2k}{a^2}\right)$$

So:

$$R = \left(\frac{6}{c^2} \left(\frac{\ddot{a}}{a} \right) + \frac{6}{c^2} \left(\frac{\dot{a}}{a} \right)^2 + \frac{6k}{a^2} \right) \tag{E.38}$$

It is recalled that the stress energy tensor is worth:

$$T_{\mu\nu} = \begin{pmatrix} \rho c^2 & 0 & 0 & 0 \\ 0 & -p & 0 & 0 \\ 0 & 0 & -p & 0 \\ 0 & 0 & 0 & -p \end{pmatrix} \tag{E.39}$$

$$R_{\mu\nu} - \frac{1}{2} g_{\mu\nu} R + \Lambda g_{\mu\nu} = -\frac{8\pi G}{c^4} T_{\mu\nu}$$

$$\begin{pmatrix} 3\left(\frac{\ddot{a}}{a}\right) & 0 & 0 & 0 \\ 0 & -\frac{2k}{1-kr^2} - \frac{\ddot{a}a}{c^2(1-kr^2)} - \frac{2\dot{a}^2}{c^2(1-kr^2)} & 0 & 0 \\ 0 & 0 & -\frac{2r^2\left(\dot{a}\right)^2}{c^2} - 2kr^2 - \frac{r^2 a\ddot{a}}{c^2} & 0 \\ 0 & 0 & 0 & -2\left(\frac{(\dot{a})^2 r^2 \sin^2\theta}{c^2}\right) - 2kr^2\sin^2\theta - \frac{\left(\ddot{a}ar^2\sin^2\theta\right)}{c^2} \end{pmatrix}$$

$$-\frac{1}{2} \begin{pmatrix} c^2 & 0 & 0 & 0 \\ 0 & -\frac{a^2}{1-kr^2} & 0 & 0 \\ 0 & 0 & -a^2r^2 & 0 \\ 0 & 0 & 0 & -a^2r^2\sin^2\theta \end{pmatrix} \left(\frac{6}{c^2}\left(\frac{\ddot{a}}{a}\right) + \frac{6}{c^2}\left(\frac{\dot{a}}{a}\right)^2 + \frac{6k}{a^2} \right)$$

$$+\Lambda \begin{pmatrix} c^2 & 0 & 0 & 0 \\ 0 & -\frac{a^2}{1-kr^2} & 0 & 0 \\ 0 & 0 & -a^2r^2 & 0 \\ 0 & 0 & 0 & -a^2r^2\sin^2\theta \end{pmatrix}$$

$$= -\frac{8\pi G}{c^4} \begin{pmatrix} \rho & 0 & 0 & 0 \\ 0 & -p & 0 & 0 \\ 0 & 0 & -p & 0 \\ 0 & 0 & 0 & -p \end{pmatrix} \begin{pmatrix} c^2 & 0 & 0 & 0 \\ 0 & -\frac{a^2}{1-kr^2} & 0 & 0 \\ 0 & 0 & -a^2r^2 & 0 \\ 0 & 0 & 0 & -a^2r^2\sin^2\theta \end{pmatrix}$$

$$ds^2 = c^2 dt^2 - a^2(t) \left\{ \frac{dr^2}{1 - kr^2} + r^2 \left(d\theta^2 + \sin^2\theta d\varphi^2 \right) \right\} \tag{E.40}$$

E.5 Steps 4 Expression of Einstein's Equation of General Relativity and Deduction of Friedmann–Lemaitre's Equations [52] and [53]

E.5.1 Reminder of Einstein's Equation

$$R_{\mu\nu} - \frac{1}{2} g_{\mu\nu} R + \Lambda g_{\mu\nu} = -\frac{8\pi G}{c^4} T_{\mu\nu} \tag{E.41}$$

We have to solve:

$$\begin{bmatrix} 3\left(\frac{\ddot{a}}{a}\right) & 0 & 0 & 0 \\ 0 & -\frac{2k}{1-kr^2} - \frac{\ddot{a}a}{c^2(1-kr^2)} - \frac{2\dot{a}^2}{c^2(1-kr^2)} & 0 & 0 \\ 0 & 0 & -\frac{2r^2\dot{a}^2}{c^2} - 2kr^2 - \frac{r^2 a\ddot{a}}{c^2} & 0 \\ 0 & 0 & 0 & -2\left(\frac{\dot{a}^2 r^2 \sin^2\theta}{c^2}\right) - 2kr^2\sin^2\theta - \frac{\ddot{a}ar^2\sin^2\theta}{c^2} \end{bmatrix}$$

$$-\frac{1}{2}\begin{bmatrix} c^2 & 0 & 0 & 0 \\ 0 & -\frac{a^2}{1-kr^2} & 0 & 0 \\ 0 & 0 & -a^2r^2 & 0 \\ 0 & 0 & 0 & -a^2r^2\sin^2\theta \end{bmatrix}\left\{\frac{6}{c^2}\left(\frac{\ddot{a}}{a}\right) + \frac{6}{c^2}\left(\frac{\dot{a}}{a}\right)^2 + \frac{6k}{a^2}\right\} + \Lambda\begin{bmatrix} c^2 & 0 & 0 & 0 \\ 0 & -\frac{a^2}{1-kr^2} & 0 & 0 \\ 0 & 0 & -a^2r^2 & 0 \\ 0 & 0 & 0 & -a^2r^2\sin^2\theta \end{bmatrix}$$

$$= -\frac{8\pi G}{c^4}\begin{bmatrix} \rho c^2 & 0 & 0 & 0 \\ 0 & -p & 0 & 0 \\ 0 & 0 & -p & 0 \\ 0 & 0 & 0 & -p \end{bmatrix}\begin{bmatrix} c^2 & 0 & 0 & 0 \\ 0 & -\frac{a^2}{1-kr^2} & 0 & 0 \\ 0 & 0 & -a^2r^2 & 0 \\ 0 & 0 & 0 & -a^2r^2\sin^2\theta \end{bmatrix}$$

E.5.2 Time-Related Equation No. 1 (μ and t)

$$R_{tt} - \frac{1}{2} g_{tt} R + \Lambda g_{tt} = -\frac{8\pi G}{c^4} T_{tt}$$

where

$$R_{tt} = 3\left(\frac{\ddot{a}}{a}\right)$$

$$T_{tt} = \rho c^2 g_{tt}$$

$$g_{tt} = c^2$$

$$R = \left(\frac{6}{c^2}\left(\frac{\ddot{a}}{a}\right) + \frac{6}{c^2}\left(\frac{\dot{a}}{a}\right)^2 + \frac{6k}{a^2} \right)$$

We get:

$$3\left(\frac{\ddot{a}}{a}\right) - \frac{c^2}{2}\left(\frac{6}{c^2}\left(\frac{\ddot{a}}{a}\right) + \frac{6}{c^2}\left(\frac{\dot{a}}{a}\right)^2 + \frac{6k}{a^2} \right) + \Lambda c^2 = -\frac{8\pi G}{c^4}\rho c^2 c^2$$

$$\left(-3\left(\frac{\dot{a}}{a}\right)^2 - \frac{3kc^2}{a^2} \right) + \Lambda c^2 = -\frac{8\pi G}{c^4}\rho c^4$$

$$\frac{1}{c^2}\left(-3\left(\frac{\dot{a}}{a}\right)^2 - \frac{3kc^2}{a^2} \right) + \Lambda \frac{c^2}{c^2} = -\frac{8\pi G}{c^4}\rho c^2$$

The dimensional equations give us:

$$\frac{1}{\frac{\mathrm{m}^2}{\mathrm{s}^2}} \times \left(\frac{1}{\mathrm{s}^2} + \frac{\frac{1}{\mathrm{m}^2} \times \frac{\mathrm{m}^2}{\mathrm{s}^2}}{1} + \frac{1}{\mathrm{m}^2} \times \frac{\mathrm{m}^2}{\mathrm{s}^2} \right) = \frac{\left(\frac{\mathrm{m}^3}{\mathrm{kgs}^2}\right)}{\left(\frac{\mathrm{m}}{\mathrm{s}}\right)^4} \times \frac{\mathrm{kg}}{\mathrm{m}^3} \times \frac{\mathrm{m}^2}{\mathrm{s}^2} = \frac{1}{\mathrm{m}^2}$$

$$\left(\frac{3}{c^2}\left(\frac{\dot{a}}{a}\right)^2 + \frac{3k}{a^2} \right) - \Lambda = \frac{8\pi G}{c^2}\rho$$

$$\left(\frac{3}{c^2}\left(\frac{\dot{a}}{a}\right)^2 \right) = \frac{8\pi G}{c^2}\rho - \frac{3k}{a^2} + \Lambda$$

$$\left(\left(\frac{\dot{a}}{a}\right)^2 \right) = \frac{8\pi G c^2}{3c^2}\rho - \frac{kc^2}{a^2} + \frac{\Lambda c^2}{3}$$

$\left(\frac{\dot{a}}{a}\right)^2 = \frac{8\pi G}{3}\rho + \frac{\Lambda c^2}{3} - \frac{kc^2}{a^2}$ **Friedmann–Lemaitre's first equation.**

The dimensional equation gives us:

$$\frac{1}{\mathrm{s}^2} = -\frac{\left(\frac{\mathrm{m}^3}{\mathrm{kgs}^2}\right)}{1} \times \frac{\mathrm{kg}}{\mathrm{m}^3} + \frac{1}{\mathrm{m}^2} \times \frac{\mathrm{m}^2}{\mathrm{s}^2} - \frac{1}{\mathrm{m}^2} \times \frac{\mathrm{m}^2}{\mathrm{s}^2}$$

{69} Source https://burro.cwru.edu/Academics/Astr330/Lect03/friedmann.html.

E.5.3 Space-Related Equation No. 2 (μ and t)

$$R_{\mu\nu} - \frac{1}{2} g_{\mu\nu} R + \Lambda g_{\mu\nu} = -\frac{8\pi G}{c^4} T_{\mu\nu}$$

With:

$$R_{ii} = \frac{g_{ii}}{a^2}\left(\frac{\ddot{a}\,a}{c^2} + \frac{2(\dot{a})^2}{c^2} + 2k\right)$$

$$T_{ii} = -p g_{ii}$$

$$R = \left(\frac{6}{c^2}\left(\frac{\ddot{a}}{a}\right) + \frac{6}{c^2}\left(\frac{\dot{a}}{a}\right)^2 + \frac{6k}{a^2}\right)$$

Becomes in Einstein's equation:

$$\frac{g_{ii}}{a^2}\left(\frac{\ddot{a}\,a}{c^2} + \frac{2(\dot{a})^2}{c^2} + 2k\right) - \frac{1}{2} g_{ii} R + \Lambda g_{ii} = -\frac{8\pi G}{c^4}(-p) g_{ii}$$

By simplifying by g_{ii} we get:

$$\left(\frac{\ddot{a}}{ac^2} + \frac{2(\dot{a})^2}{a^2 c^2} + \frac{2k}{a^2}\right) - \frac{1}{2} R + \Lambda = \frac{8\pi G p}{c^4}$$

Replacing R with its value:

$$\left(\frac{\ddot{a}}{ac^2} + \frac{2(\dot{a})^2}{a^2 c^2} + \frac{2k}{a^2}\right) - \frac{1}{2}\left(\frac{6}{c^2}\left(\frac{\ddot{a}}{a}\right) + \frac{6}{c^2}\left(\frac{\dot{a}}{a}\right)^2 + \frac{6k}{a^2}\right) + \Lambda = \frac{8\pi G}{c^4} p$$

$$-\left(\frac{2}{c^2}\left(\frac{\ddot{a}}{a}\right)+\frac{1}{c^2}\left(\frac{\dot{a}}{a}\right)^2+\frac{k}{a^2}\right)+\Lambda=\frac{8\pi G}{c^4}p$$

$$-\left(\frac{\ddot{a}}{a}\right)=\left(\frac{1}{2}\left(\frac{\dot{a}}{a}\right)^2+\frac{c^2k}{2a^2}\right)-\frac{\Lambda c^2}{2}+\frac{4\pi G}{c^2}p$$

$$\left(\frac{\ddot{a}}{a}\right)=\left(-\frac{1}{2}\left(\frac{\dot{a}}{a}\right)^2-\frac{c^2k}{2a^2}\right)+\frac{\Lambda c^2}{2}-\frac{4\pi G}{c^2}p$$

If we add up with Friedmann–Lemaitre's first equation.

$\left(\frac{\dot{a}}{a}\right)^2=\frac{8\pi G}{3}\rho+\frac{\Lambda c^2}{3}-\frac{kc^2}{a^2}$ Friedmann–Lemaitre's first equation.

$-\frac{8\pi G}{3}\rho-\frac{\Lambda c^2}{3}+\frac{kc^2}{a^2}+\left(\frac{\dot{a}}{a}\right)^2=0$ Friedmann–Lemaitre's first equation

$+$

$$2\left(\frac{\ddot{a}}{a}\right)=\left(-\left(\frac{\dot{a}}{a}\right)^2-\frac{c^2k}{a^2}\right)+\frac{2\Lambda c^2}{2}-\frac{8\pi G}{c^2}p$$

$$2\left(\frac{\ddot{a}}{a}\right)=\left(-\cancel{\frac{c^2k}{a^2}}\right)+\frac{2\Lambda c^2}{2}-\frac{8\pi G}{c^2}p-\frac{8\pi G}{3}\rho-\frac{\Lambda c^2}{3}+\cancel{\frac{kc^2}{a^2}}$$

$$2\left(\frac{\ddot{a}}{a}\right)=-\frac{24\pi G}{3c^2}p-\frac{8\pi Gc^2}{3c^2}\rho+\frac{2\Lambda c^2}{3}$$

$$\left(\frac{\ddot{a}}{a}\right)=-\frac{12\pi G}{3c^2}p-\frac{4\pi Gc^2}{3c^2}\rho+\frac{\Lambda c^2}{3}$$

$\left(\frac{\ddot{a}}{a}\right) = -\frac{4\pi G}{3c^2}(\rho c^2 + 3p) + \frac{\Lambda c^2}{3}$ 2nd Friedmann equation.

$$\frac{1}{\mathrm{s}^2} = \frac{\left(\frac{\mathrm{m}^3}{\mathrm{kgs}^2}\right)}{\frac{\mathrm{m}^2}{\mathrm{s}^2}} \times \left(\frac{\mathrm{kg}}{\mathrm{m}^3} \times \frac{\mathrm{m}^2}{\mathrm{s}^2} + \frac{\mathrm{kgm}}{\mathrm{s}^2\mathrm{m}^2}\right) + \frac{1}{\mathrm{m}^2} \times \frac{\mathrm{m}^2}{\mathrm{s}^2}$$

$$\frac{1}{\mathrm{s}^2} = \frac{\mathrm{m}^3\mathrm{s}^2}{\mathrm{kgs}^2\mathrm{m}^2} \times \left(\frac{\mathrm{kg}}{\mathrm{ms}^2} + \frac{\mathrm{kg}}{\mathrm{s}^2\mathrm{m}}\right) + \frac{1}{\mathrm{m}^2} \times \frac{\mathrm{m}^2}{\mathrm{s}^2}$$

$$\frac{1}{\mathrm{s}^2} = \frac{\mathrm{m}}{\mathrm{kg}} \times \left(\frac{\mathrm{kg}}{\mathrm{ms}^2} + \frac{\mathrm{kg}}{\mathrm{s}^2\mathrm{m}}\right) + \frac{1}{\mathrm{m}^2} \times \frac{\mathrm{m}^2}{\mathrm{s}^2}$$

Appendix F

Can-We Understand a Black Hole from the Strength of the Materials?

F.1 Preamble

We will show in this appendix that the plastic work of a beam in one dimension in pure bending implies a curvature tending towards infinity and a radius of curvature tending towards a singularity ($R = 0$) as in the case of a black hole in 4 dimensions.

In this appendix, the formulations of Jean-Courbon's book, "yielding applied to the calculation of structures", have been redemonstrated. They were then reformulated to highlight the notions of curvature χ and singularity ($R = 0$) as in the case of black holes. Thus, a black hole can be seen as the plasticization of the structure of space-time, overloaded by gravity during the gravitational collapse of a giant star.

We will thus study the case to one dimension to simplify the approach. We will then extend the case in 3 dimensions by generalizing the results obtained.

F.2 Determination of the Maximum Elastic Moment M_e that can Withstand a Single Bending Beam

Either a single bending beam of width b and height of $H = 2\ h$, solicited by abending moment $M = M_e$ imposing normal stresses on the extreme fibers equals to the elastic stress as σ_e defined in figure F.1.

At the elastic limit the bending moment is the following:

$$M_e = 2\left\{F \times \frac{H}{3}\right\} = 2\left\{\frac{1}{2}\left(\sigma_e \frac{H}{2} b\right) \times \frac{H}{3}\right\} = \frac{\sigma_e b H^2}{6}$$

$$\sigma = \sigma_e = \frac{Mv}{I} = \frac{M\frac{H}{2}}{\frac{bH^3}{12}} = \frac{6M}{bH^2}$$

$$M_e = \frac{\sigma_e b H^2}{6}$$

DOI: 10.1051/978-2-7598-2573-8.c916

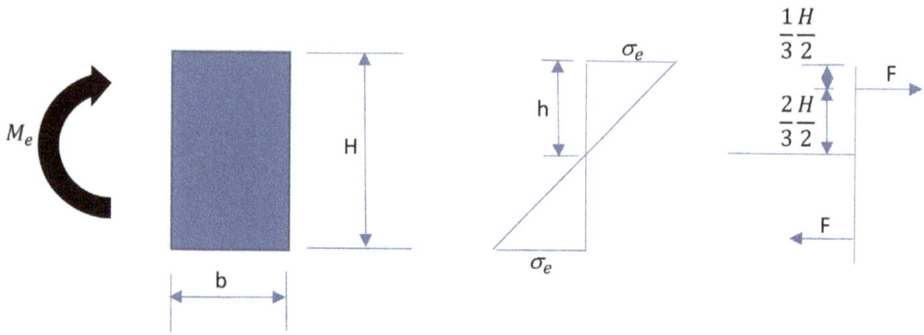

Fig. F.1 – Elasticity beam section.

F.3 Determination of the Maximum Plastic Moment M_p that can Withstand a Single Bending Beam

The same beam is this time solicited by a plastic moment $M_p > M_e$ as defined in figure F.2.

The bending moment is therefore:

$$M_p = 2\left\{F \times \frac{H}{4}\right\} = 2\left\{\left(\sigma_e \frac{H}{2} b\right) \times \frac{H}{4}\right\} = \frac{\sigma_e b H^2}{4}$$

We define:

$$\frac{H}{2} = h$$

In elastic, we have for the bending moment:

$$\boldsymbol{M_e = \frac{\sigma_e b H^2}{6} = \frac{2\sigma_e b h^2}{3}}$$

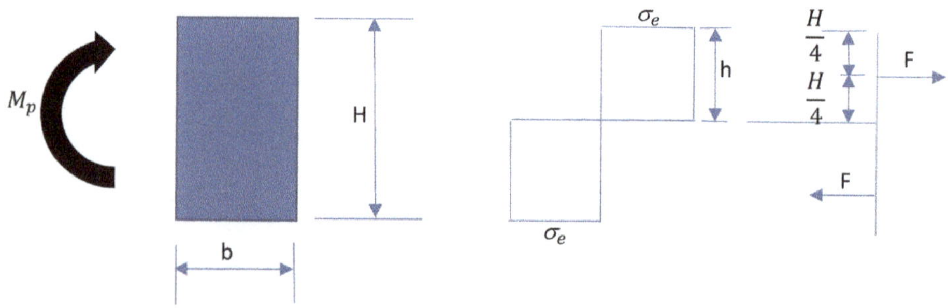

Fig. F.2 – Plastic beam section.

In total plasticity we have:

$$\boldsymbol{M_p = \frac{\sigma_e b H^2}{4} = \sigma_e b h^2}$$

And between the plastic moment and the elastic moment we have the following relationships:

$$\boldsymbol{\frac{2}{3} M_p = M_e}$$

$$\boldsymbol{M_p = \frac{3}{2} M_e}$$

F.4 Relationship Between the Curvature and Elastic Curvature of a Single Bending Beam with a Moment Less than the Elastic Moment

The relationship between the curvature and the bending moment is written (see appendix H):

$$\frac{M}{EI} = \frac{1}{R}$$

Either in pure bending and elastic limit and taking into account the value of inertia I we have:

$$M = \frac{EI}{R} = \frac{bH^3}{12}\frac{E}{R} = \frac{2}{3} bh^3 \frac{E}{R}$$

With:

$$M_e = \frac{\sigma_e b H^2}{6} = \frac{2\sigma_e b h^2}{3}$$

$$M = \frac{EI}{R} = \frac{bH^3}{12}\frac{E}{R} = \frac{2}{3} bh^3 \frac{E}{R} = \frac{M_e E h}{\sigma_e}\frac{1}{R} = \frac{M_e}{\frac{\sigma_e}{Eh}}\frac{1}{R} = M_e \frac{\chi}{\chi_e}$$

We define:

$$\boldsymbol{\frac{\sigma_e}{Eh} = \frac{1}{R_e} = \chi_e}$$

$$\frac{1}{R} = \chi$$

So, we have a curvature-moment relationship when the moment $M \leq M_e$

$$M = \frac{EI}{R} = EI\chi = \frac{M_e}{\chi_e}\chi$$

Looking for curvature χ:

$$\chi = \chi_e \frac{M}{M_e} = \frac{1}{R} = \frac{d^2y}{dx^2}$$

This gives the expression of the following curvature:

$$\frac{d^2y}{dx^2} = \frac{1}{R} = \chi_e \frac{M}{M_e} \text{ if } M \leq M_e$$

With:

$$\chi_e = \frac{\sigma_e}{Eh} = \frac{1}{R_e}$$

$$M_e = \frac{\sigma_e bH^2}{6} = \frac{2\sigma_e bh^2}{3}$$

F.5 Determination of the Elastic–Plastic Moment that can Support a Single Bending Beam

Either the same beam as above working in elastic–plastic (part of fiber are stretched to elastic stress σ_e only) as shown in figure F.3.

The bending moment is written this time:

$$M = 2(h-\psi)b\sigma_e\left(\psi + \frac{1}{2}(h-\psi)\right) + 2\frac{1}{2}(\psi)b\sigma_e\frac{2}{3}\psi$$

$$M = 2(h-\psi)\left(\frac{1}{2}\psi + \frac{h}{2}\right)b\sigma_e + \frac{2}{3}b\psi^2\sigma_e$$

$$M = (h-\psi)(\psi+h)b\sigma_e + \frac{2}{3}b\psi^2\sigma_e$$

$$M = b(h^2-\psi^2)\sigma_e + \frac{2}{3}b\psi^2\sigma_e$$

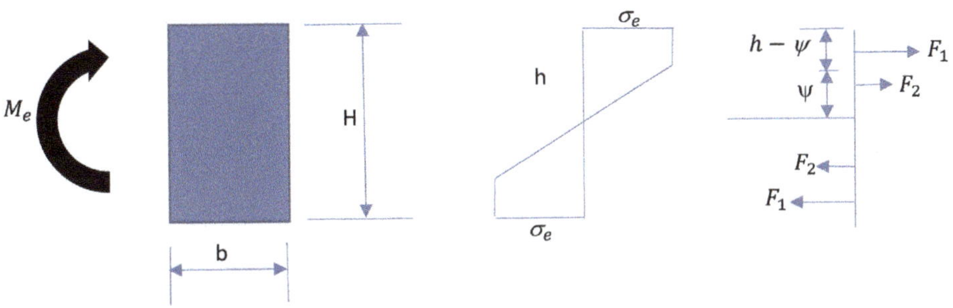

FIG. F.3 – Elasto-plasticity beam section.

$$M = bh^2\sigma_e - b\psi^2\sigma_e + \frac{2}{3}b\psi^2\sigma_e$$

$$M = bh^2\sigma_e - \frac{1}{3}b\psi^2\sigma_e$$

In addition, with the definition of the plastic moment:

$$M_p = \frac{\sigma_e bH^2}{4} = \sigma_e bh^2$$

$$\sigma_e = \frac{M_p}{bh^2}$$

That we report in the M expression above:

$$M = M_p - \frac{1}{3}\psi^2\frac{M_p}{h^2} = M_p\left(1 - \frac{1}{3}\frac{\psi^2}{h^2}\right)$$

Hence the phrase:

$$M = b(h^2 - \psi^2)\sigma_e + \frac{2}{3}b\psi^2\sigma_e = M_p\left(1 - \frac{1}{3}\frac{\psi^2}{h^2}\right)$$

F.6 Relationship Between the Curvature and the Elasto-Plastic Curvature of a Single Bending Beam with a Moment Greater or Equal to the Elastic Moment

The curvature χ is defined as following:

$$\chi = \frac{\sigma_e}{E\psi} = \frac{1}{R}$$

By introducing this formula in the bending moment M:

$$M = M_p\left(1 - \frac{1}{3}\frac{\left(\frac{\sigma_e}{E\chi}\right)^2}{h^2}\right)$$

$$M = M_p\left(1 - \frac{1}{3}\frac{(\sigma_e)^2}{E^2\chi^2 h^2}\right)$$

For memory we have:

$$\chi_e = \frac{\sigma_e}{Eh} = \frac{1}{R_e}$$

$$\chi = \frac{\sigma_e}{E\psi} = \frac{1}{R}$$

$$M = M_p\left(1 - \frac{1}{3}\frac{{\chi_e}^2}{\chi^2}\right)$$

The curvature χ of this expression can be extracted:

$$\frac{3M}{M_p} - 3 = -\frac{{\chi_e}^2}{\chi^2}$$

$$\frac{3M - 3M_p}{M_p} = -\frac{{\chi_e}^2}{\chi^2}$$

$$-\frac{M_p}{3M - 3M_p} = \frac{\chi^2}{{\chi_e}^2}$$

$$\chi^2 = {\chi_e}^2 \frac{M_p}{3M_p - 3M}$$

This gives the following expression of the curvature:

$$\frac{d^2y}{dx^2} = \frac{1}{R} = \chi_e\sqrt{\frac{M_p}{3M_p - 3M}} \text{ if } M_e \leq M < M_p$$

F.7 Interpretation of the Relationship between Curves and Plastic Moment

When the curvature $\chi = \chi_e$ reaches the elasticity limit σ_e, the bending moment M is equal to M_e or expressed depending on the plastic moment $\frac{2}{3}M_p$.

$$\boldsymbol{M = M_p\left(1 - \frac{1}{3}\frac{{\chi_e}^2}{\chi^2}\right)}$$

Indeed, when we have $\chi = \chi_e$:

$$M = M_p\left(1 - \frac{1}{3}\left(\frac{\chi_e}{\chi_e}\right)^2\right) = \frac{2}{3}M_p = M_e$$

When R tends towards 0, the curvature tends towards infinity $\lim\limits_{R\to 0}\chi = \frac{1}{R}$, and the moment is worth the maximum plastic moment M_p.

When $\chi \to \infty$ and therefore $R \to 0$, the moment tends towards the plastic moment:

$$M = M_p\left(1 - \frac{1}{3}\left(\frac{\chi_e}{\infty}\right)^2\right) = M_p$$

F.8 Energy Deformation of the Elasto-Plastic Beam

We have shown that in the post elastic field we have:

$$M = M_p\left(1 - \frac{1}{3}\frac{{\chi_e}^2}{\chi^2}\right)$$

Or:

$$\frac{M}{EI} = \frac{M_p}{EI}\left(1 - \frac{1}{3}\frac{{\chi_e}^2}{\chi^2}\right)$$

The bending energy is therefore written:

$$U = \frac{1}{2}\int_0^L \frac{M^2}{EI}dx = \frac{1}{2}\int_0^L \frac{\left(M_p\left(1 - \frac{1}{3}\left(\frac{\chi_e}{\chi}\right)^2\right)\right)^2}{EI}dx$$

As we are in simple and pure bending ($M = cte$)

$$\left(M_p\left(1 - \frac{1}{3}\frac{R^2}{{R_e}^2}\right)\right)^2 = 2EI\frac{U}{L}$$

Dividing by each member: $(EI)^2$

$$\left(\frac{M_p}{EI}\right)^2\left(1-\frac{1}{3}\frac{R^2}{R_e{}^2}\right)^2=\frac{2}{EI}\frac{U}{L}$$

If $R = R_e$, we are at the elastic limit of the section:

$$\left(\frac{2M_p}{3EI}\right)^2=\left(\frac{M_e}{EI}\right)^2=\frac{1}{R_e{}^2}=\frac{2}{EI}\frac{U}{L}$$

If the curvature χ tends towards infinity, R tends to 0 and the equation becomes:

$$\left(\frac{M_p}{EI}\right)^2=\frac{2}{EI}\frac{U}{L}$$

We define:

$$\frac{M_p}{EI}=\frac{1}{R_p}$$

$$\left(\frac{1}{R_p}\right)^2=\frac{2}{EI}\frac{U}{L}$$

When the mass becomes very large, the energy-mass and the associated strain energy U tend towards infinity, the curvature $(1/R_p)$ becomes very large, the radius R of curvature tends towards R_p which tends towards 0 (singularity). The section is laminated (the moment in the section is worth M_p).

F.9 Analogy with the General Relativity and Singularity of Black Holes

So, we found a relationship between the plastic moments and elastic curvature.

For $M \leq M_e$

$$\frac{1}{R}=\chi_e\frac{M}{M_e} \text{ or } \frac{1}{R}=\frac{1}{R_e}\frac{M}{M_e}$$

For $M_e \leq M \leq M_p$

$$M=M_p\left(1-\frac{1}{3}\frac{\chi_e{}^2}{\chi^2}\right) \text{ or } M=M_p\left(1-\frac{1}{3}\frac{R^2}{R_e{}^2}\right)$$

$$R=\sqrt{3}R_e\sqrt{\left(1-\frac{M}{M_p}\right)}$$

If we now return to the simplified two-dimensional analogy of a plate deformed by an intense point load at its center mass m, and model this plate by an assembly of pure two-dimensional bending beams as shown in figure F.4, we can visualize what a black hole is, *i.e.* a plastic deformation (Moment M tends towards the plastic moment M_p) structure of space-time. R is the curvature radius of the plate.

We find the analogy with the black hole. Indeed, under an intense mass, the plate (the assembly of beams) takes an infinite curvature $\chi = \frac{1}{R}$, the moments are plastic moments, space-time is yielded and therefore the mass concentrates in one point a singularity ($R = 0$), a singularity materialized in a way by the beam that folds on themselves.

$$\frac{d^2y}{dx^2} = \frac{1}{R} = \chi_e \sqrt{\frac{M_p}{3M_p - 3M}} \text{ if } M_e \leq M < M_p$$

It is clear that beyond the elastic moment, when the moment tends towards M_p (or R tends towards 0), the curvature of the beam tends towards infinity. This is the case for example if on a beam on two supports if you increase the F load infinitely (see figure F.4).

$$\lim_{\substack{M \to M_p \\ R \to 0}} \frac{d^2y}{dx^2} = \frac{1}{R} = \chi_e \sqrt{\frac{M_p}{3M_p - 3M}} = \infty$$

The reality of the black hole would require extending this approach to a three-dimensional assembly of beams see four-dimensional content of the 3 dimensions of space and the size of time to constitute space-time in 4 dimensions (see figure F.5).

We also notice that the overloaded beam curves more and more, the radius of curvature decreases, the fibers lengthen on the tense side and compress the other to the elastic limit. After a certain amount of load increase, the beam geometrically has nothing to do with a beam. A plastic hinge forms. The elastic strength of material no

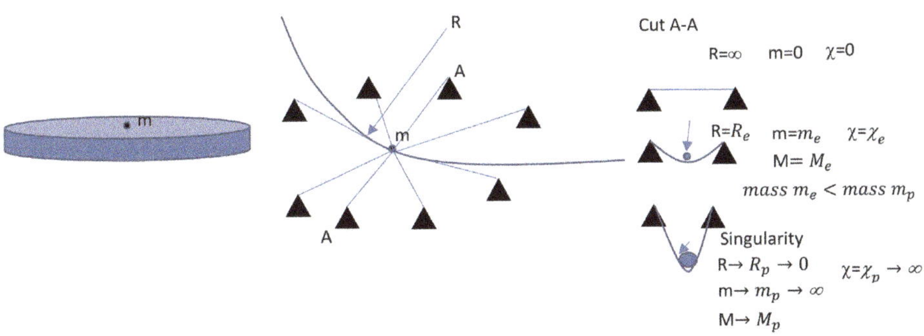

FIG. F.4 – Modeling by a 2D elastic plate of a space plane by an assembly of beams charged to yielding.

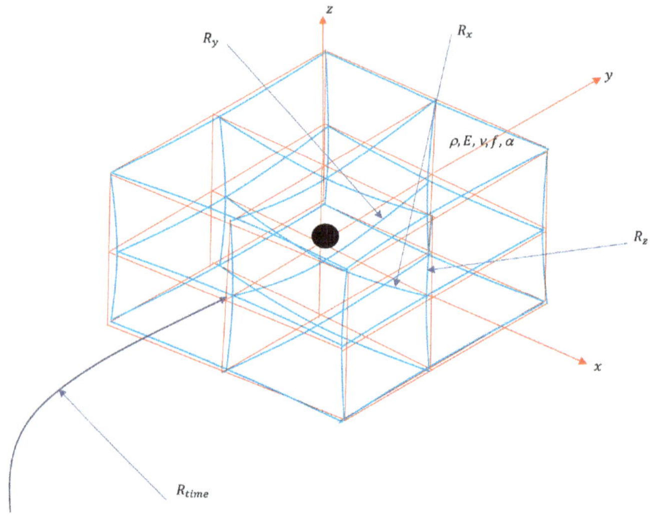

FIG. F.5 – Space-time structure modeled by as a frame of beams in bending charged by an intensive mass (black hole) imposing 4 curvatures associated with the 4 dimensions of space-time.

longer applies, the sections are yielded and laminated and stretch until all the fibers work at elastic stress, at that time the fibers stretch and compress all like a fluid, we speak of plastic flow. Continuing to increase the loading the fine structure of the beam tears. The starting beam no longer exists.

By analogy with a black hole that corresponds to a plasticization of space-time, the fabric of it follows the same path and after plasticization until its elastic limit tears in turn marking the end of the geometry of space and time.

Appendix G

Proof of the Relation Between Speed c and the Shear Modulus μ of the Elastic Medium in the Case of Gravitational Wave

G.1 Expression Issued of the General Relativity Linearized

Sources: {187, 192}.

The gravitational wave equation is:

$$\partial^{\lambda}\partial_{\lambda}\overline{h}_{\mu v} = \Box\,\overline{h}_{\mu v} = 0$$

where

$$h_{\mu v} = \overline{h}_{\mu v} - \frac{1}{2}\eta_{\mu v}\overline{h}$$

$$\overline{h}_{\mu v} = A_{\mu v}\cos(k_{\sigma}x^{\sigma})$$

$$A_{\mu v} = A_{+}\begin{bmatrix} 0 & 0 & 0 & 0 \\ 0 & +1 & 0 & 0 \\ 0 & 0 & -1 & 0 \\ 0 & 0 & 0 & 0 \end{bmatrix} + A_{\times}\begin{bmatrix} 0 & 0 & 0 & 0 \\ 0 & 0 & +1 & 0 \\ 0 & +1 & 0 & 0 \\ 0 & 0 & 0 & 0 \end{bmatrix}$$

As, h the trace of $h_{\mu v}$ (sum of terms following the diagonal) is null:

$$h_{\mu v} = \overline{h}_{\mu v}$$

The gravitational wave equation becomes:

$$\partial^{\lambda}\partial_{\lambda}h_{\mu v} = \Box\,h_{\mu v} = 0$$

DOI: 10.1051/978-2-7598-2573-8.c917

$$\frac{\partial^2 h_{\mu\nu}}{c^2\partial t^2} = \frac{\partial^2 h_{\mu\nu}}{\partial x^2} + \frac{\partial^2 h_{\mu\nu}}{\partial y^2} + \frac{\partial^2 h_{\mu\nu}}{\partial z^2} = \Delta h_{\mu\nu}$$

As we are in the 3 dimensions of space (i,j index):

$$h_{ij} = 2\varepsilon_{ij}$$

$$\frac{\partial^2 \varepsilon_{ij}}{c^2\partial t^2} = \frac{\partial^2 \varepsilon_{ij}}{\partial x^2} + \frac{\partial^2 \varepsilon_{ij}}{\partial y^2} + \frac{\partial^2 \varepsilon_{ij}}{\partial z^2} = \Delta \varepsilon_{ij}$$

So, we obtain *via* the general relativity linearized:

$$\frac{\partial^2 \varepsilon_{ij}}{c^2\partial t^2} = \Delta \varepsilon_{ij}$$

(see formula 3.9 in the paper of T.G. Tenev, M.F. Horstemeyer The Mechanics of Spacetime – A Solid Mechanics Perspective on the Theory of General Relativity).

G.2 Formula Issued of the Elastic Wave in Elasticity Theory

The fundamental dynamic equation is (force = mass × acceleration):

$$\rho \frac{\partial^2 u_i}{\partial t^2} = \frac{\partial \sigma_{ij}}{\partial x_j}$$

With for the stress tensor:

$$\sigma_{ij} = \frac{E}{(1+\nu)}\left(\varepsilon_{ij} + \frac{\nu}{(1-2\nu)}\varepsilon_{kk}\delta_{ij}\right)$$

Or with Lamé's coefficients:

$$\sigma_{ij} = 2\mu\varepsilon_{ij} + \lambda\varepsilon_{kk}\delta_{ij}$$

$$\sigma_{ij} = \frac{E}{(1+\nu)}\left(\varepsilon_{ij} + \frac{\nu}{(1-2\nu)}\varepsilon_{kk}\delta_{ij}\right)\begin{cases} \sigma_{11} = 2G\left[\varepsilon_{11} + \frac{\nu}{(1-2\nu)}(\varepsilon_{11}+\varepsilon_{22}+\varepsilon_{33})\right]; & (i=1, j=1) \\ \sigma_{22} = 2G\left[\varepsilon_{22} + \frac{\nu}{(1-2\nu)}(\varepsilon_{11}+\varepsilon_{22}+\varepsilon_{33})\right]; & (i=2, j=2) \\ \sigma_{33} = 2G\left[\varepsilon_{22} + \frac{\nu}{(1-2\nu)}(\varepsilon_{11}+\varepsilon_{22}+\varepsilon_{33})\right]; & (i=3, j=3) \\ \sigma_{13} = 2G[\varepsilon_{13}]; & (i=2, j=3) \\ \sigma_{23} = 2G[\varepsilon_{23}]; & (i=1, j=2) \\ \sigma_{12} = 2G[\varepsilon_{12}]; & (i=3, j=1) \end{cases}$$

The reciprocal equations are:

$$\varepsilon_{ij} = \frac{1}{E}\left((1+\nu)\sigma_{ij} - \nu\sigma_{kk}\delta_{ij}\right) \begin{cases} \varepsilon_{11} = \frac{1}{E}[\sigma_{11} - \nu(\sigma_{22}+\sigma_{33})]; & (i=1, j=1) \\ \varepsilon_{22} = \frac{1}{E}[\sigma_{22} - \nu(\sigma_{11}+\sigma_{33})]; & (i=2, j=2) \\ \varepsilon_{33} = \frac{1}{E}[\sigma_{33} - \nu(\sigma_{11}+\sigma_{22})]; & (i=3, j=3) \\ \varepsilon_{12} = \frac{1+\nu}{E}[\sigma_{12}]; & (i=2, j=3) \\ \varepsilon_{13} = \frac{1+\nu}{E}[\sigma_{13}]; & (i=1, j=2) \\ \varepsilon_{23} = \frac{1+\nu}{E}[\sigma_{23}]; & (i=3, j=1) \end{cases}$$

We define $\overline{\varepsilon} = \varepsilon_{11} + \varepsilon_{22} + \varepsilon_{33} = \frac{\partial u_1}{\partial x_1} + \frac{\partial u_2}{\partial x_2} + \frac{\partial u_3}{\partial x_3}$

The dynamic equation becomes with the σ_{ij} definition:

$$\rho\frac{\partial^2 u_i}{\partial t^2} = \frac{\partial}{\partial x_j}\left\{2\mu\varepsilon_{ij} + \lambda\varepsilon_{kk}\delta_{ij}\right\}$$

We obtain for example with $i = 1$ with the summation on j on ε_{ij} and $\frac{\partial}{\partial x^j}$ and with δ_{ij} the Kronecker symbol:

$$\rho\frac{\partial^2 u_1}{\partial t^2} = \frac{\partial}{\partial x_j}\left\{2\mu\varepsilon_{1j} + \lambda\varepsilon_{kk}\delta_{1j}\right\}$$

$$\rho\frac{\partial^2 u_1}{\partial t^2} = 2\mu\frac{\partial\varepsilon_{11}}{\partial x_1} + 2\mu\frac{\partial\varepsilon_{12}}{\partial x_2} + 2\mu\frac{\partial\varepsilon_{13}}{\partial x_3} + \lambda\frac{\partial(\varepsilon_{11}+\varepsilon_{22}+\varepsilon_{33})}{\partial x_1}$$

With for the strain tensor:

$$\varepsilon_{ij} = \frac{1}{2}\left(\frac{\partial u_i}{\partial x_j} + \frac{\partial u_j}{\partial x_i}\right)$$

$$\varepsilon_{11} = \left(\frac{\partial u_1}{\partial x_1}\right); \varepsilon_{22} = \left(\frac{\partial u_2}{\partial x_2}\right); \varepsilon_{33} = \left(\frac{\partial u_3}{\partial x_3}\right)$$

$$\varepsilon_{12} = \frac{1}{2}\left(\frac{\partial u_1}{\partial x_2} + \frac{\partial u_2}{\partial x_1}\right)$$

$$\varepsilon_{13} = \frac{1}{2}\left(\frac{\partial u_1}{\partial x_3} + \frac{\partial u_3}{\partial x_1}\right)$$

We can write a new version of the dynamic equation function of the displacement u_1:

$$\rho\frac{\partial^2 u_1}{\partial t^2} = 2\mu\frac{\partial\left(\frac{\partial u_1}{\partial x_1}\right)}{\partial x_1} + 2\mu\frac{\partial\frac{1}{2}\left(\frac{\partial u_1}{\partial x_2} + \frac{\partial u_2}{\partial x_1}\right)}{\partial x_2} + 2\mu\frac{\partial\frac{1}{2}\left(\frac{\partial u_1}{\partial x_3} + \frac{\partial u_3}{\partial x_1}\right)}{\partial x_3} + \lambda\frac{\partial(\varepsilon_{11}+\varepsilon_{22}+\varepsilon_{33})}{\partial x_1}$$

$$\rho \frac{\partial^2 u_1}{\partial t^2} = 2\mu \frac{\partial^2 u_1}{\partial x_1^2} + \mu \frac{\partial^2 u_1}{\partial x_2^2} + \mu \frac{\partial^2 u_2}{\partial x_1 \partial x_2} + \mu \frac{\partial^2 u_1}{\partial x_3^2} + \mu \frac{\partial^2 u_3}{\partial x_1 \partial x_3} + \lambda \frac{\partial(\varepsilon_{11} + \varepsilon_{22} + \varepsilon_{33})}{\partial x_1}$$

We define the Laplacian:

$$\Delta u_1 = \nabla^2 u_1 = \frac{\partial^2 u_1}{\partial x_1^2} + \frac{\partial^2 u_1}{\partial x_2^2} + \frac{\partial^2 u_1}{\partial x_3^2}$$

$$\rho \frac{\partial^2 u_1}{\partial t^2} = \mu \frac{\partial^2 u_1}{\partial x_1^2} + \mu \nabla^2 u_1 + \mu \frac{\partial^2 u_2}{\partial x_1 \partial x_2} + \mu \frac{\partial^2 u_3}{\partial x_1 \partial x_3} + \lambda \frac{\partial \left(\frac{\partial u_1}{\partial x_1} + \frac{\partial u_2}{\partial x_2} + \frac{\partial u_3}{\partial x_3} \right)}{\partial x_1}$$

$$\rho \frac{\partial^2 u_1}{\partial t^2} = (\lambda + \mu) \frac{\partial^2 u_1}{\partial x_1 \partial x_1} + \mu \nabla^2 u_1 + (\lambda + \mu) \frac{\partial^2 u_2}{\partial x_1 \partial x_2} + (\lambda + \mu) \frac{\partial^2 u_3}{\partial x_1 \partial x_3}$$

$$\rho \frac{\partial^2 u_1}{\partial t^2} = \mu \nabla^2 u_1 + (\lambda + \mu) \frac{\partial}{\partial x_1} \left\{ \frac{\partial u_1}{\partial x_1} + \frac{\partial u_2}{\partial x_2} + \frac{\partial u_3}{\partial x_3} \right\}$$

$$\rho \frac{\partial^2 u_1}{\partial t^2} = \nabla^2 u_1 + (\lambda + \mu) \frac{\partial}{\partial x_1} \{ \varepsilon_{11} + \varepsilon_{22} + \varepsilon_{33} \}$$

We obtain for all the direction ($i = 1, 2, 3$) with the 3 terms A, B, C below:

$$(A) \rightarrow \rho \frac{\partial^2 u_1}{\partial t^2} = \mu \nabla^2 u_1 + (\lambda + \mu) \frac{\partial}{\partial x_1} \left\{ \frac{\partial u_1}{\partial x_1} + \frac{\partial u_2}{\partial x_2} + \frac{\partial u_3}{\partial x_3} \right\}$$

$$(B) \rightarrow \rho \frac{\partial^2 u_2}{\partial t^2} = \mu \nabla^2 u_2 + (\lambda + \mu) \frac{\partial}{\partial x_2} \left\{ \frac{\partial u_1}{\partial x_1} + \frac{\partial u_2}{\partial x_2} + \frac{\partial u_3}{\partial x_3} \right\}$$

$$(C) \rightarrow \rho \frac{\partial^2 u_3}{\partial t^2} = \mu \nabla^2 u_3 + (\lambda + \mu) \frac{\partial}{\partial x_3} \left\{ \frac{\partial u_1}{\partial x_1} + \frac{\partial u_2}{\partial x_2} + \frac{\partial u_3}{\partial x_3} \right\}$$

That can be written with the differential operators and the vectorial notations:

$$\rho \frac{\partial^2 \overrightarrow{u}}{\partial t^2} = \mu \overrightarrow{\Delta} (\overrightarrow{u}) + (\lambda + \mu) \overrightarrow{\nabla} (\mathrm{div}(\overrightarrow{u}))$$

where:

$$\vec{\nabla} = \begin{Bmatrix} \dfrac{\partial}{\partial x_1} \\ \dfrac{\partial}{\partial x_2} \\ \dfrac{\partial}{\partial x_3} \end{Bmatrix}$$

$$\operatorname{div}(\vec{u}) = \frac{\partial u_1}{\partial x_1} + \frac{\partial u_2}{\partial x_2} + \frac{\partial u_3}{\partial x_3}$$

We can calculate now $\frac{\partial A}{\partial x_1} + \frac{\partial B}{\partial x_2} + \frac{\partial C}{\partial x_3}$:

$$\frac{\partial}{\partial x_1}\left(\mu\nabla^2 u_1 + (\lambda+\mu)\frac{\partial}{\partial x_1}\bar{\varepsilon}\right) + \frac{\partial}{\partial x_2}\left(\mu\nabla^2 u_2 + (\lambda+\mu)\frac{\partial}{\partial x_2}\bar{\varepsilon}\right)$$
$$+ \frac{\partial}{\partial x_3}\left(\mu\nabla^2 u_3 + (\lambda+\mu)\frac{\partial}{\partial x_3}\bar{\varepsilon}\right) = \frac{\partial}{\partial x_1}\left(\rho\frac{\partial^2 u_1}{\partial t^2}\right)$$
$$+ \frac{\partial}{\partial x_2}\left(\rho\frac{\partial^2 u_2}{\partial t^2}\right) + \frac{\partial}{\partial x_3}\left(\rho\frac{\partial^2 u_3}{\partial t^2}\right)$$

With the definition of the strains we obtain now:

$$\left(\mu\nabla^2\varepsilon_1 + (\lambda+\mu)\frac{\partial^2}{\partial x_1}\bar{\varepsilon}\right) + \left(\mu\nabla^2\varepsilon_2 + (\lambda+\mu)\frac{\partial^2}{\partial x_2}\bar{\varepsilon}\right) + \left(\mu\nabla^2\varepsilon_3 + (\lambda+\mu)\frac{\partial^2}{\partial x_3}\bar{\varepsilon}\right)$$
$$= \left(\rho\frac{\partial^2\varepsilon_1}{\partial t^2}\right) + \left(\rho\frac{\partial^2\varepsilon_2}{\partial t^2}\right) + \left(\rho\frac{\partial^2\varepsilon_3}{\partial t^2}\right)$$

$$\mu\nabla^2(\varepsilon_1+\varepsilon_2+\varepsilon_3) + (\lambda+\mu)\nabla^2\bar{\varepsilon} = \rho\frac{\partial^2}{\partial t^2}(\varepsilon_1+\varepsilon_2+\varepsilon_3)$$

$$\boldsymbol{(\lambda+2\mu)\nabla^2\bar{\varepsilon} = \rho\frac{\partial^2}{\partial t^2}(\bar{\varepsilon})}$$

Or:

$$\boldsymbol{\frac{\partial^2}{\partial t^2}(\bar{\varepsilon}) = \frac{(\lambda+2\mu)}{\rho}\nabla^2\bar{\varepsilon}}$$

The dimensional equation of this term is:

$$\frac{\frac{\text{kgm}}{\text{s}^2\text{m}^2}}{\frac{\text{kg}}{\text{m}^3}} = \frac{\text{m}^2}{\text{s}^2}$$

We have longitudinal wave of speed:

$$\boldsymbol{v_{\text{longitudinal}} = \sqrt{\frac{(\lambda+2\mu)}{\rho}}}$$

If we have shear wave, the longitudinal strains are null:

$$\rho\frac{\partial^2 u_1}{\partial t^2} = \mu\nabla^2 u_1 + (\lambda+\mu)\frac{\partial}{\partial x_1}\left\{\frac{\partial u_1}{\partial x_1}+\frac{\partial u_2}{\partial x_2}+\frac{\partial u_3}{\partial x_3}\right\}$$

$$\rho\frac{\partial^2 u_2}{\partial t^2} = \mu\nabla^2 u_2 + (\lambda+\mu)\frac{\partial}{\partial x_2}\left\{\frac{\partial u_1}{\partial x_1}+\frac{\partial u_2}{\partial x_2}+\frac{\partial u_3}{\partial x_3}\right\}$$

$$\rho\frac{\partial^2 u_3}{\partial t^2} = \mu\nabla^2 u_3 + (\lambda+\mu)\frac{\partial}{\partial x_3}\left\{\frac{\partial u_1}{\partial x_1}+\frac{\partial u_2}{\partial x_2}+\frac{\partial u_3}{\partial x_3}\right\}$$

Becomes:

$$\rho\frac{\partial^2 u_1}{\partial t^2} = \mu\nabla^2 u_1$$

$$\rho\frac{\partial^2 u_2}{\partial t^2} = \mu\nabla^2 u_2$$

$$\rho\frac{\partial^2 u_3}{\partial t^2} = \mu\nabla^2 u_3$$

$$\boldsymbol{\rho\frac{\partial^2 u_i}{\partial t^2} = \mu\nabla^2 u_i}$$

And the speed of the shear wave becomes:

$$\boldsymbol{v_{\text{shear}} = \sqrt{\frac{\mu}{\rho}}}$$

That is the formula 3.11 of the paper "Mechanics of spacetime — A Solid Mechanics perspective on the theory of General Relativity".

With:

$$\varepsilon_{ij} = \frac{1}{2}\left(\frac{\partial u_i}{\partial x_j}+\frac{\partial u_j}{\partial x_i}\right)$$

And with for the gravitational wave $\varepsilon_{kk} = \frac{\partial u_k}{\partial x_k} = 0$ (plane deformations)

$$\rho\frac{\partial^2 \frac{\partial u_i}{\partial x_j}}{\partial t^2} = \mu\nabla^2\frac{\partial u_i}{\partial x_j}$$

$$\rho \frac{\partial^2 \frac{\partial u_j}{\partial x_i}}{\partial t^2} = \mu \nabla^2 \frac{\partial u_j}{\partial x_i}$$

We do the sum of these two terms:

$$\rho \frac{\partial^2 \left(\frac{\partial u_i}{\partial x_j} + \frac{\partial u_j}{\partial x_i}\right)}{\partial t^2} = \mu \nabla^2 \left(\frac{\partial u_i}{\partial x_j} + \frac{\partial u_j}{\partial x_i}\right)$$

So, we obtain (formula 3.11 of the paper Mechanics of spacetime — A Solid Mechanics perspective on the theory of General Relativity):

$$\rho \frac{\partial^2 \varepsilon_{ij}}{\partial t^2} = \mu \nabla^2 \varepsilon_{ij}$$

G.3 Comparison of the Results Between General Relativity and Elasticity Theory

We obtain in general relativity linearized basing on $h_{\mu\nu} = 2\varepsilon_{\mu\nu}$:

$$\frac{\partial^2 \varepsilon_{ij}}{c^2 \partial t^2} = \Delta \varepsilon_{ij}$$

Or:

$$\frac{\partial^2 \varepsilon_{ij}}{\partial t^2} = c^2 \Delta \varepsilon_{ij}$$

We obtain according to the theory of elasticity in plane strains for a homogeneous and isotropic elastic medium:

$$\rho \frac{\partial^2 \varepsilon_{ij}}{\partial t^2} = \mu \nabla^2 \varepsilon_{ij}$$

Or:

$$\frac{\partial^2}{\partial t^2} \varepsilon_{ij} = \frac{\mu}{\rho} \Delta \varepsilon_{ij}$$

We deduct well from the parallelism of the two equations that the gravitational waves are shear wave of speed:

$$c_{\text{shear}} = \sqrt{\frac{\mu}{\rho}}$$

As we are in pure shear, we have so:

$$\varepsilon_{11} = -\varepsilon_{22}$$

$$\varepsilon_{12} = \varepsilon_{21}$$

This deformation state is compatible with the 2 polarizations of the gravitational wave.

Conclusion

T. Damour is therefore quite right when he says [5] I quote **«the speed of sound for the elastic deformation of space in Einstein's theory is the speed of light; gravitational waves are waves of elastic deformations of space»**.

Appendix H

Proof of Curvature in Beam Theory

H.1 General Geometric Curvature in One Dimension

We are going to determine firstly the general expression of the curvature in elastic theory applied at a beam.

(Sources see [2, 3, 29, 30] {1} à {17}).

Let be the average fiber of an elastic body as defined in figure H.1 source see {16, 17}.

We have so between the curve line ds and the angle $d\varphi$:

$$Rd\varphi = ds$$

$$R = \frac{ds}{d\varphi}$$

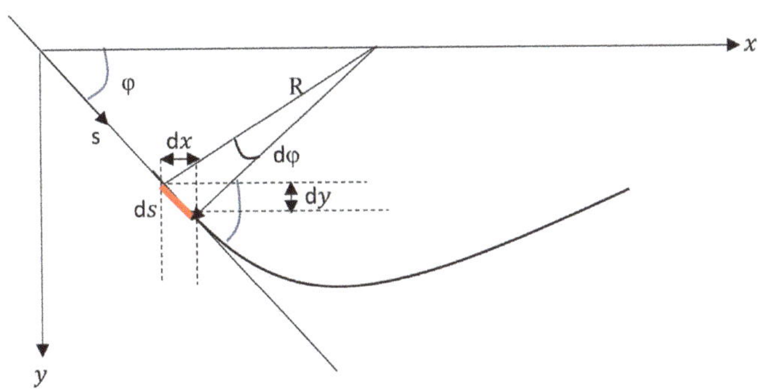

FIG. H.1 – Medium fiber of an elastic body.

DOI: 10.1051/978-2-7598-2573-8.c918

$$\frac{1}{R} = \frac{d\varphi}{ds}$$

$$\int d\varphi = \int \frac{ds}{R}$$

The differential approach allows to write:

$$\frac{d\varphi}{ds} \times \frac{ds}{dx} = \frac{d\varphi}{dx}$$

We define $\gamma = \frac{ds}{dx}$ that we introduce in the above equation:

$$\frac{d\varphi}{ds} \times \gamma = \frac{d\varphi}{dx}$$

$$\frac{d\varphi}{ds} = \frac{d\varphi}{dx} / \gamma$$

Following Pythagoras, we have for the metric:

$$ds^2 = dx^2 + dy^2$$

$$ds = \sqrt{dx^2 + dy^2}$$

$$\frac{ds}{dx} = \sqrt{\left(1 + \frac{dy^2}{dx^2}\right)}$$

In addition, we have by definition of the tangent of the angle φ:

$$\tan\varphi = \frac{dy}{dx}$$

$$1 + \tan^2\varphi = 1 + \left(\frac{dy}{dx}\right)^2$$

With:

$$\frac{ds}{dx} = \sqrt{\left(1 + \frac{dy^2}{dx^2}\right)}$$

$$\left(\frac{ds}{dx}\right)^2 = \left(1 + \frac{dy^2}{dx^2}\right)$$

So, we get:

$$1+\tan^2\varphi = 1+\frac{\sin^2\varphi}{\cos^2\varphi} = 1+\left(\frac{dy}{dx}\right)^2 = \left(\frac{ds}{dx}\right)^2 = \left(1+\frac{dy^2}{dx^2}\right) = \frac{\cos^2\varphi+\sin^2\varphi}{\cos^2\varphi}$$
$$= \frac{1}{\cos^2\varphi}$$

In addition, we can calculate the derivative in respect to x of the tangent:

$$\tan\varphi = \frac{dy}{dx}$$

$$\frac{d}{dx}\tan\varphi(x) = \frac{d^2y}{dx^2}$$

The calculation is the following:

$$u = \sin U(x) \quad u' = U'\cos U(x)$$

$$v = \cos U(x) \quad v' = -U'\sin U(x)$$

$$\left(\frac{u}{v}\right)' = \frac{u'v - v'u}{v} = \frac{U'\cos(U(x))\cos(U(x)) + U'\sin(U(x))\sin(U(x))}{\cos^2(U(x))}$$

$$\left(\frac{u}{v}\right)' = \frac{u'v - v'u}{v} = \frac{U'\cos^2(U(x)) + U'\sin^2(U(x))}{\cos^2(U(x))}$$

$$\left(\frac{u}{v}\right)' = \frac{u'v - v'u}{v} = \frac{U'}{\cos^2(U(x))} = \frac{d\varphi}{dx}\times\frac{1}{\cos^2\varphi(x)} = \frac{d^2y}{dx^2}$$

So, we obtain:

$$\frac{d\varphi}{dx} = \frac{\frac{d^2y}{dx^2}}{\frac{1}{\cos^2\varphi(x)}}$$

With:

$$\left(1+\frac{d^2y}{dx^2}\right) = \frac{\cos^2\varphi+\sin^2\varphi}{\cos^2\varphi} = \frac{1}{\cos^2\varphi}$$

$$\frac{d\varphi}{dx} = \frac{\frac{d^2y}{dx^2}}{\left(1+\frac{dy^2}{dx^2}\right)}$$

$$\frac{d\varphi}{ds} \times \frac{ds}{dx} = \frac{d\varphi}{dx}$$

With the curvature definition:

$$\frac{1}{R} = \frac{d\varphi}{ds}$$

$$\frac{1}{R} \times \frac{ds}{dx} = \frac{d\varphi}{dx}$$

$$\frac{ds}{dx} = \sqrt{\left(1 + \frac{dy^2}{dx^2}\right)}$$

We can write a new expression of the angle $d\varphi$:

$$\frac{d\varphi}{dx} = \frac{\frac{d^2y}{dx^2}}{\left(1 + \frac{d^2y}{dx^2}\right)} = \frac{1}{R} \times \frac{ds}{dx}$$

$$\frac{d\varphi}{dx} = \frac{\dfrac{d^2y}{dx^2}}{\left(1 + \dfrac{dy^2}{dx^2}\right)} = \frac{1}{R} \times \sqrt{\left(1 + \frac{dy^2}{dx^2}\right)}$$

$$\frac{\frac{d^2y}{dx^2}}{\left(1 + \frac{dy^2}{dx^2}\right)\sqrt{\left(1 + \frac{dy^2}{dx^2}\right)}} = \frac{1}{R}$$

And we obtain the final version of the curvature:

$$\frac{1}{R} = \frac{\frac{d^2y}{dx^2}}{\left(1 + \frac{dy^2}{dx^2}\right)^{3/2}} = \frac{\frac{d^2y}{dx^2}}{\left(1 + \left(\frac{dy}{dx}\right)^2\right)^{3/2}} = \frac{\frac{d^2y}{dx^2}}{\left(\sqrt{1 + \left(\frac{dy}{dx}\right)^2}\right)^3}$$

H.2 Case of the Beam (Sources see [1] à [3])

The term $\left(\sqrt{1 + \left(\frac{dy}{dx}\right)^2}\right)^3$ tends to 1, so $\frac{dy}{dx} \lll\lll 1$ tangent almost horizontal so, it remains:

$$\frac{1}{R} = \frac{d^2y}{dx^2}$$

H.3 Relation Between the Bending Moment and the Curvature in One Dimension

We will show here what is the relation between $1/R$ and the bending moment $M_{(x)}$:

For that, we will establish a relation between the bending moment and the normal stress of bending in a beam (see figure H.2).

We have by definition of the bending moment:

$$M_{(x)} = \int_{-h/2}^{h/2} z\sigma b dz$$

We are in elastic, the straight sections before deformations remain straight after deformations.

$$\sigma = kz$$

By deferring this expression of the stress in the expression of the bending moment, we obtain:

$$M_{(x)} = \int_{-h/2}^{h/2} kz^2 b dz$$

We define the inertia of the beam with respect to the axis z:

$$I_{zz} = \int_{-h/2}^{h/2} z^2 b dz$$

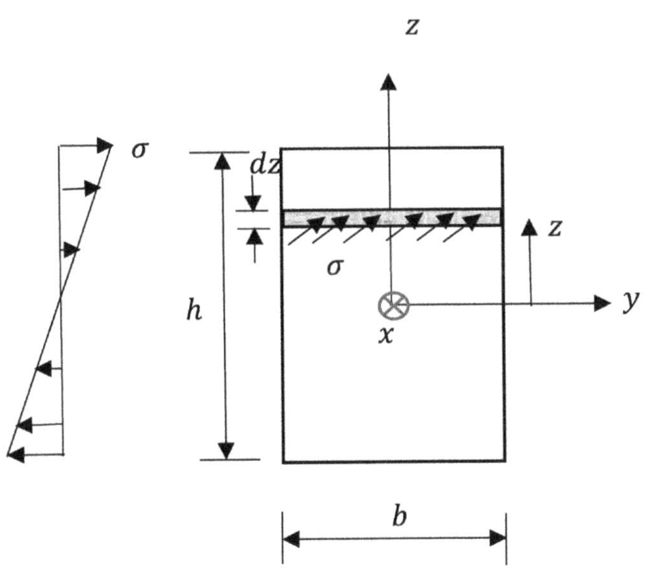

FIG. H.2 – Diagram of normal stresses in a bending beam.

And we obtain:

$$M_{(x)} = kI_{zz}$$

Or in other way:

$$\frac{\sigma}{z} = k$$

$$M_{(x)} = \frac{\sigma}{z} I_{zz}$$

So, we obtain:

$$\sigma_{xx} = \frac{M_{(x)} z}{I_{zz}}$$

Now let us look at the relationship between moment and curvature (see figures H.3 and H.4).

$$\sigma_{\text{sup}} = \frac{M\left(\frac{h}{2}\right)}{I} = \varepsilon E = E\left(\frac{L_{\text{sup}} - dx}{dx}\right)$$

$$\sigma_{\text{inf}} = -\frac{M\left(\frac{h}{2}\right)}{I} = \varepsilon E = E\left(\frac{L_{\text{inf}} - dx}{dx}\right)$$

The tensile stresses are considered negative.

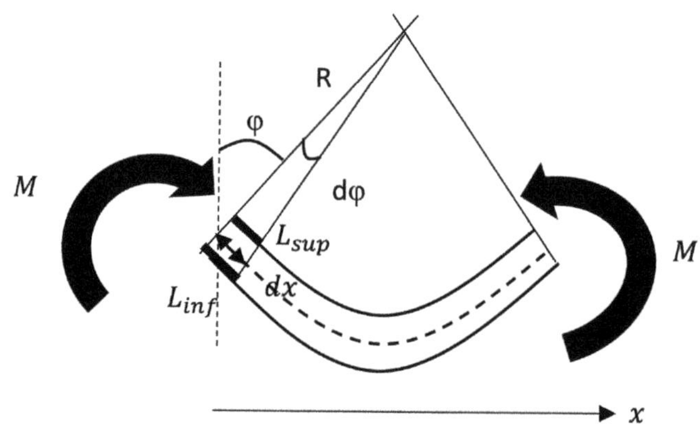

FIG. H.3 – Bending beam stressed by two bending moments.

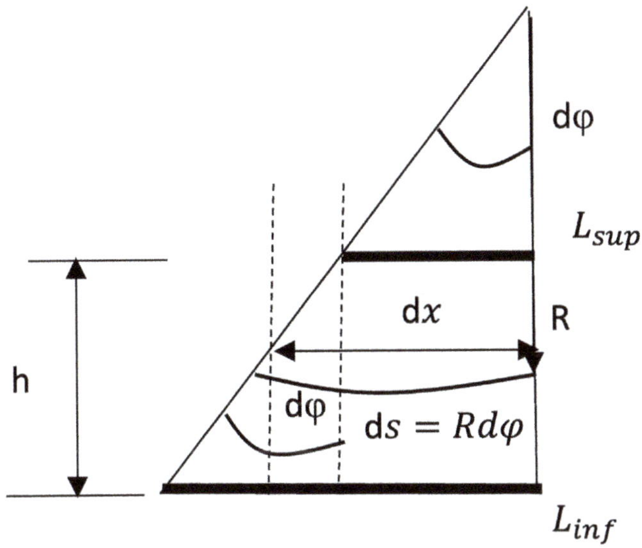

FIG. H.4 – Elongation and shortening of the fibers in bending of the beam.

From these equations we get:

$$L_{\text{sup}} = \frac{M\left(\frac{h}{2}\right)}{EI}dx + dx$$

$$L_{\text{inf}} = -\frac{M\left(\frac{h}{2}\right)}{EI}dx + dx$$

The definition of the angle $d\varphi$ allows to write:

$$\text{tang}(d\varphi) = \frac{L_{\text{inf}} - L_{\text{sup}}}{h} = \frac{dx}{R}$$

So:

$$\text{tang}(d\varphi) = \frac{\left(-\frac{M\left(\frac{h}{2}\right)}{EI}dx + dx\right) - \left(\frac{M\left(\frac{h}{2}\right)}{EI}dx + dx\right)}{h}$$

$$\text{tang}(d\varphi) = -\frac{\left(2\frac{M\left(\frac{h}{2}\right)}{EI}dx\right)}{h} = \frac{dx}{R}$$

Hence the final expression relating the curvature $\frac{1}{R}$ to the bending moment M.

$$\frac{d\varphi}{dx} = -\frac{M}{EI} = \frac{1}{R}$$

Appendix I

Proof of Quantum Value of Young's Modulus of Space Space-Time Obtained in Tables 16.1 and 16.2

I.1 Introduction

The purpose of this appendix is to verify the results and the formulas obtained by iterations in tables 16.1 and 16.2 of chapter 16 of this book from the value of Young's modulus given in the publication of T.G. Tenev and M.F. Horstemeyer The Mechanics of Spacetime. A Solid Mechanics Perspective on the Theory of General Relativity chapter 4 Summary and Conclusion.

In this publication, the quantum Young's modulus of space-time takes the following expression:

$$Y = \frac{6c^7}{2\pi\hbar G^2}$$

The numerical application gives:

$$Y = \frac{6(299\,792\,458)^7}{2\pi\left(\frac{6.62607015\times10^{-34}}{2\pi}\right)\times(6.6742\times10^{-11})^2} = 4.42\times10^{113}\,\text{Pa}$$

I.2 Determination of Young's Modulus Formula from the Expressions of the Table 16.1 of Chapter 16

We have demonstrated in chapter 16 the following expression:

$$\rho\, Vc^2 = \frac{Gh^2}{4\pi\, Vc^2}$$

DOI: 10.1051/978-2-7598-2573-8.c919

Or multiplying each side by Vc^2, we get:

$$\rho V^2 c^4 = \frac{Gh^2}{4\pi}$$

We have shown in the numerical application of table 16.1 that the length of the spatio-temporal fibers is of the order of magnitude of the Planck length therefore $k \times L_p$. We therefore replace V in the previous formula by the expression of the volume of a sphere of radius $r = k \times L_p$ and we obtain:

$$\rho\left(\frac{4}{3}\pi(kL_p)^3\right)^2 c^4 = \frac{Gh^2}{4\pi}$$

Replacing L_p by its expression, we get:

$$\rho\left[\frac{4}{3}\pi k^3\left(\sqrt{\frac{hG}{2\pi c^3}}\right)^3\right]^2 c^4 = \frac{Gh^2}{4\pi}$$

So:

$$\rho\left[\frac{16}{9}\pi^2 k^6\left(\frac{hG}{2\pi c^3}\right)^3\right] c^4 = \frac{Gh^2}{4\pi}$$

After simplification we get:

$$\rho = \frac{9c^5}{8k^6 hG^2}$$

We are in the first case (compression wave in the arms of the interferometer), we can replace density ρ by its expression as a function of Young's modulus:

$$E = Y = \rho c^2$$

So, we get:

$$Y = \frac{9c^7}{8k^6 hG^2}$$

Replacing Planck's constant by the reduced Planck's constant we get:

$$\hbar = \frac{h}{2\pi}$$

We obtain Young's modulus final expression:

$$\boldsymbol{Y = \frac{9c^7}{16k^6\pi\hbar G^2}}$$

We obtained by iterations $r = 1.566 \times 10^{-35}$ m in table 16.1 that we compare with the exact value of Planck's length $1.616125518 \times 10^{-35}$ m and therefore we obtain $k = 0.96875656$.

Introducing k in the previous formula, we obtain so:

$$Y = \frac{9c^7}{13.2253729\pi\hbar G^2}$$

$$\mathbf{Y = \frac{6c^7}{8.81691525\pi\hbar G^2}}$$

The numerical application gives:

$$Y = \frac{6(299\,792\,458)^7}{8.81691525\pi \times \left(\frac{6.62607015\times10^{-34}}{2\pi}\right) \times \left(6.6742 \times 10^{-11}\right)^2} = 1.00 \times 10^{113}\,\text{Pa}$$

This value is well close to the value of T.G. Tenev and M.F. Horstemeyer with $\nu = 1$:

$$\mathbf{Y = \frac{6c^7}{2\pi\hbar G^2}}$$

I.3 Determination of Young's Modulus Formula from the Expression of the Table 16.2 of Chapter 16

We demonstrated in §I.2 that

$$\rho = \frac{9c^5}{8k^6 hG^2}$$

This time we consider the shear wave approach as done in chapter 16, table 16.2.

The formula for Young's modulus as a function of density ρ and Poisson's ratio ν is this time, based on the shear wave speed $c_{\text{shear}} = \sqrt{\frac{\mu}{\rho}}$:

$$E = Y = 2\rho c^2(1+\nu)$$

We extract ρ from this formula and we get:

$$\rho = \frac{Y}{2c^2(1+\nu)}$$

We introduce this expression for ρ in the previous formula for this value and we get:

$$\frac{Y}{2c^2(1+\nu)} = \frac{9c^5}{8k^6 hG^2}$$

We obtain the following expression of Young's modulus:

$$\mathrm{Y} = \frac{18c^7(1+v)}{8k^6\hbar G^2}$$

We take into account of the value of Planck's reduced constant:

$$\hbar = \frac{h}{2\pi}$$

We get so:

$$Y = \frac{18c^7(1+v)}{16k^6\pi\hbar G^2}$$

And finally:

$$\boldsymbol{Y = \frac{9c^7(1+v)}{8k^6\pi\hbar G^2}}$$

Taking into account of Poisson's ratio ($v = \mathbf{1}$), we get:

$$\mathrm{Y} = \frac{1.5 \times 6c^7}{4k^6\pi\hbar G^2}$$

Replacing k by its value (see §I.2), we get by an iterative approach:

$$\boldsymbol{Y = \frac{6c^7}{2.20422881\pi\hbar G^2}}$$

Or:

$$\boldsymbol{G = \sqrt{\frac{6c^7}{2.20422881\pi\hbar Y}}}$$

The numerical application gives:

$$Y = \frac{6(299\,792\,458)^7}{2.20422881\pi \times \left(\frac{6.62607015\times10^{-34}}{2\pi}\right) \times (6.6742 \times 10^{-11})^2} = 4.01 \times 10^{113}\,\mathrm{Pa}$$

This value is well close to the value of T.G Tenev and M.F. Horstemeyer [18] with $v = 1$:

$$\boldsymbol{Y = \frac{6c^7}{2\pi\hbar G^2}}$$

The main difference between our approach and that of T.G. Tenev and M.F. Horstemeyer is that we obtained by iterations this Planck's length (results of calculation) while they postulated this one (3.4 [18] assuming $L = L_p$).

Appendix J

Young's Modulus of the Space Time from the Energy Density of the Gravitational Wave

J.1 Origin of This Topic

This topic was discussed at the 43rd Annual Donald R. Hamilton Lecture hosted by Kip Thorne Princeton University on April 12, 2018 "Exploring the Universe with Gravitational Waves".
https://www.kaltura.com/index.php/extwidget/preview/partner_id/1449362/uiconf_id/14292362/entry_id/1_blcdwnpb/embed/auto?&flashvars%5bstreamerType%5d=auto

During the questioning session, the audience asked the order of magnitude of Young's modulus of space-time compared to a piece of steel (1 h 21 min). An answer was proposed by R. Weiss at the request of K. Thorne (approximately 10^{20} times more rigid than the steel Young's modulus!). This point and the key formulas are also stated in the article [13] by K McDonald. See also the article {215} by A.C. Melissinos. I discussed also with R. Weiss about my Pramana paper and this topic by mail the 24 August 2020.

We will try to give in this appendix the key formulas to obtain this result.

J.2 Strain Energy Density in Strength of Material

In elasticity theory, the energy density of an elastic body loaded by a stress σ which creates a strain ε is:

$$U = \int_0^{\varepsilon} \sigma \, d\varepsilon$$

DOI: 10.1051/978-2-7598-2573-8.c920

With Hooke's law:

$$\sigma = \varepsilon Y$$

That we introduce in the expression above:

$$U = \int_0^{\varepsilon} Y \varepsilon d\varepsilon$$

And we obtain so:

$$U = Y \int_0^{\varepsilon} \varepsilon d\varepsilon = Y \left[\frac{\varepsilon^2}{2} \right]_0^{\varepsilon} = \frac{Y \varepsilon \varepsilon}{2} = \frac{\sigma \varepsilon}{2} = \frac{Y \varepsilon^2}{2}$$

$$\boldsymbol{U = \frac{Y \varepsilon^2}{2}}$$

J.3 Strain Energy Density for a Gravitational Wave

The Einstein equation is:

$$T_{\mu\nu} = \frac{c^4}{8\pi G} \left(R_{\mu\nu} - \frac{1}{2} g_{\mu\nu} R \right)$$

For the gravitational wave, we have for the metric:

$$g_{\mu\nu} = \eta_{\mu\nu} + h_{\mu\nu}$$

$$\overline{h}_{\mu\nu} = h_{\mu\nu} + \frac{1}{2} \eta_{\mu\nu} \overline{h}$$

Thus, Einstein's equation becomes with an additional pseudo complementary tensor $t_{\mu\nu}$ that is interpreted as the stress energy tensor of the gravitational wave motion {219}:

$$T_{\mu\nu} = -\frac{c^4}{8\pi G} \left(R_{\mu\nu} - \frac{1}{2} g_{\mu\nu} R \right) = -\frac{c^4}{8\pi G} G_{\mu\nu}$$

With the perturbation $h_{\mu\nu}$ (see formula 4 of {219}):

$$T_{\mu\nu} + t_{\mu\nu} = -\frac{c^4}{8\pi G} G_{\mu\nu}^{(1)}$$

$G_{\mu\nu}^{(1)}$ is made of the terms of $G_{\mu\nu}$ that are linear in $h_{\mu\nu}$ (see formula 5 of {219}):

$$t_{\mu\nu} = T_{\mu\nu}^{GW} = \frac{c^4}{8\pi G} \left[G_{\mu\nu}^{(2)} + \cdots \right]$$

Using Einstein's equation and the definition of $g_{\mu\nu}$, we obtain(see formula 35.70 of {53} and 35.23 of [54]):

$$t_{\mu\nu} = T^{GW}_{\mu\nu} = \frac{c^4}{32\pi G}\left[\frac{\partial \bar{h}_{ij}}{\partial x^{\mu}}\frac{\partial \bar{h}^{ij}}{\partial x^{\nu}} - \frac{1}{2}\frac{\partial \bar{h}}{\partial x^{\mu}}\frac{\partial \bar{h}}{\partial x^{\nu}} - \frac{\partial \bar{h}_{\nu i}}{\partial x^{\mu}}\frac{\partial \bar{h}^{ij}}{\partial x^{j}} - \frac{\partial \bar{h}_{\mu i}}{\partial x^{\nu}}\frac{\partial \bar{h}^{ij}}{\partial x^{j}}\right]$$

Or according to Eq. (5) of {219} taking into account the symmetries:

$$t_{\mu\nu} = \frac{c^4}{32\pi G}\left[\frac{\partial \bar{h}_{ij}}{\partial x^{\mu}}\frac{\partial \bar{h}^{ij}}{\partial x^{\nu}} - \frac{1}{2}\frac{\partial \bar{h}}{\partial x^{\mu}}\frac{\partial \bar{h}}{\partial x^{\nu}} - \frac{\partial \bar{h}^{ji}}{\partial x^{j}}\frac{\partial \bar{h}_{i\nu}}{\partial x^{\mu}} - \frac{\partial \bar{h}^{ji}}{\partial x^{j}}\frac{\partial \bar{h}_{i\mu}}{\partial x^{\nu}}\right]$$

As the trace is null $\overline{h}_{ij} = h_{ij}$ and $h = 0$; $h = h^{\mu}_{\mu}$.
The harmonic gauge condition implies (see formula 7 of {219} and 35.24 of [54]):

$$\frac{\partial \bar{h}_{\mu\nu}}{\partial x_{\mu}} = 0$$

In link with the gauge condition and the Transverse Traceless condition (TT gauge), it remains (see formula 8 of {219}) with to designate an average of waves (see 35.23 of [54]):

$$t_{\mu\nu} = \frac{c^4}{32\pi G}\langle(\partial_{\mu} h_{ij})(\partial_{\nu} h^{ij})\rangle = \frac{c^4}{32\pi G}\left\langle\frac{\partial \bar{h}_{ij}}{\partial x^{\mu}}\frac{\partial \bar{h}^{ij}}{\partial x^{\nu}}\right\rangle$$

$$t_{\mu\nu} - \frac{c^4}{32\pi G}\langle\nabla_{\mu} h_{ij}\nabla_{\nu}\, h^{ij}\rangle$$

$$h_{xx} = -h_{yy}$$

$$h_{xy} = h_{yx} = 0$$

With the perturbation of the TT metric, for the first-time term, the energy density is then {220}:

$$t_{00} = t_{00} = \frac{c^4}{32\pi G}\langle(\dot{h}_{ij})(\dot{h}^{ij})\rangle$$

So, with two polarizations:

$$t_{00} = t_{tt} = \frac{c^4}{32\pi G}\langle(\dot{h}_{11+})(\dot{h}_{11+}) + (\dot{h}_{22+})(\dot{h}_{22+}) + (\dot{h}_{12\times})(\dot{h}_{12\times}) + (\dot{h}_{21\times})(\dot{h}_{21\times})\rangle$$

$$t_{00} = t_{tt} = \frac{c^4}{32\pi G}\langle 2\left(\dot{h}_{ij+}\right)^2 + 2\left(\dot{h}_{ij\times}\right)^2\rangle$$

$$t_{00} = t_{tt} = \frac{c^4}{16\pi G}\langle \left(\dot{h}_{ij+}\right)^2 + \left(\dot{h}_{ij\times}\right)^2\rangle$$

The point means derivative with respect to ct.

$$x = ct$$

$$dx = cdt$$

So, as it is a derivative in respect to ct (see formula 32.15 of {26}):

$$t_{00} = t_{tt} = \frac{c^4}{16\pi G}\left[\dot{h}_+^2 + \dot{h}_\times^2\right]$$

$$t_{00} = t_{tt} = \frac{c^4}{16\pi G}\left[\left(\frac{d(h_+)}{cdt}\right)^2 + \left(\frac{d(h_\times)}{cdt}\right)^2\right]$$

$$t_{00} = t_{tt} = \frac{c^2}{16\pi G}\left[\left(\frac{d(h_+)}{dt}\right)^2 + \left(\frac{d(h_\times)}{dt}\right)^2\right]$$

$$\frac{\frac{\mathrm{m}^2}{\mathrm{s}^2}}{\frac{\mathrm{m}^3}{\mathrm{kgs}^2}}\frac{1}{\mathrm{s}^2} = \frac{\mathrm{kg}}{\mathrm{ms}^2} = \frac{\frac{\mathrm{kgm}^2}{\mathrm{s}^2}}{\mathrm{m}^3} = \text{energy density}$$

Note: This expression is the formula 10.3.6 of the Weinberg book gravitation and cosmology, we have for the energy density:

$$t_{\mu v} = \frac{k_\mu k_v}{\frac{16\pi G}{c^4}}\left(\lceil e_+ \rceil^2 + \lfloor e_- \rfloor^2\right)$$

With 35.26 of [54]:

$$\begin{aligned} h_{\mu v} &= A_{\mu v}\cos(k_\sigma x^\sigma) \\ &= A_+ \begin{bmatrix} 0 & 0 & 0 & 0 \\ 0 & +1 & 0 & 0 \\ 0 & 0 & -1 & 0 \\ 0 & 0 & 0 & 0 \end{bmatrix} \cos(k_\sigma x^\sigma) + A_\times \begin{bmatrix} 0 & 0 & 0 & 0 \\ 0 & 0 & +1 & 0 \\ 0 & +1 & 0 & 0 \\ 0 & 0 & 0 & 0 \end{bmatrix} \cos(k_\sigma x^\sigma) \end{aligned}$$

$$h_{\mu\nu} = A_{\mu\nu}\cos(k_\sigma x^\sigma) = \begin{bmatrix} 0 & 0 & 0 & 0 \\ 0 & A_+ & A_\times & 0 \\ 0 & A_\times & -A_+ & 0 \\ 0 & 0 & 0 & 0 \end{bmatrix}\cos(k_\sigma x^\sigma)$$

$$k_\mu = k_\nu = \tfrac{\omega}{c} = \tfrac{1}{\lambda}$$

We consider the following two polarizations to then take an average over a period for a time T {221}:

$$h_+ = A_+\cos\Big(\frac{\omega}{c}(ct-z)\Big); \dot{h}_+ = -A_+\frac{\omega}{c}\sin\Big(\frac{\omega}{c}(ct-z)\Big)$$

The h point want to say a derivative in respect to ct here.

$$h_\times = A_\times\cos\Big(\frac{\omega}{c}(ct-z)\Big); \dot{h}_\times = -A_\times\frac{\omega}{c}\sin\Big(\frac{\omega}{c}(ct-z)\Big)$$

$$\int_0^T t_{00}\,dt = \frac{c^4}{16\pi G}\frac{\omega^2}{c^2}\left\langle\left[A_+^2\int_0^{T=\frac{2\pi}{\omega}}\sin^2\Big(\frac{\omega}{c}(ct-z)\Big)dt + A_\times^2\int_0^{T=\frac{2\pi}{\omega}}\sin^2\Big(\frac{\omega}{c}(ct-z)\Big)dt\right]\right\rangle$$

$$t_{00}\,T = \frac{c^4}{16\pi G}\frac{\omega^2}{c^2}\left\langle\left[A_+^2\frac{T}{2} + A_\times^2\frac{T}{2}\right]\right\rangle$$

The energy density is so with 35.27 of [54]:

$$t_{00} = \frac{c^4}{32\pi G}\frac{\omega^2}{c^2}\langle[A_+^2 + A_\times^2]\rangle$$

For the energy density, this is the term ($t_{\mu\nu}$) that must be taken into account in t, and with the adequate gauge we obtain for all the components of the gravitational waves stress energy tensor (see formula 2.24 of {217} and formulas 5.35 to 5.40 of {218} and formulas 35.23, 35.27, and 35.70 of {53}):

$$t_{00} = \frac{t_{zz}}{c^2} = -\frac{t_{0z}}{c} = \frac{c^4}{32\pi G}\frac{\omega}{c}\frac{\omega}{c}\left[A_+^2 + A_\times^2\right]$$

So, we can rewrite the expression of the energy density for the z direction (see formula 2.25 of {217} and formula 32.20 of {26} and 35.27 of [54]):

$$t_{00} = \frac{t_{zz}}{c^2} = -\frac{t_{0z}}{c} = \frac{c^2\omega^2}{32\pi G}\left[A_+^2 + A_\times^2\right]$$

With $\omega = 2\pi f$, we obtain:

$$t_{00} = \frac{4\pi^2 c^2 f^2}{32\pi G}\left[A_+^2 + A_\times^2\right]$$

$$t_{00} = \frac{\pi c^2 f^2}{8G}\left[A_+^2 + A_\times^2\right]$$

A_+ is link with h the strain h.

So, we have the strain energy of the gravitational wave as indicated in [13] and {215}:

$$t_{00} = \frac{t_{zz}}{c^2} = \frac{\pi c^2 f^2 h^2}{8G}$$

$$\frac{\mathrm{m}^2}{\mathrm{s}^2 \frac{\mathrm{m}^3}{\mathrm{kgs}^2}} \times \frac{1}{\mathrm{s}^2} = \frac{\mathrm{kg}}{\mathrm{ms}^2} = \frac{\frac{\mathrm{kgm}^2}{\mathrm{s}^2}}{\mathrm{m}^3}$$

J.4 Expression of the Young Modulus Y (See [13])

Assuming that the space is made of an elastic material, a new deformable ether, we have so:

$$U = \frac{\sigma\varepsilon}{2} = \frac{\varepsilon E\varepsilon}{2} = \frac{Yh^2}{2} = \frac{\pi c^2 f^2 h^2}{8G}$$

So, we obtain well for Young's modulus:

$$Y = E = \frac{\pi c^2 \mathrm{f}^2}{4G}$$

Or:

$$G = \frac{\pi c^2 f^2}{4Y}$$

Numerical application

The frequency f of gravitational waves measured by Ligo/Virgo is approximately $\varphi = 100$ Hz

$$Y = \frac{\pi \times 299\,792\,458\,000^2 \times 100^2}{4 \times 6.6742 \times 10^{-11}} = 1.057 \times 10^{31}\,\mathrm{Pa} = 5.0363 \times 10^{19}\, Y_{\mathrm{steel}}$$

With $Y_{\mathrm{steel}} = E = 210\,000\,000\,000\,\mathrm{Pa}$.

If we take the formulas obtained *via* the elasticity approach developed in this book (see Publication Pramana of D. Izabel):

- (case 1 and 2 one/two arms of interferometer):

$$G = \frac{\pi f^2}{\rho}; \; Y = \rho c^2$$

$$Y = \frac{\pi c^2 f^2}{G} = \frac{\pi \times 299\,792\,458\,000^2 \times 100^2}{6.6742 \times 10^{-11}} = 4.2305 \times 10^{31}\mathrm{Pa} = 2.0145 \times 10^{20}\, Y_{\mathrm{steel}}$$

- (case 3 shear by torsion of a space tube):

$$G = \frac{\pi f^2 c^2}{\mu}; \; \mu = \frac{Y}{2(1+v)}$$

$$Y = 2(1+v)\frac{\pi c^2 f^2}{G} = 2(1+1)\frac{\pi \times 299\,792\,458\,000^2 \times 100^2}{6.6742 \times 10^{-11}} = 1.692 \times 10^{32}\,\mathrm{Pa}$$
$$= 8.058 \times 10^{20}\, Y_{\mathrm{steel}}$$

R. Weiss' response to the Hamilton conference in 2018 and in his Nobel lecture p. 71 I quote "The power per area in the wave is proportional to the square of the rate of change of the strain times a gigantic factor which tells that a small amount of strain in space is accompanied by a huge amount of energy. In other words, it takes enormous amounts of energy to distort space. **One way to say it is, the stiffness (Young's modulus) of space at a distortion frequency of 100 Hz is 10^{20} larger than steel" is therefore redemonstrated.**

$$S_g = \frac{dE}{dA\,dt} = \frac{\frac{\mathrm{kgm}^2}{\mathrm{s}^2}}{\mathrm{m}^3}\frac{\mathrm{m}}{\mathrm{s}} = ct_{00} = ct_{tt} = \frac{c^3}{16\pi G}\left[\left(\frac{d(h_+)}{dt}\right)^2 + \left(\frac{d(h_\times)}{dt}\right)^2\right]$$

With $\frac{c^3}{16\pi G} = 7.8 \times 10^{36}\,\mathrm{ergsec/cm}^2$.

See formulas 4.6–4.9 of A. Buonanno "GRAVITATIONAL WAVES arXiv.org > gr-qc > arXiv:0709.4682v1 and Weiss Nobel lecture.

References from the Author and from the Article Published at PRAMANA on August 13, 2020

[1] D. Izabel (2015) Does the sequences of random numbers follows a log periodic law? Has memory? *Am. Rev. Math. Stat.* **3**, 1.

[2] D. Izabel (2020) Mechanical conversion of the gravitational Einstein constant κ, *Pramana* **94**, 119.

[3] D. Izabel (2007) *Formulaire de résistance des matériaux* **2**, 1.

[4] T. Damour (2000) *Conf. n°183 la Relativité Générale* **1**, 1.

[5] T. Damour (2019) *Conf. Enigmatiques Trous Noirs Conférence Organisée Par Les Amis De l'IHES Slide 8 21ème Minute et 44ème minute* **1**, 1.

[6] A. Aspect (2019) *Conf. Le Photon Onde Ou Particule l'étrangeté Quantique Mise En Lumière Organisée Par Les Amis De l'IHES* **1**, 1.

[7] K. Thorne (2019) *Conf. the warped side of the universe? Cardiff University* **1**, 1.

[8] A. Einstein (1915) *Phys. Math.* **1**, 844.

[9] A. Einstein (1916) *Phys. Math.* **1**, 688.

[10] A. Einstein (1934) *Querido Verlag* **6**, 220.

[11] W. Dyson, A.S. Eddington, and C. Davidson (1920) *Philos. Trans. R. Soc. London* A **220**, 291.

[12] A. Einstein (1915) *Sitz. Preus. Akad. Wiss.* **1**, 831.

[13] K.T. McDonald (2018) *J. Henry. Lab. Princ. Univ.* NJ08544 what is stiffness of space time, 1.

[14] R. Chaubey and A.K. Shukla (2017) *Pramana – J. Phys.* **88**, 65.

[15] T. Damour (2016) *Conf. Mardi. Esp. Des Sci.* **1**, 1.

[16] T. Damour (2017) *Conf. Assos. Amis. Lab, Arago.* **1**, 1.

[17] R. Feynman, F. Morinigo, and W. Wagner, *Lectures on gravitation* (CRC Press, T and F Group lecture 3 chapter 3.4 figures 3.3 and 3.4 1962–1963) **1**, 53.

[18] T.G. Tenev and M.F. Horstemeyer, *Int. J. Mod. Phys. D* **27**, 1850083.

[19] K. Thorne (2018) Conference, Irvine, UCI, school, physical, sciences (Reines lectures exploring the universe with gravitation waves from big bang to black holes) **1**, 1.

[20] K. Thorne (2019) Conference, Prague, learned, society, Czech, republic, and, faculty, mathematics, and, physics, Charles, university (Geometrodynamics the Nonlinear Dynamics of Curved space time) **1**, 1.

[21] E. Hubble (1929) *Astronmy* **15**, 168.

[22] C.W.F. Everitt (2011) *Phys. Rev. Lett.* **106**, 221101.

[23] Collective LIGO Scientific Collaboration and Virgo Collaboration (2016) *Phys. Rev. Lett.* **116**, 061102.

[24] Collective LIGO Scientific Collaboration and Virgo Collaboration (2017) *Phys. Rev. Lett.* **119**, 161101.

[25] A. Einstein (1918) *Sitz. König. Preuß. Akad. Wiss.* **1**, 154.

[26] H.I. Ringermacher and L.R. Mead (2015) *R. Astron.* **149**, 137.

[27] A. Einstein (1916) *Einstein Arch. Univ. Héb. J.* **16**, 453.

[28] A. Einstein (1921) *Univ. Leyde. Gauthier. Villars. Cie.* **1**, 1.

[29] S. Timoshenko and J.N. Goodier (1951) *Theory of elasticity* (McGraw Hill, New York) **1**, 146.
[30] S. Timoshenko (1930) *Strength of material* (Van Nostrand) **1**, 91.
[31] A.D. Sakharov (1968) *Sov. Phys.* **12**, 1040.
[32] J.L.A. Synge (1959) *Math. Zeitschrift.* **72**, 82.
[33] C.B. Rayner (1963) *Proc. R. Soc. A. Math. Phys. Eng.* **272**, 44.
[34] K. Kondo (1964) *Int. J. Eng. Sci.* **2**, 219.
[35] C. Truesdell (1966) *The elements of continuum mechanics* (Springer-Verlag, New York) **1**, 1.
[36] R. Grot and A. Eringen (1966) *Int. J. Eng. Sci.* **2**, 1.
[37] J.E. Marsden and T.J. Hughes (1983) *Mathematical foundations of elasticity* (Prentice-Hall, Englewoods Cliffs, NJ) **1**, 1.
[38] U.H. Gerlach and J.F. Scott (1986) *Phys. Rev.* **34**, 3638.
[39] G.A. Maugin (1971) *An. Inst. Henri. Poincare* **4**, 275.
[40] J. Kijowski (1994) *Leonardo Centre Phys. Th. CNRS Luminy*, ArXiv. 9411212. **CPT-94**, 3102.
[41] A. Tartaglia (1994) *Front. Fund. Phys.* **1**, 147.
[42] S. Antoci and L. Mihich (1999) *Nuovo Cimento* **1**, 8.
[43] M. Karlovini and L. Samuelsson (2002) *Quantum. Grav.* **20**, 3613.
[44] T. Padmanabhan (2004) *Int. J. Mod. Phys.* **13**, 2293.
[45] M. Rangamani (2009) *Quantum Grav.* **26**, 224003.
[46] M. Beau (2015) *Fondation Louis de Broglie* **40**, 1.
[47] C.G. Böhmer, N. Tamanini, and M. Wright (2018) *Int. J. Mod. Phys.* **27**, 1850007.
[48] W. Hehl, C. Kiefer, and G. Rel (2017) *Gravitation* **8**, 50.
[49] A. Tartaglia and N. Radicella (2019) arXiv:9034096.
[50] M. Beau (2018) arXiv:180503020v2.
[51] T. Tenev and M.F. Horstemeyer (2018) *Adv. Phys. Sci.* **2**, 1.
[52] T. Tenev and M.F. Horstemeyer (2019) *Int. J. Mod. Phys. D* **28**, 1.
[53] T. Tenev and M.F. Horstemeyer (2019) *Int. J. Mod. Phys. D* **28**, 9.
[54] C.W. Misner, K.S. Thorne, and J.A. Wheeler (1970) *Gravitation* (W.H. Freeman and Company San Francisco) **1**, 941.
[55] G. Mihai (2012) *The bases of the unifying theory of physics* (University of Craiova, Faculty of Electrotechnics, Romania) **1**, 12.
[56] Collectif (2019) *Cours De Physique Gratuit.* **1**, 1.
[57] R. Taillet (2013) Université, Savoie, Mont, blanc, LAPTH, Laboratoire, d'Annecy-le-Vieux, Physique, Théorique, e-lecture, podcast (Introduction à la relativité générale) **24**, 1.
[58] E. Ventsel and T. Krauthammer (2001) *Thin plates and shells theory and applications* (Deckker, New York) **1**, 347.
[59] S. Timoshenko and S. Woinowsky Krieger (1961) *Theory of plate and shell* (McGraw Hill) **1**, 432.
[60] J. Pascal (2008–2009) *Vibration et acoustique 2* (Ensim, 2A) **1**, 14.
[61] G. Lhermet and G. Vessiere (2008–2009) *Module F312 Elasticité linéaire* (problèmes, et, corrigés, semestre, 3) **1**, 3.
[62] H. Casimir (1948) *R. Netherlands Academy Arts Sciences* **1**, 793.
[63] A. Michelson and W. Morley (1887) *A. J. Sci.* **1**, 1.
[64] Collectif (2006) mms2.ENSMP.fr.mmc.Paris.annales.corrige. dim.disc.helicop **1**, 1.
[65] R. Owen, J. Brink, Y. Chen, J. Kaplan, G. Lovelace, K. Matthew, D. Nichols, M. Scheel, F. Zhang, A. Zimmerman, and K. S. Thorne (2011) *Phys. Rev. Lett.* **106**, 151101.
[66] Damien André (2020) *Résistance des matériaux* (Ensil-ensci, Université de Limoge cours TD TP Module, 425) **1**, 31.
[67] Collectif Kigo (2019) *Tests of General Relativity with the Binary Black Hole Signals from the LIGO-Virgo Catalog GWTC-1 arXiv:1903.04467v3.*
[68] L. Bernus, O. Minazzoli, A. Fienga, M. Gastineau, J. Laskar, and P. Dera (2019) Constraining the mass of the graviton with the planetary ephemeris INPOP, *Phys. Rev. Lett.* **123**, 161103.
[69] T. Damour (2005) *Note Einstein et la physique du vingtième siècle* **1**, 1.

[70] T. Damour (2012) si Einstein m'était conté de la relativité à la théorie des cordes (la loi Einstiennenne d'élasticité de l'espace-temps édition cherche, midi) **1**, 1.
[71] F. Dournac, Dournac.org (Note T. Damour, Einstein n'a jamais utilisé l'expression de la loi d'élasticité de l'espace-temps) https://dournac.org/sciences/general_relativity/node31.html.
[72] R. Taillet (2015) Université, Savoie, Mont, blanc, LAPTH, Laboratoire, d'Annecy-le-Vieux, Physique, Théorique, e-lecture, podcast (cours de cosmologie) **1**, 1.

References Bibliographies – General Culture and Sources of this Book Noted {xx} in the Body of the Text

For the Mechanics of Structures

Strength of Material of Beams

[1] S. Timoshenko (1949) «Strength of material» Translated from English 2e edition by Ch. Laffitte.
[2] D. Izabel (2007) «Formulaire de résistance des matériaux tome 1 à 5» Sebtp.
[3] S. Timoshenko (1968) «Strength of material» tome 1 et 2.
[4] S. Timoshenko (1933–1951) Theory of elasticity introduction chapter 1 introduction.

Strength of Material Plate

[5] S. Timoshenko (1951) «Théorie des plaques et des coques, Traduit de l'anglais par L. Vial Reliure inconnue» p. 46, chap. 12 and p. 37, chap. 10.
[6] Pa Kelly Auckland university «lecture note section 6.9 solid mécanique part II».

Matrix Calculations of Structures

[7] Michel Cazenave «Méthode des éléments finis 2ème édition Approche pratique en mécanique des structures Collection: Technique et Ingénierie» Dunod.
[8] Alain Chateauneuf «structures comprendre les éléments finis» élipse technosup.
[9] J.M. Aribert «Calculs matriciels des structures» Cours INSA Rennes.
[10] Uday S. (2007) «Dixit Finite Elements Methods in engineering (Lecture Notes)» Department of Mechanical Engineering Indian Institute of Technology Guwahati Guwahati – 781039, Assam, India.

Eigen Period of String Vibration, Beams and Plates

[11] George Rawitscher and Jakob Liss (2010) «The vibrating inhomogeneous string» Department of Physics, University of Connecticut, Storrs, Connecticut, 06269.
[12] Ménad CHENAF Jean-Vivien HECK (2007) «Compléments technique relatif à l'effet de surpression Recommandations et précautions en vue de réduire les risques – rapport d'étude» CSTB Guide PPRT.
[13] S. Timoshenko and D/H Young (1948) «advanced dynamique».
[14] Jean-claude Pascal «vibration et acoustique 2 Ecole nationale d'ingénieur du Man».
[15] Dominique Nizerhy «mécanique et dynamique de structure».

[16] Engineering Mathematics: «Open Learning Unit Level 113.4: Tangents and Normals».
[17] E.L. Lady «Curvature».

For General Relativity

[18] Tenev T.G. and Horstemeyer M.F. (2016) «The Mechanics of Space time. A Solid Mechanics Perspective on the Theory of General» Foundations of Physics manuscript arXiv:1603.07655v3» p. 24.
[19] M.R. Beau (2014) «Théorie des champs des contraintes et des déformations en relativité générale et expansion cosmologique» – Foundations of Physics manuscript arXiv:1209.0611v2, p. 4 and Annales de la Fondation Louis de Broglie, Volume 40, 2015.
[20] T. Damour (2006) «La Relativité Générale aujourd'hui» Institut des Hautes Etudes Scientifiques Bures-sur-Yvette, France – Séminaire Poincaré, pp. 1–40.
[21] T. Damour (2005) «Einstein et la physique du vingtième siècle» Membre de l'Académie des sciences, p. 4.
[22] M.R. Beau (2014) «Sur l'hypothèse concernant l'existence d'un champ de gravité de type vectoriel couplant avec l'accélération des masses» Hal archive ouvertes 00683021v2 point c, p. 13.
[23] Collective (2016) «Observation of Gravitational Waves from a Binary Black Hole Merger» Physical Review Letter.
[24] Collectif Cours de physique «tenseur énergie impulsion» http://physique.coursgratuits.net/relativite-generale/tenseur-d-energie-impulsion.php.
[25] Richard Taillet «Cours en ligne initiation à la relativité générale, cosmologie et relativité restreinte» teacher-researcher at Savoie Mont blanc university.
[26] Moore «Relativité Générale» édition Boeck.
[27] DR physics A «Einstein field equation for beginner».
[28] David O'Neill «Derivation of the Riemannian Curvature Tensor» https://www.youtube.com/watch?v=Dcxo6AhQABo.
[29] T. Damour (2000) «conférence Relativité générale» université de tous les savoir.
[30] sienceclic. https://www.youtube.com/user/ScienceClic.
[31] T. Damour «si Einstein m'était conté...» champs sciences Flammarion.

For Quantum Mechanics

Quantum Curvature

[32] C. De Beule (2016) «Graphical interpretation of the Schrodinger equation» University Antwerpen.
[33] Michael Fowler (2007) «Schrödinger's Equation in 1-D: Some Examples Curvature of Wave Functions».
[34] Steven Errede (2016) «UIUC Physics 406 Acoustical Physics of Music» Department of Physics, University of Illinois at Urbana-Champaign, IL.
[35] Xinzhong Wu (2015) «Probability and Curvature in Physics School of History and Culture of Science» Shanghai Jiaotong University, Shanghai, China Journal of Modern Physics, 2191–2197.
[36] Kansas state university «The shape of the wave function».

General Lectures in Quantum Mechanics

[37] J. Dalibard «Online e-lecture of quantum mechanics X polytechnic 1st year 2013».
[38] Online courses: «variational calculation and special relativity X polytechnic».
[39] Paul D'Alessandris (2003) «Spiral Modern Physics» NY 14623.

Quantum Mechanics and Schrodinger in 1–3 Dimensions and Bohr Atom

[40] Pr. M. ABD-LEFDIL Université Mohammed V – Agdal Faculté des Sciences Département de Physique Année universitaire 07-08Filières SM-SMI – «mécanique quantique».
[41] Gary Larson «Chapter 6 very brief introduction to quantum mechanics».
[42] T. Brierre «cour L1 CHIM 110 atomes et molécules USTHB info».

Spontaneous Symmetry Break

[43] Jean Iliopoulos «Introduction to the origins of mass elementary particles and fundamental symmetries» OUP Oxford.
[44] Perez Carles Pujol «Quantum foundations and applications» de Boeck.
[45] Ph. Grangier/J. Dalibard «1st and 2nd year quantum mechanics course» X Polytechnique.
[46] Jean Jacques Samuleli «The standard model of electron particle physics at the Higgs boson».

Particles and Waves

[47] Alice Sinatra (2008) «Introduction à la mécanique quantique».
[48] Richard Taillet «initiation à l'électromagnétisme» – teacher-researcher at Savoie Mont blanc university.

For the Waves

[49] P-PAM-307B (2017) «Ondes et Acoustique dans les Fluides Notes de cours» Université Paris Saclay.
[50] Autiwa (2009) «PHQ952 Mécanique Quantique Approfondie».
[51] Jae Yu (2007) «PHYS 5326 – Lecture 12, 13, 14».

For All Subjects

[52] R. Feynman (1963) «The Feynman Lectures on Physics» Addison-Wesley, vol. II, pp. 38–10.
[53] C.W. Misner, K.S. Thorne, and J.A. Wheeler (1973) «Gravitation» W.H. Freeman; x17.5, pp. 426–428.
[54] Paty Michel (2005) HAL «L'espace-temps de la théorie de la relativité».
[55] Bernard Guy (2016) «Relier la mécanique et la relativité générale? Réflexions et propositions».
[56] Etienne Klein (2014) «collections les grandes idées de la science».

For the Author David Izabel Hal Publications

[57] David Izabel (2017) Hal-01535955v1 «Can we understand the fundamentals of general relativity and quantum mechanics from the concepts of material resistance?» HAL archives ouvertes.

[58] David Izabel (2017) Hal 01546414v1 «Taking into account the rigidity of space-time in general relativity -Understanding by a simple analogy based on the concepts of material resistance» HAL archives ouvertes.

[59] David Izabel (2017) Hal 01558504v1 «Can we estimate the value of Young's space-time module from simple analogies based on the concepts of material resistance? Reflections and proposals» HAL archives ouvertes.

For Casimir's Force

[60] Gert-Ludwig Ingold (2014) «Casimir effect from a scattering approach» Gert-Ludwig Ingold Institut fur Physik, Universitat Augsburg, Universitatsstrae 1, D-86135 Augsburg, Germany arXiv:1404.6919v1 [quant-ph] 28 Apr 2014, p. 5, chapter 5.

[61] G. Bressi, G. Carugno, R. Onofrio, and G. Ruoso (2002) «Measurement of the Casimir Force between Parallel Metallic Surfaces» Volume 88, Number 4 Physical Review Letters, p. 3.

[62] Mubassira Nawaz, Remco J. Wiegerink, Theo S.J. Lammerink, Miko Elwenspoek «Parallel Plate Structures For Optical Modulation and Casimir Force Measurment» Paper ID: 142.

[63] Matthew Schwartz (2008) «Lecture 14 the Casimir effect».

[64] Joseph P. Straley (University of Kentucky) Luke S. Langsjoen (University of Virginia), Hussain Zaidi (University of Virginia) «Casimir effect due to a single boundary as a manifestation of the Weyl problem» arXiv:1002.1762v1.

[65] http://physiquereussite.fr/energie-du-vide/.

[66] Bruce C. Denardo, Joshua J. Puda, and Andres Larraza, Am. J. Physics (2009) 77 (12) «A water wave analog of the Casimir effect».

For Demonstration of Friedmann's Equations

[67] Joan Arnau Romeu (2014) «Derivation of Friedman equations» Facultat de Fsica, Universitat de Barcelona, Diagonal 645, 08028 Barcelona, Spain Barcelona.

[68] Richard Taillet «cours en ligne de cosmologie» teacher-researcher at Savoie Mont blanc university.

[69] Source http://burro.cwru.edu/Academics/Astr330/Lect03/friedmann.html.

For Special Relativity

[70] Richard Taillet «cours en ligne Initiation à la relativité restreinte» teacher-researcher at Savoie Mont blanc university.

For the Concept of Eigen Value and Eigen Mode of Space-Time

[71] E. Schrodinger (1939) «The proper vibrations of the expanding universe» Physica 6,899.

For Quantum Field Theory Klein–Gordon–Dirac

[72] http://www.diffusion.ens.fr/vip/pageC00.html.
[73] Pierre Salati (2017) «Parcours physique subatomique et cosmologie» PSC Universitées de Grenoble-Alpes & de Savoie Mont-Blanc Master 2ème année de physique.
[74] David London (2013) «Introduction à la théorie quantique des champs PHY6611A: Le Modèle Standard» Université de Montréal.
[75] http://www.sciences.ch/htmlfr/physatomique/physatommecquantchamps01.php.
[76] http://www.sciences.ch/htmlfr/physatomique/physatommecphyspart01.php.
[77] Bertrand Delamotte (2005) «Introduction à la théorie quantique des champs».
[78] http://cordier-phychi.toile-libre.org/phy/TQC.html.
[79] Autiwa (2009) «PHQ952 Mécanique Quantique Approfondie».
[80] Alain Comtet (2006) «équation de Dirac» DEA<cel-00092970>.
[81] Pascal Début (2015) «L'équation de Dirac Le mariage de la relativité et de la mécanique quantique et la naissance de l'antimatière Notes de 25ème festival d'astronomie de Fleurance».
[82] Mossadek Talby (2017) «Master 1 Physique 1 Physique des Particules Chapitre II Equation de Dirac».
[83] Ilarion Pavel ingénieur en chef des mines chercheur au Laboratoire de Physique Théorique École Normale Supérieure «La prédiction de l'antimatière par Dirac».
[84] Dr. Prasanta Tripathy «Cour en ligne Quantum Field Theory» Department of Physics univ de Madras.
[85] Wikipédia.
[86] http://www.sciences.ch/htmlfr/physatomique/physatommecphyspart01.php.
[87] François Delduc «DEA de Physique Théorique Rhône-Alpin TC3, Introduction à la théorie des champs».
[88] «voyage vers l'infiniment petit (portail)» http://www.diffusion.ens.fr/vip/pageC00.html.
[89] Mabrouk Benhamou (2007) «Méthodes fonctionnelles en Mécanique Quantique: Une présentation simple HAL Id: cel-00140579» https://cel.archives-ouvertes.fr/cel-00140579.
[90] Laura Messio (2012) «Notes de cours du Professeur Shaposhnikov L'équation de Dirac».
[91] Autiwa (2009) «PHQ952 Mécanique Quantique Approfondie 11 partie 5».
[92] Etienne (2017) «Klein:la masse ? le boson de Higgs? IFG parenthèse culture» http://www.youtube.com/watch?v=hbyrxG4xih8.

For Gravitational Waves

[93] Richard Taillet «Cours en ligne d'initiation à la relativité générale» teacher-researcher at Savoie Mont blanc university.
[94] Cascina (2007) «Second VESF School on Gravitational Waves».
[95] B.P. Abbott *et al.* (2016) «Observation of Gravitational waves from a binary black holes merger» – LIGO scientific collaboration and Virgo collaboration physical review letter.

For Laplacian's Own values in Hyper-Spherical Coordinates

[96] Vernon D. Sandbergbl (1978) «Tensor spherical harmonics on S 2 and S 3 as eigenvalue problem» sa> Center for Relativity, University of Texas, Austin 78712 and California Institute of Technology, Pasadena, California 91125.
[97] David Skinner «Mathematical Methods University of Cambridge Part IB Mathematical Tripos».

[98] N. Yakubu, S.X.K. Howusu, W.L. Lumbi, and N. Ibrahim (2016) «Field Equation of a Static Homogeneous Oblate Spheroidal Massive Body» International Journal of Theoretical and Mathematical Physics 2016, 6(3): 104–109. DOI: 10.5923/j.ijtmp.20160603.03 Solution of Newton's Gravitational.

Overview of the Topics

[99] Joe Polchinski «Gravity and Quantum Mechanics – The Quest for Unification» https://www.youtube.com/watch?v=FqwXeeXnDZg#t=175.7710337.

Vibrations of the Universe and Space-Time

[100] https://www.sciencesetavenir.fr/espace/univers/les-sept-contractions-de-l-univers_101886.
[101] https://www.youtube.com/watch?v=9j6W8jCna9c.
[102] H.I. Ringermacher and L.R. Mead Submit. AJ (2015) «Observation of discrete oscillations in a model – independent plot of cosmological scale Factor VS Looback time and scalar field model» Dept. of Physics and Astronomy, U. of Southern Mississippi, Hattiesburg, MS 39406, USA.
[103] L. De Broglie (1927) «L'univers à 5 dimensions et la mécanique ondulatoire» le journal de physique et radium tome VIII.

Duality Waves Particles

[104] Antonin Eddi (2011) «Marcheurs, dualité onde-particule et Mémoire de chemin. Dynamique des Fluides» [physics.u-dyn]. Université Paris-Diderot – Paris VII, 2011. Français. HAL Id: tel-00575626 https://tel.archives-ouvertes.fr/tel-00575626.

Acoustic Sound Energy

[105] http://www.formules-physique.com/categorie/437.

Yukawa's Potential

[106] Yukawa see Search Utah State University. http://www.physics.usu.edu/Wheeler/Classical Mechanics/CMYukawaPotential.pdf.

String Theory and Quantum Qravity

[107] C. Bachas «String Theory and Quantum Gravity, Theoretical Physics, UMR 8549 CNRS/ENS Paris».
[108] Florian Perrin (2003) «Introduction to String Theory Commission of Cosmology of the Astronomical Society of France».
[109] Patricia Schwarz «theory of the strings» http://www.superstringtheory.com/basics/basic4a.html.

[110] String Theory. http://www.sciences.ch/htmlfr/cosmologie/cosmothcordes01.php.
[111] CNRS University Southern Paris «Chapter 3 Equation of Propagation and Energy» https://www.equipes.lps.u-psud.fr/gr_15/SMPEHTML/polypdf/ondchap3.pdf.
[112] Jean-Jacques LABARTHE (2005) «Corde Vibrante and Acoustique Laboratory» Aimé-Cotton.
[113] Dr. David Tong String «Theory University of Cambridge Part III Mathematical Tripos».
[114] Alexandre Depire «The cordes theory» Institut Point Carré.

Cosmological Diffuse Background

[115] Joey Shapiro Key and Neil J. Cornish Department of Physics, Montana State University, Bozeman, MT 59717 David N. Spergel Department of Astrophysical Sciences, Princeton University, Princeton, NJ 08544Glenn D. Starkman Center for Education and Research in Cosmology and Astrophysics, Department of Physics, Case Western Reserve University, Cleveland, OH 44106–7079 and Astrophysics Department, University of Oxford, Oxford OX1 3RH, UK «Extending the WMAP Bound on the Size of the Universe».
[116] Angélica de Oliveira-Costa & Max Tegmark MIT Kavli Institute for Astrophysics and Space Research, Cambridge, MA 02139; (2008) «CMB multipole measurements in the presence of foregrounds» to be submitted to Phys. Rev. D.
[117] Max Tegmark1, Angelica de Oliveira-Costa1 & Andrew J. S. Hamilton2. 1Department of Physics & Astronomy, University of Pennsylvania, Philadelphia, PA 19104, USA, and 2JILA and Dept. of Astrophysical and Planetary Sciences, U. Colorado, Boulder, CO 80309, USA (2003) «A high resolution foreground cleaned CMB map from WMAP» Submitted to Phys. Rev. D.
[118] Marc Betoule (2009) «Analyse des données du fond diffus cosmologique: simulation et séparation de composantes» Direction: Jacques Delabrouille (APC) et Marc-Antoine Miville-Deschênes (IAS) Soutenance de thèse.
[119] Fredéric Guilloux (2008) «Analyse harmonique et Estimation spectrale sur la Sphère. Applications à l'étude du Fond diffus cosmologique. Mathématiques» Université Paris Diderot – ParisVII.
[120] Frédéric GUILLOUX (2014) «Statistiques pour le fond diffus cosmologique» Université Pierre et Marie Curie Laboratoire de Statistique Théorique et Appliquée Travaux en collaboration avec Marc BETOULE, Jean-François CARDOSO, Jacques DELABROUILLE, Jacques Gilles FAŸ et Maude LE JEUNE (Université Paris-Diderot, Laboratoire Astro-Particule et Cosmologie) Mathématiques en mouvement.
[121] David skinner «Laplace's equation».
[122] Pierre Amiot (2011) «Le Laplacien et les Harmoniques sphériques» Source Département de Physique, Université Laval, Québec. http://physicsgirl.com/version9_002.htm.
[123] Mathieu Langer et Hervé Dole «Base de la cosmologie moderne notes préliminaires cours 8 et 9 sur l'univers perturbé» Master 2 NPAC.

Higgs Boson

[124] Christophe Roland «Vous avez dit boson de Higgs?».
[125] Ivo van Vulpen Assistant: Ivan Angelozzi (2014) «The Standard Model Higgs Boson Part of the Lecture Particle Physics II» UvA Particle Physics Master Université Paris Sud.
[126] Elias KHAN «Notions sur le modèle standard: du lagrangien au mécanisme de Higgs Institut Universitaire de France Institut de Physique Nucléaire» – Orsay. IN2P3 – CNRS.
[127] Bruno Machet (2007) «Rapport de stage Brisure spontanée de symétrie et mécanisme de Higgs dans le modele standard des interactions électrofaibles stage au Laboratoire de Physique Théorique et Hautes Energies».

[128] Abdelhak Djouadi «Le boson de Higgs au LHC: la quête de l'origine de la masse» Article proposé par: Laboratoire de Physique Théorique d'Orsay, UMR 8627, CNRS/Université Paris-Sud 11, Orsay.
[129] Diego Bettoni (2011) «The Higgs Mechanism Elementary Particle Physics Strong Interaction Fenomenology».
[130] Jean Iliopoulos (2014) "The origin of the mass elementary particles and fundamental symmetries". Oxford University Press.

Ether

[131] https://www.youtube.com/watch?v=k6gd3bQLiFc.
[132] Frederic Boissac. https://www.youtube.com/watch?v=x7A9ylimxiQ.
[133] Jean-Pierre Luminet (2011) «Histoire de l'Ether, d'Aristote au vide quantique, conférence centre Pompidou sur l'Ether» https://www.youtube.com/watch?v=au1sD-cbrC8.
[134] Einstein (1916) «Lettre à Lorentz 17 juin 1916 archive Albert Einstein Université Hébraïque de Jérusalem 16-453».
[135] Einstein (1920) «L'éther et la théorie de la relativité la géométrie et l'expérience» Conférence d'Einstein à l'université de Leyde le 5 mai 1920. http://quanthomme.free.fr/energielibre/energie/Einstein.htm.
[136] T. Damour (2016) «si Einstein m'était conté» édition Flammarion révisée.
[137] Ludwik Kostro «Albert Einstein's new ether and his general relativity».
[138] J. Einsenstaedt and A.J. Kox «Studies in the story of the general relativity» edited by bikhauser.
[139] Galina Granek Einstein's Ether: F. (2001) «Why did Einstein Come Back to the Ether?» Department of Philosophy, Haifa University Mount Carmel, Haifa 31905, Israel Apeiron, Vol. 8, No. 3.
[140] A. Michelson and E. Morley (1887) «On the relative motion of the earth and the luminifer Ether» American journal of sciences N°203.

Expansion of Space-Time

[141] Edwin Hubble (1929) «A relation between distance and radial velocity among extra-galactic Nebulae» Mount Wilson observatory, Carnegie Institution of Washington.

Distortion of Space-Time by the Rotation of the Earth – Lense Thirring Effect

[142] C.W.F. Everitt *et al.* (2011) «Gravity Probe B: Final Results of a Space Experiment to Test General Relativity» Phys. Rev. Lett. 106, 221101 and arXiv:1105.3456v1 [gr-qc] 17 May 2011.
[143] Wikipédia Effet Lense–Thirring.
[144] Wikipedia «Satellite Gravity probe B».
[145] O.I. Chashchina, L. Iorio, and Z.K. Silagadze (2009) «Elementary derivation of the Lense-Thirring precession» Cornell University library.
[146] Dr. Philippe Jetzer typeset and revision: Felix Hahl (2011) «Astrophysical and Cosmological Applications of General Relativity» ETH Zurich and University of Zurich FS.
[147] Richard Taillet (2017) «Introduction à la relativité générale» teacher-researcher at Savoie Mont blanc university.
[148] Vasco Miguel rold ao mantegas physics engineering Gravito-electromagnétisme «The Lense–Thirring Effect».

[149] Chandrachur chakraborty «Lense–Thirring precession in strong gravitational field» Institute fundamental de recherche de Mumbai India.
[150] Saul A. Teukolski «The Kerr metric» Cornel university USA.
[151] Klaus L'ammerzahl and Gernot Neugebauer «The Lense–Thirring Effect:From the Basic Notions to the Observed Effects» 1 Fakultat fur Physik, Universitat Konstanz, 78457 Konstanz, Germany 2 Institut fur Theoretische Physik, Friedrich–Schiller–Universitat Jena, Max–Wien–Platz 1, 07743 Jena, Germany.

Moment of Inertia and Kinetic Energy of Rotating Bodies

[152] Marc Séguin «Physics XXI Volume A Page 1 Course note written by: Simon Vézina Chapter 4.4 – The Moment of Inertia and Kinetic Rotation Energy».
[153] https://www.youtube.com/watch?v=i0dTMI1N9Co.
[154] https://www.youtube.com/watch?v=ToPJB1V_xKA.
[155] Marc Séguin «Physical XXI Volume A course note written by Simon Vezina Chapter 3.2 elastic energy of an ideal spring».

Einstein Cartan Kibble Sciama Theory

[156] Steuard Jensen (2005) «General Relativity with Torsion: Extending Wald's Chapter on Curvature» Enrico Fermi Institute and Department of Physics University of Chicago 5640 S. Ellis Avenue, Chicago IL 60637, USA.
[157] Jean-Philippe Bruneton «Théories alternatives de la gravitation et applications».
[158] Andrew Culham (1993) «Black hole solution with torsion» Shaun university of British Columbia Vancouver.
[159] Nikodem J. Poplawski c Elsevier B. V. (2010) «Cosmology with torsion an alternative to cosmic inflation» Department of Physics, Indiana University, Swain Hall West, Physics Letters B, Vol. 694, No. 3, pp. 181–185.
[160] R.J. Petti «Derivation of Einstein–Cartan theory from general relativity» 146 Gray Street, Arlington, Massachusetts 02476 U.S.A.
[161] F. Chambat (2015) «Mécanique globale de la Terre» ENS Lyon.
[162] Andrzej Trautman «Einstein–Cartan Theory» Warsaw University, Warsaw, Poland Published with typographical changes in: Encyclopedia of Mathematical Physics edited by J.-P. Francoise, G.L. Naber and Tsou S.T. Oxford: Elsevier (2006) vol. 2, pages 189–195.
[163] Nikodem J. Popławski «Einstein–Cartan Gravity in Particle Physics and Cosmology» Department of Mathematics and Physics University of New Haven West Haven, CT, USA.
[164] Nikodem J. Popławski (2010) «Cosmology with torsion: an alternative at the cosmic inflation» Physical Letter B 694 (2010) 181–185.
[165] H. Hadi, Y. Heydarzade, M. Hashemi, and F. Darabi «Emergent cosmos in Einstein–Cartan theory» Department of Physics, Azarbaijan Shahid Madani University, Tabriz 53714-161 European physical journal.
[166] L. N. Katkar «Einstein–Cartan Theory of Gravitation Kinematical parameters and Maxwell equations».
[167] Myron W. Evans «Chapter 2 A Review of Einstein–Cartan–Evans (ECE) Field Theory by Alpha Institute for Advance Study» (AIAS).

Twist in RDM

[168] S. Baraka et A. Bureau «Calcul des contraintes de torsion dans un éléments soumis à de la torsion» Revue construction métallique n°1 de 2000.

[169] Jean Azoulet «formulaire de RDM» Dunod.
[170] Access steel «NCCI Torsion SN007b-FR-EU».

Superfluid Space-Time Without Viscosity

[171] Stefano Liberati and Luca Maccione «Astrophysical constraints on Planck scale dissipative phenomena SISSA» Via Bonomea 265, 34136, Trieste, Italy and INFN, Sezione di Trieste Arnold Sommerfeld Center, Ludwig-Maximilians-Universitat, Theresienstra_e 37, 80333 Munchen, Germany 3Max-Planck-Institut fur Physik (Werner-Heisenberg-Institut), Fohringer Ring 6, 80805 Munchen, Germany.
[172] Marco Fedi «Special and general relativity according to superfluid quantum space».

Stiffness of the Space Time

[173] Andrew Ochadlick (2016) «Space time Stiffness and Rigidity Approximation» https://www.youtube.com/watch?v=vT_zkTu6wZk.
[174] Kirk T. McDonald (2018) «What is the Stiffness of Space time?» Joseph Henry Laboratories, Princeton University, Princeton, NJ 08544.
[175] Mathieu R. Beau (2018) «On the nature of space-time, cosmological inflation, and expansion of the universe» Department of Physics, University of Massachusetts, Boston, MA 02125, USA.
[176] T.G. Tenev and M.F. Horstemeyer (2018) «Mechanics of space time – A Solid Mechanics perspective on the theory of General Relativity» International Journal of Modern Physics D, Vol. 27, No. 08, 1850083 (2018).
[177] S.R. Hwang (2020) «Estimation of space time stiffness based on LIGO observations».
[178] D. Izabel (2020) «Mechanical conversion of the Einstein's constant Kappa» Pramana.

Global Warming of the Universe

[179] Yi-Kuan Chiang, Ryu Makiya, Brice Ménard, and Eiichiro Komatsu (2020) «The Cosmic Thermal History Probed by Sunyaev–Zeldovich Effect Tomography» Astrophysical journal volume 902 n°1 -American Astronomy so.
[180] Françoise Combes «observatoire de Paris Dark energy and new physics (slide 36) "The case particule/horizon applies to dS, the intrication entropy produces states of thermal excitation responsible of dark energy. Dark energy and accelerated expansion are due to the slow thermalisation of the emergence of space-time"».

Number of Dimensions in the Universe (4) According to Gravitational Waves

[181] Kris Pardo, Maya Fishbach, Daniel E. Holz, David N. Spergel (2018) «Limits on the number of space time dimensions from GW170817» Journal of Cosmology and Astroparticle Physics 2018 (07).

Black Hole

[182] K. Schwarzschild (1916) «Uber das gravitationfeld eines massenpunktes nach der einsteinschen Theorie».

Wave Particle Duality – Thermal Fluctuation

[183] Einstein (1909) «On the present status of the radiation problem 15 march 1909 Physikalische Zeitschrift 185–193)».
[184] Alexei Kojevnikov Einstein's «Fluctuation Formula and the Wave-Particle Duality».
[185] Sándor Varró «Einstein's Fluctuation Formula. A Historical Overview».
[186] Alain Aspect (2019) «The photon wave or particle quantum strangeness brought to light conference Le photon onde ou particule? L'étrangeté quantique mise en lumière» les amis de l'IHES».

Wave Equations in Elasticity

[187] Prof. Prasenjit Khanikar (2019) «NPTEL IIT Guwahati department of mechanical Engineering IIT Guwahati dynamic behaviour of material Lecture 6: Propagation of Elastic Waves in Continuum. https://www.youtube.com/watch?v=dyoizjrJNkU».
[188] T. Tenev and M.F. Horstemeyer (2018) «In their publication formula 3.9 at 3.14 of Mechanics of space time – A Solid Mechanics perspective on the theory of General Relativity».
[189] Marc François (2009) «Introduction à la théorie des ondes élastiques».
[190] Tony Valier-Brasier (2005) «UPMC Notes de cour Ondes élastiques dans les solides isotropes».
[191] Bernard Giroux (2011) «Institut National de la Recherche Scientifique Centre Eau Terre Environnement Méthode sismique les ondes sismiques».
[192] Prof. Matteo Ciccotti, révisé par Bruno Bresson (2018) «ESPCI Paris – Laboratoire de Science et Ingénierie de la Matière Molle Mécanique des Solides et des Matériaux Elasticité-Viscoélasticité Plasticité-Rupture».

Human Brain Frame and Frame of the Univers

[193] F. Vazza and A. Feletti, Frontière of physique (2020) «The Quantitative Comparison Between the Neuronal Network and the Cosmic Web».

About the Expression of G as a Function of c, f and Y

[194] T. Young (1800) «Outlines of Experiments and Inquiries respecting Sound and Light, Phil. Trans. Roy. Soc. London 90, 106».
[195] C. Huygens (1912) «Traite de la Lumiere (Leiden, 1690), Treatise on Light, trans. by S.P. Thompson (Macmillan)».
[196] A. Einstein (1915) «Die Feldgleichungen der Gravitation, Sitzb. Konig. Preuß. Akad. Wissen. (Berlin) 844, The field Equations of Gravitation».

[197] B.P. Abbott *et al.* (2016) «Observation of Gravitational Waves from a Binary Black Hole Merger, Phys. Rev. Lett. 116, 061102».

[198] B.P. Abbott *et al.*, GW150914: (2016) «The Advanced LIGO Detectors in the Era of First Discoveries, Phys. Rev. Lett. 116, 131103».

[199] B.P. Abbott *et al.*, GW151226: (2016) «Observation of Gravitational Waves from a 22-SolarMass Binary Black Hole Coalescence, Phys. Rev. Lett. 116, 241103».

[200] L.D. Landau and E.M. Lifshitz (1975) «The Classical Theory of Fields, 4th ed. (ButterworthHeinemann, Oxford)».

[201] S. Weinberg (1972) «Gravitation and Cosmology (J. Wiley and Sons, New York)».

[202] A. Melissinos and F. Lobkowicz (1975) «Physics for Scientists and Engineers (W.B. Saunders Company, Philadelphia)».

[203] M. Maggiore (2008) «Gravitational Waves» (Oxford University Press).

[204] C.W. Misner, K.S. Thorne and J.A. Wheeler (1973) «Gravitation (W.H. Freeman and Co)».

[205] T.G. Tenev and M.F. Horstemeyer (2018) «Mechanics of space time "A Solid Mechanics perspective on the theory of General Relativity", Int. J. Mod. Phys. D 27, 1850083».

[206] D. Izabel (2017) «Can we estimate the Youngs modulus of the space-time from simple analogies based on the concepts of the strength of materials? Reflection and proposals».

[207] B.P. Abbott *et al.* (2017) «GW170104: Observation of a 50-Solar-Mass Binary Black Hole Coalescence at Redshift 0.2, Phys. Rev. Lett. 118, 2221101».

[208] B.P. Abbott *et al.* (2017) «GW170814: A Three-Detector Observation of Gravitational Waves from a Binary Black Hole Coalescence, Phys. Rev. Lett. 119, 141101».

[209] B.P. Abbott *et al.* (2017) «GW170817: Observation of Gravitational Waves from a Binary Neutron Star Inspiral, Phys. Rev. Lett. 119, 161101».

[210] B.P. Abbott *et al.* (2017) «Gravitational Waves and Gamma-Rays from a Binary Neutron StarMerger: GW170817 and GRB 170817A, Ap. J. Lett. 848, L13».

[211] S.P. Timoshenko and J.N. Goodier (1970) «Theory of Elasticity, 3rd ed. (McGraw-Hill)».

[212] U.H. Gerlach and J.F. Scott (1986) «Metric elasticity in a collapsing star: Gravitational radiation coupled to torsional motion, Phys. Rev. D 34, 3638».

[213] M.R. Beau (2018) «On the acceleration of the expansion of a cosmological medium».

[214] A. Einstein (1917) «Kosmologische Betrachtungen zur allgemeinen Relativit¨ atstheorie, Sitzb.König. Preuß. Akad. Wissen. (Berlin), 142 Cosmological Considerations in the General Theory of Relativity».

[215] A.C. Melissinos (2018) «Upper limit on the Stiffness of space-time (Apr. 27, 2018)».

[216] Hamilton lecture (2018) «Hamilton Lecture at Princeton "Exploring the Universe with Gravitational Waves Kip Thorne" the question arose as to what is the Young's modulus Y of space-time».

[217] Kostas D. Kokkotas (2002) «Gravitationnal Wave Physics Department of Physics, Aristotle University of Thessaloniki Thessaloniki 54124, Greece».

[218] Eanna E. Flanagan and Scott A. Hughes (2005) «New journal of physics, the basics of gravitational wave theory New Journal of Physics 7 (2005) 204 formula 5.35 to 5.40».

[219] Steven A. Balbus (2016) «Simplified derivation of the gravitational wave stresstensor from the linearized Einstein field equations».

[220] Carlos F. Sopuerta (2016) «Gravitational wave theory 2nd ICE summer school gravitational wave astronomy IEEC institute of space science».

[221] J.W. Maluf and F.F. Faria (2003) «The gravitational energy momentum flux foumulas 32 and 33».

Terms and Definitions

General relativity: Theory developed by A. Einstein which makes it possible to study gravitation, cosmology, the infinitely large and whose principle is that the force of gravity is not a force but a deformation of space-time resulting from mass-energy applied to it.

$$\begin{array}{c}\text{Mean}\\ \text{spatial}\\ \text{curvature}\end{array} = \frac{8\pi G}{c^4} \times \frac{\text{ENERGY}}{\text{VOLUME}} = \frac{8\pi G}{c^2} \times \frac{\text{MASS}}{\text{VOLUME}}$$

Quantum field theory: Studying the interactions between particles modeled as a field at the scale of the infinitely small.

Standard model: Standard model: Nature consists of leptons $(e^-(electrons); \mu^-(Muon); \tau(\text{tau}); (neutrinos)v_e; v_\mu; v_\tau)$ and quarks (u (up), c (charm), t (top), d (down), s (strange), b (bottom)) We also find hadrons (baryons based on quarks (qqq) and mesons (q and q^-)). According to this theory in the infinitely small there are 3 interaction forces (electromagnetic force, weak force and strong force) that interact at the levels of quantum particles. The mediators of its forces are bosons of gauges respectively the photons, the bosons W-, W+ and Z0, the gluons g. The origin of the mass of its particles relates to the Higgs boson (H).

Continuum mechanics: Branch of physics concerning mechanics intended to study the balance of a system, its motions, displacements, strains and stresses within a medium (solid or liquid or fluid or otherwise) created by surface, electromagnetic, temperature or volume forces applied to it.

Theory of elasticity: Theory defining the behavior laws between stresses and strains arising from the continuum Mechanics assuming that the body returns to its original form when we remove the load that initially deformed it (Hook's law). This results in the properties of elastic materials, Young's modulus Y (or E), Poisson's ratio v, shear modulus G.

Strength of material: Branch of mechanics that is a simplification of elasticity theory, aimed at establishing external support reactions, internal force (M bending moment, T twist, shear force (Q or V) N normal effort), stresses, deformations and deflections of elastic solid (beams, plates, shells,...) in static and dynamics.

Beam theory: Branch of the strength of material developed notably by S. Timoshenko applied to linear elastic solids deformed simultaneously in traction and compression by bending moments. The beam, when it curves, has a radius of curvature R and a curvature $1/R$. The fundamental relationship of the beams, as in general relativity, connects the curvature to the second derivative of the deflection at the external moments applied. ($y_{(x)}$ deflection, $\theta_{(x)}$ the rotation, EI the bending rigidity, $M_{(x)}$ the bending moment).

$$\frac{d^2 y_{(x)}}{dx^2} = \frac{d\left(\frac{dy_{(x)}}{dx}\right)}{dx} = \frac{d\theta_{(x)}}{dx} = \frac{M_{(x)}}{EI} = \frac{1}{R}$$

Longitudinal deformation: Relative length variation to an initial length:

$$\varepsilon = \frac{\Delta L}{L_{\text{initial}}} = \frac{L_{\text{final}} - L_{\text{initial}}}{L_{\text{initial}}}$$

Shear deformation: Angular deformation γ of a parallelogram under shear force action.

Stress: Force applied to a surface.

Normal stress: Stress perpendicular to a surface. $\sigma = \frac{N}{S}$.

Shear stress: Stress applied in the plane of a surface. $\tau = \frac{V}{S}$.

Strength of material Curvature: Reverse of curvature radius R: curvature $= 1/R$.

Curvature in the general relativity sense: Angle/surface.

$$\text{Curvature} = \frac{\alpha + \beta + \gamma - 180°}{\text{Surface}}$$

Tensor: A Mathematical object whose explicit form is a matrix, or a letter provided with indices and exponents, characterizing a physical object, making it possible to express mathematical laws independently of the frame of reference in which we place ourselve sex: σ_{ij} stress tensor, ε_{ij} strain tensor, $G_{\mu\nu}$ Einstein's tensor, $R^{\alpha}_{\mu\nu\beta}$ curvature tensor or Riemann tensor, $R_{\mu\nu}$ Ricci's tensor, $g_{\mu\nu}$ metric tensor. Stress energy tensor $T_{\mu\nu}$.

Gravitational wave: Elastic deformation of space-time that propagates as a wave within the structure of it.

Space-time: Elastic medium with 3 dimensions of space and one of time.

Black hole: Singularity, tears of space-time or nothing even light can come out of it. A star/singularity made up of the distorted texture of space-time resulting from the gravitational collapse of a giant star. Local yielding of space-time has a gigantic excess load concentrated in a small volume. Locally and intensely distorted space-time.

Interferometer: Device consisting of two vacuum arms traveled by a laser beam between two mirrors to measure the space strains during the passage of a gravitational wave.3 interferometers exist (2 in the USA Ligo, 1 in Europe Virgo) several are under construction especially in India and Japan. Gigantic high-precision deformation gauge that measures longitudinal compression/traction deformation simultaneously and opposite in each 3 or 4 km arm in the order of $(\varepsilon_{xx} = -\varepsilon_{yy} = \pm\varepsilon = \pm 10^{-21})$.

Vacuum: Space environment filled with virtual particles appearing and disappearing in very short times thanks to the Heisenberg's principle. Medium filled by the Higgs scalar field at the time of the big bang. The medium where Casimir's force manifests itself. Deformable elastic space medium according to Einstein.

Dark matter: Matter of unknown origin undergoing gravity but not emitting light. Supposed to exist to explain in particular why stars on the periphery of galaxies have velocities that do not decrease with the radius of the galactic disk or why the structure of the universe on a large scale is in the form of clusters of galaxies following a wired texture. Would have influenced the formation of galaxies since the big bang. Dark matter accounts for 26.8% of matter.

Neutron star: A star made principally of neutrons resulting from the collapse of a medium-sized star.

Dark energy: Energy of unknown origin that opposes gravitation and accelerates the expansion of the universe. It would be associated with the cosmological constant Λ. It represents 68.3% of what constitutes the universe.

Ordinary material: Matter made of particles defined in the standard model. It represents 4.9% of the content of the universe.

Einstein's constant: $\kappa = \frac{8\pi G}{c^4}$: A constant of proportionality between the curvature tensor and the stress-energy tensor.

Luminerous ether: Medium supposed to be vector of the propagation of light by its own vibrations. Michelson and Morley's experience failed to detect it.

New ether: Elastic medium filling the empty space distorted by the presence of mass/energy within it.

Graviton: Hypothetical spin 2 boson of theoretical mass null (speed of gravitational wave = c at a precision of $1/10^{15}$) supposed to transmit gravitational force according to quantum field theory.

Spin: Ownership of quantum objects, a kind of magnetic moment attached to each particle.

Quantum object: An infinitely small object, which manifests itself as a wave or a particle following the experiment carried out.

Quantum loop gravity: Theory under construction attempting to quantify space-time and gravity. If confirmed, space-time would be quantified to one dimension of the Planck scale ("space atom" of Planck length and volume connected together by loops).

String theory: Theory under construction aimed at unifying all the forces of nature (electromagnetic, weak, strong and gravitation). The elementary element of nature would be Planck-length strings, open or closed, vibrating in different ways to characterize different particles including the graviton.

Definition of the Main General Relativity Terms (Source e-Lecture of General Relativity Initiation of Richard Taillet Teacher-Researcher at Savoie Mont Blanc University)

Metric tensor: (geometric unknown to be determined which characterizes the gravitation)

$$g_{\mu v} = \eta_{\alpha\beta} \frac{\partial \xi^{\alpha}}{\partial x^{\mu}} \frac{\partial \xi^{\beta}}{\partial x^{v}}$$

Minkowski's metric:

$$\eta_{\alpha\beta} = \begin{bmatrix} 1 & 0 & 0 & 0 \\ 0 & -1 & 0 & 0 \\ 0 & 0 & -1 & 0 \\ 0 & 0 & 0 & -1 \end{bmatrix}$$

The interval (m^2):

$$ds^2 = c^2\,dt^2 - dx^2 - dy^2 - dz^2$$

$$ds^2 = \eta_{\alpha\beta}\, d\xi^{\alpha}\, d\xi^{\beta}$$

We note the coordinates of an inertial frame in free fall (which does not feel gravitation).

$$\xi^{\alpha} \text{ such as: } \xi^0 = ct;\ \xi^1 = x;\ \xi^2 = y;\ \xi^3 = z$$

We note the coordinates of a Galilean frame that is not in free fall (which is subject to gravitation).

$$x^{\mu} \text{ such as: } x^0;\ x^1;\ x^2;\ x^3$$

The geodesic equation: to calculate the movement of objects within warped space-time):

$$\frac{d^2 x^{\beta}}{d\tau^2} + \Gamma^{\beta}_{\mu v} \times \frac{dx^{v}}{d\tau} \times \frac{dx^{\mu}}{d\tau} = 0$$

$$\frac{d^2x^\beta}{d\tau^2} = -\Gamma^\beta_{\mu\nu} \times \frac{dx^\nu}{d\tau} \times \frac{dx^\mu}{d\tau}$$

$$\Gamma^\beta_{\mu\nu} = \frac{\partial x^\beta}{\partial \xi^\alpha} \times \frac{\partial^2 \xi^\alpha}{\partial x^\mu \partial x^\nu}$$

Christoffel coefficients based on the metric tensor:

$$\boldsymbol{\Gamma^a_{bc(x)} = \frac{1}{2} g^{ad} \left\{ \frac{\partial g_{dc}}{\partial x^b} + \frac{\partial g_{db}}{\partial x^c} - \frac{\partial g_{bc}}{\partial x^d} \right\}}$$

$$\Gamma^\lambda_{\mu\nu} \text{ Knowing } \Gamma^\lambda_{\mu\nu} = \frac{1}{2} g^{\lambda\rho}(g_{\mu\rho,\nu} + g_{\nu\rho,\mu} - g_{\mu\nu,\rho}) = \frac{1}{2} g^{\lambda\rho}\left(\frac{\partial g_{\mu\rho}}{\partial x^\nu} + \frac{\partial g_{\nu\rho}}{\partial x^\mu} - \frac{\partial g_{\mu\nu}}{\partial x^\rho}\right)$$

Riemann's curvature tensor functions with Christoffel coefficients:

$$\boldsymbol{R^\alpha_{\beta\mu\nu} = \Gamma^\alpha_{\beta\nu,\mu} - \Gamma^\alpha_{\beta\mu,\nu} + \Gamma^\alpha_{\mu\eta}\Gamma^\eta_{\beta\nu} - \Gamma^\alpha_{\nu\eta}\Gamma^\eta_{\beta\mu}}$$

$$R^\lambda_{\mu\nu\alpha} = \Gamma^\lambda_{\mu\alpha,\nu} - \Gamma^\lambda_{\mu\nu,\alpha} + \Gamma^\lambda_{\nu\eta}\Gamma^\eta_{\mu\alpha} - \Gamma^\lambda_{\alpha\eta}\Gamma^\eta_{\mu\nu}$$

Ricci's tensor:

$\boldsymbol{R_{\mu\nu} = R^\lambda_{\mu\nu\lambda}}$ with a contraction on λ: The scalar curvature:

$$R = g^{\mu\nu}R_{\mu\nu} = g^{tt}R_{tt} + g^{rr}R_{rr} + g^{\theta\theta}R_{\theta\theta} + g^{\varphi\varphi}R_{\varphi\varphi}$$

The stress energy tensor:

$$T^{\mu\nu} - -pg^{\mu\nu} + (\rho + p)u^\mu u^\nu$$

$$T^{\mu\nu} = \begin{bmatrix} \rho c^2 & 0 & 0 & 0 \\ 0 & -p & 0 & 0 \\ 0 & 0 & -p & 0 \\ 0 & 0 & 0 & -p \end{bmatrix} g^{\mu\nu}$$

Einstein's equation without cosmological constant:

$$\boldsymbol{R^{\mu\nu} - \frac{1}{2} g^{\mu\nu} R = -\frac{8\pi G}{c^4} T^{\mu\nu}}$$

Einstein's equation linearized:

$$\boldsymbol{\frac{\partial^2 \bar{h}_{\mu\nu}}{\partial t^2} - c\Delta \bar{h}_{\mu\nu} = -\frac{16\pi G}{c^4} T^{\mu\nu}}$$

$$g_{\mu\nu} = \eta_{\mu\nu} + h_{\mu\nu}$$

This corresponds to almost the flat metric $\eta_{\mu\nu}$ plus a complementary term $h_{\mu\nu}$ called perturbation.

Metric in cosmology leading to Friedman Lemaitre's equations:

$$ds^2 = c^2 dt^2 - a^2(t)\left\{\frac{dr^2}{1 - kr^2} + r^2(d\Omega^2)\right\}$$

$$ds^2 = c^2 dt^2 - a^2(t)\left\{\frac{dr^2}{1 - kr^2} + r^2(d\theta^2 + \sin^2\theta d\varphi^2)\right\}$$

In this expression:
c is the speed of light,

$a_{(t)} = \frac{r_{(t)}}{r_0}$ is the scale factor.
k is the curvature factor in ($1/m^2$) (1 spherical shape, 0 flat universe - 1 hyperbolic shape).
t, r, θ, φ the 4 space-time coordinates considered (spherical coordinates).

$$ds^2 = g_{\mu\nu} dx^\mu dx^\nu$$

$$ds^2 = \sum_{\mu=0}^{3}\sum_{\nu=0}^{3} g_{\mu\nu} dx^\mu dx^\nu$$

μ and ν vary from 0 to 4 with 0 for time and 1,2,3 for space.
The development of $g_{\mu\nu}$ is given below:

$$\begin{aligned} ds^2 &= g_{tt} dt^2 + 2g_{rt} dr dt + 2g_{\theta t} d\theta dt + 2g_{\varphi t} d\varphi dt \\ &+ g_{rr} dr^2 + 2g_{\theta r} d\theta dr + 2g_{\varphi r} d\varphi dr \\ &+ g_{\theta\theta} d\theta^2 + 2g_{\varphi\theta} d\varphi d\theta \\ &+ g_{\varphi\varphi} d\varphi^2 \end{aligned}$$

Note that there are ten different terms, hence the 10 independent equations to solve.

In the case of the universe, Friedmann–Lemaitre's metric is a solution in force today. We only have the following terms, all the others are zero:

$$g_{tt} = c^2$$

$$g_{rr} = -\frac{a^2(t)}{1-kr^2}$$

$$g_{\theta\theta} = -a^2(t) r^2$$

$$g_{\varphi\varphi} = -a^2(t)\sin\theta$$

$$g_{\mu\nu} = \begin{bmatrix} c^2 & 0 & 0 & 0 \\ 0 & -\frac{a^2(t)}{1-kr^2} & 0 & 0 \\ 0 & 0 & -a^2(t) r^2 & 0 \\ 0 & 0 & 0 & -a^2(t) r^2 \sin^2\theta \end{bmatrix}$$

Synthesis of Formula Links in Einstein's Equation

$$R^{\mu\nu} - \frac{1}{2} g^{\mu\nu} R + \Lambda g^{\mu\nu} = -\frac{8\pi G}{c^4} T^{\mu\nu}$$

κ

$$\rho_v c^2 = \frac{c^4 \Lambda}{8\pi G}$$

$$R_{\mu\nu} = R^{\lambda}_{\mu\nu\lambda}$$

$$R = g^{\mu\nu} R_{\mu\nu} = g^{tt} R_{tt} + g^{rr} R_{rr} + g^{\theta\theta} R_{\theta\theta} + g^{\varphi\varphi} R_{\varphi\varphi}$$

$$T^{\mu\nu} = -p g^{\mu\nu} + (\rho + p) u^{\mu} u^{\nu}$$

$$R^{\lambda}_{\mu\nu\alpha} = \Gamma^{\lambda}_{\mu\alpha,\nu} - \Gamma^{\lambda}_{\mu\nu,\alpha} + \Gamma^{\lambda}_{\nu\eta}\Gamma^{\eta}_{\mu\alpha} - \Gamma^{\lambda}_{\alpha\eta}\Gamma^{\eta}_{\mu\nu}$$

$$\Gamma^{\lambda}_{\mu\nu} = \frac{1}{2} g^{\lambda\rho} (g_{\mu\rho,\nu} + g_{\nu\rho,\mu} - g_{\mu\nu,\rho})$$

$$\Gamma^{\lambda}_{\mu\nu} = \frac{1}{2} g^{\lambda\rho} \left(\frac{\partial g_{\mu\rho}}{\partial x^{\nu}} + \frac{\partial g_{\nu\rho}}{\partial x^{\mu}} - \frac{\partial g_{\mu\nu}}{\partial x^{\rho}}\right)$$

$$g_{\mu\nu} = \eta_{\alpha\beta} \frac{\partial \xi^{\alpha}}{\partial x^{\mu}} \frac{\partial \xi^{\beta}}{\partial x^{\nu}}$$

$$\eta_{\alpha\beta} = \begin{bmatrix} 1 & 0 & 0 & 0 \\ 0 & -1 & 0 & 0 \\ 0 & 0 & -1 & 0 \\ 0 & 0 & 0 & -1 \end{bmatrix}$$

ξ^{α} such as: $\xi^0 = ct$: $\xi^1 = x$; $\xi^2 = y$; $\xi^3 = z$

We note the coordinates of a **Galilean** frame that is not in free fall (which is subject to gravitation).

x^{μ} such as: x^0; x^1; x^2; x^3

Detail of the curvature tensor consisting of the second derivatives of the components of the metric tensor:

$R^{\alpha}_{\beta\mu v} = \Gamma^{\alpha}_{\beta v,\mu} - \Gamma^{\alpha}_{\beta\mu,v} + \Gamma^{\alpha}_{\mu\eta}\Gamma^{\eta}_{\beta v} - \Gamma^{\alpha}_{v\eta}\Gamma^{\eta}_{\beta\mu}$ Is the curvature tensor

With:

$$\Gamma^{\alpha}_{\beta v,\mu} = \frac{\partial\Gamma^{\alpha}_{\beta v}}{\partial x^{\mu}} = \frac{\partial\left(\frac{1}{2}g^{\alpha\psi}\left(\frac{\partial g_{\beta\psi}}{\partial x^{v}} + \frac{\partial g_{v\psi}}{\partial x^{\beta}} - \frac{\partial g_{\beta v}}{\partial x^{\psi}}\right)\right)}{\partial x^{\mu}}$$

With: $\Gamma^{\alpha}_{\beta v} = \frac{1}{2}g^{\alpha\psi}\left(g_{\beta\psi,v} + g_{v\psi,\beta} - g_{\beta v,\psi}\right)$

$$\Gamma^{\alpha}_{\beta\mu,v} = \frac{\partial\Gamma^{\alpha}_{\beta\mu}}{\partial x^{v}} = \frac{\partial\left(\frac{1}{2}g^{\alpha\chi}\left(\frac{\partial g_{\beta\chi}}{\partial x^{\mu}} + \frac{\partial g_{\mu\chi}}{\partial x^{\beta}} - \frac{\partial g_{\beta\mu}}{\partial x^{\chi}}\right)\right)}{\partial x^{v}}$$

With: $\Gamma^{\alpha}_{\beta\mu} = \frac{1}{2}g^{\alpha\chi}\left(g_{\beta\chi,\mu} + g_{\mu\chi,\beta} - g_{\beta\mu,\chi}\right)$

$$\Gamma^{\alpha}_{\mu\eta} = \frac{1}{2}g^{\alpha\lambda}\left(g_{\mu\lambda,\eta} + g_{\eta\lambda,\mu} - g_{\mu\eta,\lambda}\right) = \frac{1}{2}g^{\alpha\lambda}\left(\frac{\partial g_{\mu\lambda}}{\partial x^{\eta}} + \frac{\partial g_{\eta\lambda}}{\partial x^{\mu}} - \frac{\partial g_{\mu\eta}}{\partial x^{\lambda}}\right)$$

$$\Gamma^{\eta}_{\beta v} = \frac{1}{2}g^{\eta\alpha}\left(g_{\beta\alpha,v} + g_{v\alpha,\beta} - g_{\beta v,\alpha}\right) = \frac{1}{2}g^{\eta\alpha}\left(\frac{\partial g_{\beta\alpha}}{\partial x^{v}} + \frac{\partial g_{v\alpha}}{\partial x^{\beta}} - \frac{\partial g_{\beta v}}{\partial x^{\alpha}}\right)$$

$$\Gamma^{\alpha}_{v\eta} = \frac{1}{2}g^{\alpha\gamma}\left(g_{v\gamma,\eta} + g_{\eta\gamma,v} - g_{v\eta,\gamma}\right) = \frac{1}{2}g^{\alpha\gamma}\left(\frac{\partial g_{v\gamma}}{\partial x^{\eta}} + \frac{\partial g_{\eta\gamma}}{\partial x^{v}} - \frac{\partial g_{v\eta}}{\partial x^{\gamma}}\right)$$

$$\Gamma^{\eta}_{\beta\mu} = \frac{1}{2}g^{\eta\rho}\left(g_{\beta\rho,\mu} + g_{\mu\rho,\beta} - g_{\beta\mu,\rho}\right) = \frac{1}{2}g^{\eta\rho}\left(\frac{\partial g_{\beta\rho}}{\partial x^{\mu}} + \frac{\partial g_{\mu\rho}}{\partial x^{\beta}} - \frac{\partial g_{\beta\mu}}{\partial x^{\rho}}\right)$$

$\boxed{g_{\mu v} = \eta_{\alpha\beta}\frac{\partial\xi^{\alpha}}{\partial x^{\mu}}\frac{\partial\xi^{\beta}}{\partial x^{v}}}$ Is the metric tensor such as $\boxed{ds^2 = d\tau^2 = g_{\mu v}dx^{\mu}dx^{v}}$.

With $\boxed{\eta_{\alpha\beta} = \begin{pmatrix} 1 & 0 & 0 & 0 \\ 0 & -1 & 0 & 0 \\ 0 & 0 & -1 & 0 \\ 0 & 0 & 0 & -1 \end{pmatrix}}$ the inverse of $g_{\alpha\rho}$ is $g^{\alpha\rho}$.

ξ^{α} such as:

$$\xi^0 = c \times t;$$

$$\xi^1 = x;$$

$$\xi^2 = y;$$

$$\xi^3 = z;$$

Key Formula of this Book

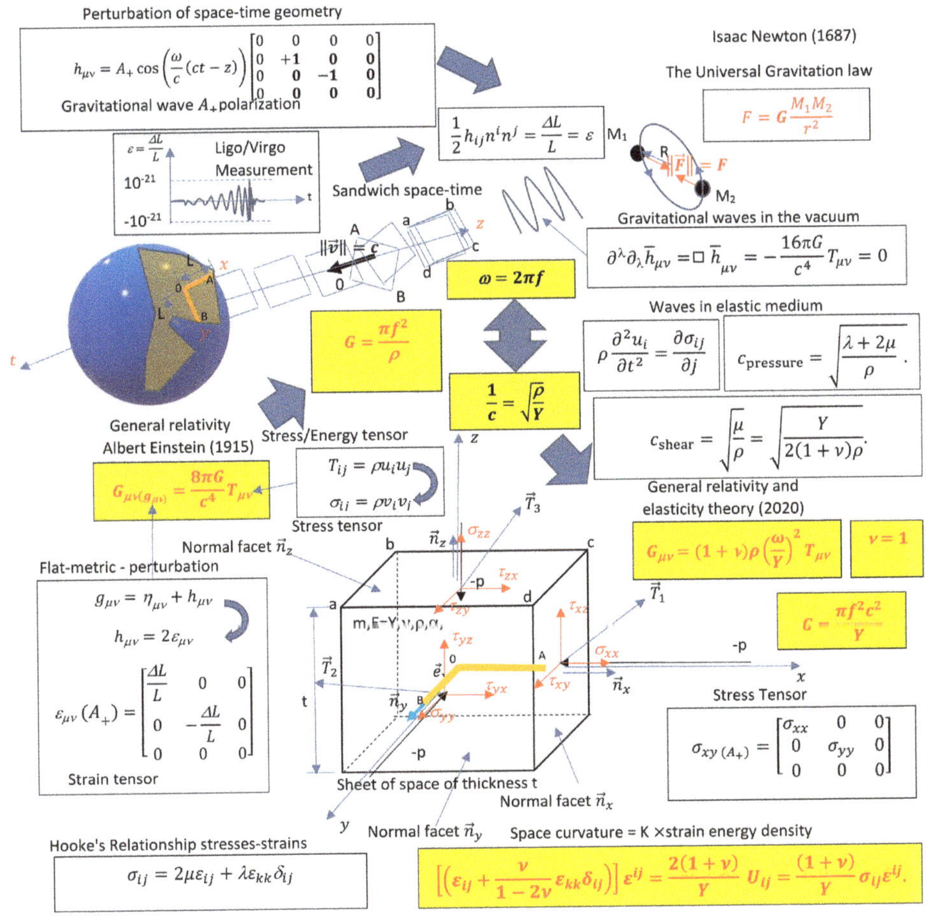

About the Author

David Izabel was born on November 1, 1969 in Troyes in the Aube department of France. He did brilliant engineering studies at INSA Rennes in mechanics where he studied in particular the theory of continuum mechanics and elasticity with Jean-Marie Aribert, the strength of materials by Timoshenko with A. Lachal. He graduated major from his class in 1992. Always eager to understand in detail how the efforts progress within all natural or artificial structures, he specializes in the study of the active and resistant forces within human constructions (bridge structures, building frames, gothic and roman structures with his father...) at the Center des Hautes Etudes de la Construction and at the University of Paris VI Pierre Et Marie Curie where he obtained a DESS in structure and method in 1993.

Passionate about mechanics, in 2007 he published a "Formular of strength of Materials" in 5 volumes on the theory of sandwich beams, an extension of the formulations of the classical strength of material taking into account the shear force strains in addition of the bending strains.

Having a strong potential for abstraction and passionate about the fractal aspect of nature, he demonstrates in mathematics that randomness games and numbers like π and gold follow a fractal law. He published his work in the American Journal of Mathematics and Statistics in June 2015. "Does the Sequences of Random Numbers Follows a Log Periodic Law? Randomness has Memory?" DOI: 10.15640/arms. v3n1a9.

Eager to understand the structure of the universe, he self-taught physics of the 20th and 21st centuries. From 2013 to 2016 he took online quantum mechanics lectures (class X of the polytechnic school with J. Dalibard), general relativity, special relativity, electromagnetism, cosmology with online lectures by professor – researcher R. Taillet at the University of Savoie Mont Blanc, gravitational waves and black holes with conferences by T. Damour, K. Thorne and R. Weiss and the lecture by R. Taillet, relativistic quantum mechanics and quantum fields theory with the online lectures of Prasanta Tripathy at the University of Madras, Bell's equations with A. Aspect conferences, quantum gravity with the C. Rovelli's lectures available on the net. He also studied the philosophy and history of science and physics with the E. Klein's conferences.

General relativity explains the elastic deformations of space-time but says nothing about the mechanical nature in the behavioral sense of the material constituting the physical space-time object. He therefore decided in 2011 to reconstruct Einstein's formula for general relativity $G_{\mu\nu} = \kappa T_{\mu\nu}$ and in particular its proportionality constant κ through the theory of elasticity and not through Newton's mechanics making intervene a formulation and a universal gravitational constant G dating from the 17th century which turns out to be only a simplification of the exact theory established by Einstein in 1915.

This leads him to a new formulation of Einstein's equation *via* a new expression of the gravitational constant G based on the mechanical parameters of the quantum vacuum that he publishes on August 13, 2020 in the Indian peer-reviewed scientific journal PRAMANA in connection with the Indian Academy of Sciences. "Mechanical conversion of the gravitational Einstein constant" κ art 119. doi.org/10.1007/s12043-020-01954-5.

Having become an expert in structures and envelopes in cold formed thin metal members and sheeting, the most sophisticated because involving all the phenomena of strength of material and becoming a professor at CHEC in 2017, he then decides to explain in detail the intellectual development that he led to this publication.

Therefore, to help understand general relativity in a didactic way, he decided to develop in this book the path which led him to discover general relativity through the elasticity theory. To help understand Einstein's theory explicitly, he decided to give explicit and very detailed examples of application of this theory such as calculating the scalar curvature R of a sphere or applying Einstein's gravity field formula to the universe by a detailed demonstration of Friedmann–Lemaitre equations. Finally, to position this theory in the current context, he created at the beginning of this book a summary of the physics of the last 3 centuries.

Summary

In this book, author David Izabel tries to answer the question: how space-time is made. After a rapid synthesis of the current situation of physics, where its two pillars beautifully demonstrated general relativity and quantum field theory are for the moment irreconcilable, he proposes to bring general relativity of quantum field theory closer together using an intermediary; the elasticity theory. For this purpose, he uses the analogy between Einstein's gravitational geometric theory in one and two-dimensional linear deformations and the elasticity theory *via* a possible spatial material based on strains measurements ($\frac{\Delta L}{L} = \varepsilon = 1 \times 10^{-21}$) performed on the Ligo or Virgo interferometer during the passage of gravitational waves. He demonstrates the fundamental principle "space curvature $= K \times$ energy density" based on the general relativity and Timoshenko's strength of materials and then generalizes this analogy using the concepts of the elasticity theory (strain and stress tensor, Hooke's law generalized). He thus draws an analogy between Einstein's gravitational constant κ and Young's modulus E (or Y) and Poisson's ratio ν of an elastic material that can constitute the space fabric, in the context of the propagation of weak gravitational waves measured by Ligo and Virgo. Thus, he demonstrates that the two polarizations A_+ and $A_\times$ of the gravitational waves correspond exactly to two strain tensors of an elastic medium twisted by the rotation and coalescences of two black holes *via* the expression $h_{\mu\nu} = 2\varepsilon_{\mu\nu}$. The junction with quantum field theory and the hypothetical spin 2 graviton then appears naturally *via* an image of the elasticity theory that is Mohr's circle. It follows from all this analogy that the space seems to consist of as very thins multi-sandwich sheets of elastic space (plane stresses) whose elastic microstructure of grain is 1.566 x 10^{-35} m as proposed in string theory or loop quantum gravity. The most extraordinary is that from this analogy *via* the strength of materials and then elasticity is that appears in the equations naturally a factor $\pi\frac{f^2}{\rho}$ that has the dimensions and structure of the G gravitational constant. Newton therefore reappeared but calibrated on the mechanical characteristics of the empty space that is no longer empty, namely its f frequency and its density ρ rather than in an absolute and universal way as fixed forever. From this new definition of G it is then possible to generalize Einstein's field equation which, as in the theory of elasticity, is now a coefficient of proportionality κ no longer determined by Newton's passage to limits in low fields, but depending on the mechanical characteristics of a material constituting space. Time also becomes elastic and quantified; it corresponds to the time needed to transmit information from one space sheet to another at the speed of light. The speed of light then becomes what it is given the density of the material constituting the space. In summary, the approach of general relativity and quantum field theory *via* the theory of elasticity seems to indicate, that we must give up a fixed G gravitational constant such as Newton's pose according to the data he had at the time. Since general relativity has dethroned Newton by demonstrating that space-time is deformable and elastic (a sort of new deformable ether) like any material in elasticity theory, on the one hand, and since it is possible to reconstruct $\kappa = \frac{8\pi G}{c^4}$ without going to the limits by the approach of Newton (calibration of the temporal components of general relativity), but through the spatial components of general relativity on the other hand, why hesitate on this new definition of the gravitational constant straight out of the unification of the general relativity theory with the elasticity theory? This book is therefore a way to understand simply the dynamics of general relativity from the strength of materials and the theory of elasticity. The reader will then make his own opinion. Good travel in this exciting quest for the truth about our world and universe.

David Izabel, December 22, 2020

www.ingramcontent.com/pod-product-compliance
Ingram Content Group UK Ltd.
Pitfield, Milton Keynes, MK11 3LW, UK
UKHW020705041026
15590UKWH00001B/472

9 782759 825738